Climate Change in Asia and Africa - Examining the Biophysical and Social Consequences, and Society's Responses

Edited by John P. Tiefenbacher

Published in London, United Kingdom

IntechOpen

Supporting open minds since 2005

Climate Change in Asia and Africa - Examining the Biophysical and Social Consequences, and Society's Responses
http://dx.doi.org/10.5772/intechopen.77537
Edited by John P. Tiefenbacher

Contributors
Md. Golam Rabbani, Md. Nasir Uddin, Sirazoom Munira, Shyamli Singh, Ovamani Olive Kagweza, Mohit Arora, Kalyan De, Nandini Ray Chaudhury, Mandar Nanajkar, Prakash Chauhan, Brijendra Pateriya, Daniel Kipkosgei Murgor, Khalid Turk, Faisal Zeineldin, Abdulrahman Alghannam, Fibor J. Tan, Narendran Kodandapani, Ramesh Prasad Bhatt, Nqobizitha Dube, Gun Mardiatmoko, Snigdha Chatterjee, Biswabara Sahu, Ruby Patel, Abdelkrim Benaradj, Hafidha Boucherit, Abdelkader Bouderbala, Okkacha Hasnaoui, Hicham Salhi, Akanwa Angela Oyilieze, Ngozi N. Joe-Ikechebelu, Kenebechukwu J. Okafor, Ijeoma N. Okedo-Alex, Fred A. Omoruyi, Jennifer Okeke, Sophia N. Amobi, Angela C. Enweruzor, Chinonye E. Obioma, Princess I. Izunobi, Theresa O. Nwakacha, Chinenye B. Oranu, Nora I. Anazodo, Chiamaka A. Okeke, Uwa-Abasi E. Ugwuoke, Uche M. Umeh, Emmanuel O. Ogbuefi, Sylvia T. Echendu

Notice
Statements and opinions expressed in the chapters are these of the individual contributors and not necessarily those of the editors or publisher. No responsibility is accepted for the accuracy of information contained in the published chapters. The publisher assumes no responsibility for any damage or injury to persons or property arising out of the use of any materials, instructions, methods or ideas contained in the book.

First published in London, United Kingdom, 2022 by IntechOpen
IntechOpen is the global imprint of INTECHOPEN LIMITED, registered in England and Wales, registration number: 11086078, 5 Princes Gate Court, London, SW7 2QJ, United Kingdom
Printed in Croatia

British Library Cataloguing-in-Publication Data
A catalogue record for this book is available from the British Library

Additional hard and PDF copies can be obtained from orders@intechopen.com

Climate Change in Asia and Africa - Examining the Biophysical and Social Consequences, and Society's Responses
Edited by John P. Tiefenbacher
p. cm.
Print ISBN 978-1-83962-629-6
Online ISBN 978-1-83962-630-2
eBook (PDF) ISBN 978-1-83962-631-9

We are IntechOpen,
the world's leading publisher of
Open Access books
Built by scientists, for scientists

6,000+
Open access books available

146,000+
International authors and editors

185M+
Downloads

156
Countries delivered to

Our authors are among the
Top 1%
most cited scientists

12.2%
Contributors from top 500 universities

Interested in publishing with us?
Contact book.department@intechopen.com

Numbers displayed above are based on latest data collected.
For more information visit www.intechopen.com

Meet the editor

Dr. John P. Tiefenbacher (Ph.D., Rutgers, 1992) is a Professor of Geography and Environmental Studies, at Texas State University. His research has focused on various aspects of hazards and environmental management. Dr. Tiefenbacher has published on a diverse array of topics that examine perception and behaviors with regard to the application of pesticides, releases of toxic chemicals, environments of the US–Mexico borderlands, wildlife hazards, and the geography of wine. More recently his work pertains to adaptation to climate change, spatial responses of wine growing to climate change, the geographies of viticulture and wine, and artificial intelligence and machine learning to predict patterns of natural processes.

Contents

Preface

The idea of climate, scientifically speaking, was formulated to represent the stable meteorological patterns of the terrestrial surface of Earth with a classification that could reflect the extant conditions. Each climate was based on the numerical tabulation of data for 30-year periods of meteorological observations of temperature, precipitation, and sometimes other phenomena, and the annual and seasonal patterns of means and extremes. Every year, the climate could be recalculated based on the 30 previous years. As long as there were minor changes, the patterns could be regarded as stable. Climates could therefore be used to forecast the distribution of vegetation, resources (like water and fertile soils), and hazards. Climate classifications were extremely useful for planning, particularly for agricultural, land use, and residential adaptations to stable environments. Our past activities could be continued on with little reason to fear that they would no longer be supported by the climatological conditions of a place.

Climatologically speaking, climate change is instability: the (increasingly significant) departure from past climates. Climate change is not "real" in the sense that we can empirically observe it; it is statistical and comparative over a longer time period than most people could recall in detail. The change signified is neither good nor bad, it is just changing. It can be a subtle change or it could be a dramatic change. Though rare, the changes taking place within a climate might be beneficial to the local community, as when arid climates become more humid or when cold or warm seasons become more temperate. The change of any climate (i.e., of a region of Earth), however, is the byproduct of global heating's influence on atmospheric and oceanic circulations. Although some might see a silver lining in the warming of Earth's atmosphere, there will likely be greater negative experiences somewhere else around the world or for people living sometime in the future.

The changes in the climates of the world are more important than just a matter of comfortable temperatures or warm sunny days. The results of changing daily, seasonal, or annual thermal and hydrological regimes yield fundamental consequences for the biological and ecological conditions in a location and may undermine human economic activities or even present threats to the lives and properties of those present. In fact, the most important consequence of changing climates is that patterns that may have long been relatively static are no longer predictable. The most important element of climate change is that it makes the future more precarious for both nature and humans. Natural organisms have more limited adaptability than people; their environment might evolve beyond their tolerances. People have a greater capacity to adapt, change, persevere, or migrate.

This volume contains studies that consider the implications of changing climates in Asia and Africa. Combined, these two regions contain the majority of the population of Earth's lesser developed countries whose survival lacks the kinds of cushion or insulation provided by advanced technologies or economic safety nets enjoyed by people in more developed regions. Africa and Asia comprise environments that have significant limitations for resource development, particularly food production. There are regions that are too dry, too wet, too hot, and too cold, and some might even experience two or more of these conditions from one season to the next. Consideration of the quite varied experiences across these regions allows us to examine the consequences of changing climates for the natural and human environments in this context.

This book is organized into three parts. The first includes four studies that focus on the biophysical consequences that climate change has on evapotranspiration rates, precipitation extremes, and coral bleaching in the Indian Ocean. The three chapters of the second section regard the social consequences of climate change in the patterns of humid-region flood risk and hazards in Asia and the implications of climate change for Zimbabwe's horticultural sector. The third section contains seven studies of adaptation to climate change. These chapters examine agricultural vulnerability in Uganda, mitigation and adaptation of palm-oil plantations in Indonesia, farmer's localized knowledge for mitigating precipitation variability in eastern and southern Africa, sustainable carbon management in paddy rice-growing regions, adaptation to changing patterns of hazards in India, river flooding and temporary displacement of women and children of Nigerian villages, and management and mitigation of ecological impacts and diversity in Nepal.

The challenges posed to these regions by global warming are often complicated by the paucity of data and the unavailability of relatively common technological systems to enable monitoring and field work. The scholars of these regions often work on the margins of global science because of the costs of access to scholarly literature and the expense of publication in the most prestigious journals. It is my hope that the exposure afforded to these scholars published in the volume will enable greater international support and collaboration with them.

John P. Tiefenbacher
Department of Geography and Environmental Studies,
Texas State University,
San Marcos, Texas, USA

Section 1

Biophysical Consequences
of Climate Change

Chapter 1

Biophysical Effects of Evapotranspiration on Steppe Areas: A Case Study in Naâma Region (Algeria)

Abdelkrim Benaradj, Hafidha Boucherit,
Abdelkader Bouderbala and Okkacha Hasnaoui

Abstract

The Algerian steppe is of great interest in terms of vegetation, mainly in the Naâma region. This steppe vegetation is generally composed of annual and perennial grasses and other herbaceous plants, as well as, bushes and small trees. It is characterized by an arid Mediterranean climate where the average annual precipitation (100 to 250 mm) is insufficient to ensure the maintenance of the vegetation, in which the potential evaporation always exceeds the precipitations. This aridity has strong hydrological effect and edaphic implications from which it is inseparable. Water losses are great than gains due to the evaporation and transpiration from plants (evapotranspiration). The wind moves soils for one location to another, and causes a strong evapotranspiration of the plants, which is explained by a strong chronic water deficit of climatic origin of these compared to the potential evapotranspiration, opposed to a humid climate. Evapotranspiration is certainly closely linked to climate factors (solar radiation, temperature, wind, etc.), but it also depends on the natural environment of the studied region. Potential evapotranspiration (PET) data estimated from Thornthwaite's method for the three stations (Mécheria, Naâma and Ainsefra). The average annual value of potential evapotranspiration is of the order of 807 mm in Mécheria, of 795 mm in Naâma de and in Ainsefra of 847 mm. It is more than 3 times greater than the value of the rainfall received. This propels it globally in the aridity of the region and from which the water balance of plants is in deficit. The potential evapotranspiration of vegetation in arid areas is very important due to high temperature and sunshine. During the cold season, precipitation covers the needs of the potential evapotranspiration and allows the formation of the useful reserve from which the emergence of vegetation. From the month of April there is an exhaustion of the useful reserve which results of progressive deficit of vegetation. Faced with this phenomenon of evatranspiration, the steppe vegetation of the region then invests in "survival" by reducing the phenomena of evapotranspiration, photosynthetic leaf surfaces, in times of drought. These ecophysiological relationships can largely explain the adaptation of steppe species (low woody and herbaceous plants) to the arid Mediterranean climate. Mechanisms and diverse modalities were allowing them to effectively resist for this phenomenon. The adaptation of the steppe vegetation by the presence of a root system with vertical or horizontal growth or both and seems to depend on the environmental conditions, and by the reduction of the surface of transpiration, and by the fall or the rolling up of the leaves, and by a seasonal reduction of transpiration surface of the plant to reduce water losses during the dry season (more than 6 months) of the year. Some xerophytes produce "rain roots" below the soil surface, following light precipitation or during dew formation. Other persistent sclerophyllous species by which decreases transpiration by the hardness of the leaves often coated with a thick layer of wax or cutin.

Keywords: evapotranspiration, climate, factors, Naâma, vegetation, Algeria

1. Introduction

The South Oranian steppe of Naâma is characterized by sparse vegetation and only drought tolerant plants that can live there. The steppe vegetation is therefore made up of open grassy formations, revealing bare soil between the plants and the amount of existing plant matter per unit area and roughly proportional to the precipitation received [1]. They are plant formations of a steppe character, primary or secondary, low and open in their typical physiognomy, and mainly subservient to arid and desert areas (rainfall <350 mm) [2].

Soil, vegetation and atmosphere form a single continuous system in which water circulates at a negative energy gradient, water moves throw the soil and then absorbed by roots, and from branches to leaves and then evaporated into intercellular cavities of leaves, and then diffused through the stomata to the layer of calm air in contact with the surface of leaves and finally to the outside atmosphere [3]. In the water cycle, water inputs correspond to precipitation and water losses are due to evapotranspiration, runoff and infiltration. Therefore, part of the precipitation can be intercepted by vegetation and returned to the atmosphere by evapotranspiration (ET) or sublimation [3].

Evapotranspiration is defined as "the response of vegetation to natural climatic conditions in relation to the physiological properties of the plant and its water resources. It is a complex climatic parameter, knowledge of which has significant practical interest at the moment of the estimation of water reserve of soils and of water requirements of crops and vegetation, and in the estimation of the volumes of water necessary for the development of plants. Evapotranspiration is also an indicator of interest in studies concerning climate change [4]. Evapotranspiration can act on the water balance and modify its various components through vegetation. Also, evapotranspiration plays a key role in the evaluation of the climatic capacities of a given region and is considered to be the indicator of optimal vegetation development [5].

This study aims to estimate the evapotranspiration in this region, and evaluate its impact on vegetation, in order to better understand the adaptation of this vegetation on this arid climate. In this chapter, we begin to present the basic concepts relating to the notions of water balance, the phenomenon of evapotranspiration and the methods of its evaluation. Next, we will describe the general characteristics of the steppe environment of the Naâma region (geographic location, climate, natural resources, etc.), then we will present the methodology adopted to achieve the study objectives, as well as, the impact of the evapotranspiration on vegetation. Finally, we end this work with a conclusion and some recommendations.

1.1 Theoretical framework on evapotranspiration

1.1.1 Notions of evapotranspiration

Evapotranspiration can be defined as the loss of water through soil and plant surface, usually expressed in mm/day. Indeed, the term "evapotranspiration" (ET)

designates the water losses of a plant cover depending on the soil water reserve, the stage of vegetation development and the atmospheric environment [6]. Evapotranspiration is a combination of two terms, namely, evaporation (from a surface, from a body of water), and transpiration (from plants). It constitutes a fundamental characteristic of the climate, represents the cumulative evaporation of the soil and the transpiration of plants [7].

Evapotranspiration is a combination of two processes:

- **Evaporation**, is the passage of water from the liquid state to the gaseous state in the form of vapor, or more precisely direct evaporation, refers to the water that evaporates from a soil or from water body to the atmosphere [8].

- **Transpiration**, is the loss of water in the form of vapor by the plant. At the leaf scale, it is caused by the difference in vapor pressure between the intercellular spaces of the leaves and the surrounding air [9, 10]. Transpiration is the driving force behind the transfer of water through the plant: a difference in water potential is created between the leaves and roots, which are at the origin of the absorption flow (a hydraulic suction pump transferring the water from the soil to the atmosphere, i.e. the soil–plant-atmosphere continuum) [11].

1.2 Different types of evapotranspiration

- **Potential evapotranspiration (PET)** is the maximum water consumption of an active vegetation cover, dense and spread over a large area and well supplied with water. It is considered to be the upper limit of evapotranspiration for a crop in a given time [12]. The evapotranspiration of a plant cover is said to be potential when the energy available for vaporization is the only factor limiting this first. It is also called the reference evapotranspiration (ET_0). The evapotranspiration of a low, contained and homogeneous plant cover whose water supply is not limiting and which is not subject to any nutritional, physiological or pathological limitation [13].

- **Real evapotranspiration (RET)** is a key term in the water balance [14]. It is linked to maximum evapotranspiration by a stress coefficient. It designates the amount of water actually lost in the form of water vapor by the plant cover when the water supply is not optimally assured. It corresponds to the effective evapotranspiration of a plant cover when the water supply is not optimally assured. This means that the limiting factor can be of a climatic nature (insufficient rainfall for example), or of a pedological nature (rapid exhaustion of the easily usable water reserve in the soil), or of a physiological nature (for example, plant incapable of ensuring, from the soil to the leaves, a sufficient flow of water to use all the energy that might be available for vaporization [15].

- **Maximum evapotranspiration (MET)** is linked to PET by a crop coefficient (Kc). This is the amount of water lost by vegetation enjoying an optimal water supply (good soil fertility, good sanitary condition, etc.). The maximum evapotranspiration varies during the development of an annual crop, increases gradually with the rate of soil cover by the plant to reach PET and decreases at the end of the vegetative cycle [16]. This form of evapotranspiration (MET) is used by agronomists to determine the water requirements of plants [17–19].

1.3 Climatic factors influencing evapotranspiration

According to Ferchichi (1996) [7], evapotranspiration is certainly closely linked to climatic factors, but it also depends on the natural environment of the region studied, the plant species concerned and soil properties. Evapotranspiration strongly depends on the availability of two factors: abiotic (climatic, geographic: topographic and orographic, hydrological and edaphic: soil) and biotic (biological: vegetation).

Evapotranspiration occurs under the influence of solar radiation, which is the source of energy that allows water to change from liquid to vapor. It depends on two elements: the heat supplied by solar radiation and the quantity of water available in the ground [20]. Evapotranspiration is certainly closely linked to climatic factors (evaporating power): air temperature, temperature of the earth's surface, wind speed and turbulence, duration of sunstroke or solar radiation, precipitation, relative air humidity and atmospheric pressure.

Indeed, the transpiration process depends on the following parameters: solar radiation, temperature, humidity, wind speed, the water vapor concentration gradient and therefore the water vapor pressure between the spaces substomatics of the leaf and the atmosphere, physiological mechanisms and metabolic activity of the plant, density of the root system, type of plant cover (structure, size, leaf area, presence or absence of leaves, nature of pigmentations, etc.). Finally, the process of evaporation depends on temperature, precipitation, air humidity and plant cover... (**Figure 1**).

1.4 Methods used for the estimation of evapotranspiration

Evapotranspiration is an important component of the water balance, involving both physical and biological processes. Many methods used for the estimation of evapotranspiration, which have been proposed by different authors. Each method is distinguished by the parameters taken into consideration, by the climatic conditions in which it was developed and by its application limits. Evapotranspiration can be evaluated by several methods (direct and indirect), which take into account

Figure 1.
Different streams of the water balance.

different meteorological parameters (climatic, energetic) in relation to soil and vegetation parameters. In the meteorological stations of Naâma region, the evapo-transpiration is measured directly with the lysimeter and evaporation pan and is estimated indirectly by using methods (equations) such as those of Thornthwaite, (1944), Penman (1948), Turkish (1961), etc.

2. Material and methods

2.1 Presentation of the study area

2.1.1 Geographical location

The wilaya (province) of Naâma is part of the southern high plains of Algeria; it extends between latitude 32° 08′45″ and 34° 22′13″ North, and longitude from 0° 36′45″ to 0° 46′05″ west. It covers an area of 3 million hectares. It is occupied by a population located along the Oran-Bechar road axis, i.e. 37% of the total area, which translates poor use of space. The region of Naâma has a large set of ecosystems and biological diversity. Administratively, the province of Naâma is limited (**Figure 2**):

- In the north by the provinces of Tlemcen and Sidi Bel Abbes,

- In the east by the province of El-Bayadh,

- In the south by the province of Béchar,

- In the west by the Algerian-Moroccan border of 275 km long.

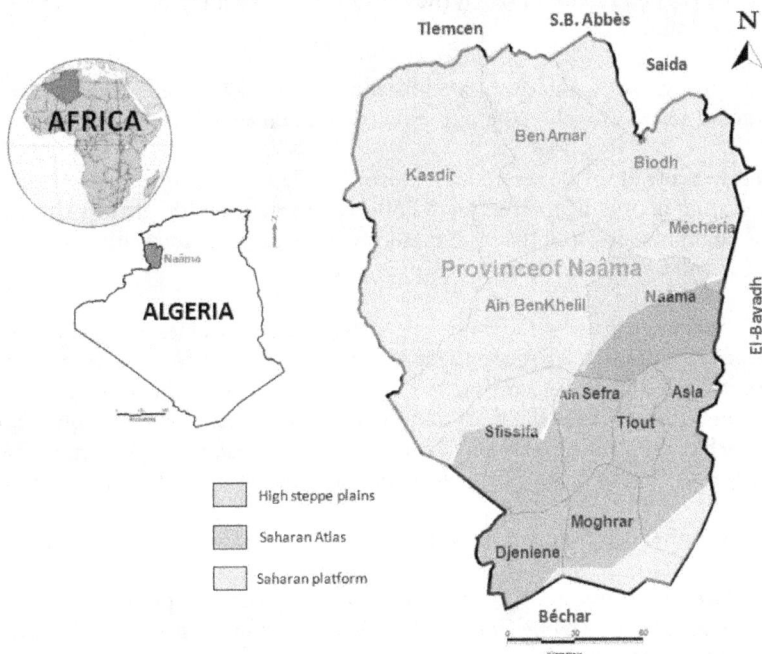

Figure 2.
Geographic location of the study region (Naâma, Algeria).

2.1.2 Biophysical characterization

The analysis of the biophysical environment will be done on the basis of the analysis of factors (geographic, hydrographic, pedological, climatic, etc.), in order to allow us to identify and characterize the potentialities and physical constraints as well as their interaction.

2.1.2.1 Geomorphological framework

From a geomorphological point of view, the territory of the Naâma regionis formed on immense depressed plain located between the two Atlas (Tellien and Saharan) There are three homogeneous geographic areas (**Figure 2**):

- Steppe space (high steppe plains) shows a vast plain (74% of the territory of the province) whose altitude increases significantly towards the south (1000 to 1300 m). It is characterized by the predominance of pastoral activity.

- Mountainous area is located in the southwest region reaching an altitude of 2000 meters and occupying 12% of the territory of the province. It is a part of the Ksours Mountains and the foothills of the Saharan Atlas. It is characterized by oasis-type agriculture.

- A pre-Saharan area covers an area of around 14% of the total area of the province.

2.1.2.2 Soil framework

According to the work of (Pouget [21]; Djebaili et al. [22], Halitim [23]; Haddouche [24] and Bensaid [25]), the soils of the Naâma region are generally classified as follows:

- Raw mineral soils are represented by raw mineral soils from erosion, raw mineral soils from alluvial input and raw mineral soils from wind input.

- Poorly evolved soils: Present an AC profile, a low degree of evolution and poor alteration in organic matter devoid of clay-humic complex, they are located at the edges of wadis and they surround raw mineral soils, they cover the glacis of the North South-East plain of the province.

- Calcimagnesium soils (Calcimorphic soils) called rendzines are located on the slopes of montains. They are the most common type of soil in this area, and occupy vast surface. According to the geomorphological characteristics, the Halomorphic soils are the most dominant, mainly located in the Chott-Chergui, the sebkhas and the Mekmen. These soils support halophyte vegetation based on *Atriplex halimus, Atriplex glauca, Frankenia thymifolia, Salsola sieberi*.

2.1.2.3 Biogeographical framework

The North African steppes in general and the Algerian steppes in particular are part of the Mauretano-steppe floristic domain defined by Maire [26]. This area belongs to the Mediterranean floristic region, therefore to the Holarctic Empire.

Biogeographically, the study region belongs to the Mediterranean area, to the high-lands sector and to the Saharan Atlas sector according to the Quézel and Santa [27].

2.1.2.4 Biological framework

The steppe space in the Naâma region is characterized mainly by plant formations of herbaceous, sub-shrub and shrub types. The steppe vegetation is characterized by the abundance either of cespitose grasses (*Stipa tenacissima, Lygeum
spartum, Stipagrostis pungens*) or of chamaephytes (*Artemisia herba-alba, Hammada
scoparia*), but also by the frequency of annual species. For the shrub layer is quite
common in arid and pre-Saharan areas (*Ephedra alata, Retama retam, Salsola
vermiculata, Thymellaea microphyla*).

For the shrub layer is characterized by the presence of forest species (*Quecus
ilex, Pinus halepensis, Juniperus phoenicea, Juniperus oxycedrus, Pistacia atlantica*) in
the mountains such as the case in Djebels (mountains) of Aissa, Merghad, Mekther,
etc. **Figure 3** below shows a map of the main plant formations in the study region.

2.1.2.5 Climate framework

The climate is Mediterranean with a bioclimatic gradient decreasing from North
to South, ranging from semi-arid to lower arid and pre-Saharan. Rainfall is low and
irregular, varying from 190 to 250 mm/year. The frequency of drought seems to be
increasing in recent decades.

2.1.3 Methodological approach

2.1.3.1 Bioclimatic study

As part of our study, we took into consideration, as climatic parameters: rainfall
and temperature because they represent the essential element of plant growth, soil

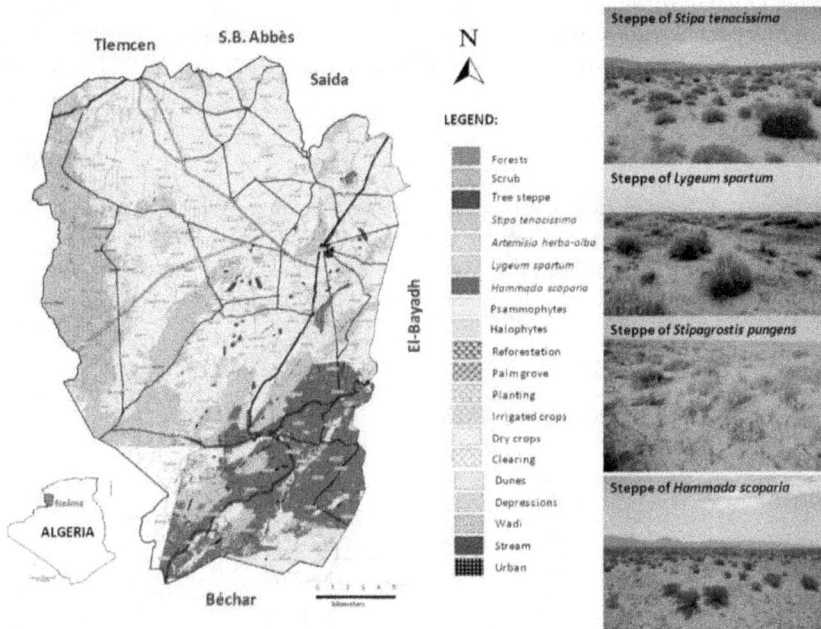

Figure 3.
Map of the main plant formations in the Naâma region [28].

formation and evolution. Several studies were carried out about the climate of steppe regions in Algeria [29–36].

The study of the climate and bioclimate is based on the automated processing of old meteorological data [29], taken over 25 years (1913–1938) and from the recent period (1990–2014). All data are collected from the National Meteorological Office (NMO) [37].

- Presentation of weather stations

From a climatic point of view, the Naâma region inscribes its territorial limits on three distinct geographic natural domains:

- Steppe area (High Steppe Plains): it is covered by 2 meteorological stations: Naâma and Mécheria.

- Atlas area (Saharan Atlas): it is covered by a single meteorological station of Aïn Sefra (**Table 1**).

- Presentation of climate data

Temperature and precipitation values are synthesized to determine climatic parameters for the entire study region by extrapolation. Data are listed in **Table 2**.

Rainfall: The geographical location of the study region shows decreasing rainfall gradient from North to South. The distribution of the average monthly rainfall during the periods 1913–1938 and 1990 to 2014 is presented as follows.

Temperatures: Temperatures are an important component of plant life, especially the two extremes: The lowest average temperatures are in January for the three stations, while the highest averages are in July for the three stations (**Table 3**) according to data from NMO (**Table 4**) [37].

Calculation of different climatic parameters

De Martonne aridity index: The aridity index is a quantitative indicator of the degree of water scarcity present in a given location. The De Martonne index is given by the formula below [38].

$$I = R/T + 10 \tag{1}$$

Where: R = Average annual rainfall in mm and T = Annual average temperature in °C. When the index is low, the climate is more arid, and vice versa (**Table 5**).

Calculation of thermal continentality: The thermal continentality is given by the Debrach method [39]. It is distinguish four types of climates:

- Islander climate: M-m < 15°C,

- Coastal climate: 15 0C < M-m < 25°C,

Station	Latitude	Longitude	Altitude
Mécheria	33° 31' N	00° 17' W	1149 m
Naâma	33° 16' N	00° 18' W	1166 m
Aïn Sefra	32° 45' N	00° 36' W	1058 m

Table 1.
Main reference weather stations in the study region [36].

Station	Rainfall	J	F	M	A	M	J	Jt	A	S	O	N	D	Annual
Méchéria	P1 (1913–1938)	21,00	24,00	32,00	29,00	25,00	14,00	5,00	8,00	34,00	29,00	43,00	29,00	293,00
	P2 (1990–2014)	1849	17,88	27,95	26,24	21,05	10,60	5,28	10,14	26,34	35,52	26,98	16,62	243,11
Naâma	P1 (1913–1938)	—	—	—	—	—	—	—	—	—	—	—	—	—
	P2 (1990–2014)	13,57	15,37	25,47	18,15	18,05	14,18	5,64	14,61	23,51	31,32	26,68	12,14	218,75
Ain sefra	P1 (1913–1938)	10,00	10,00	14,00	9,00	15,00	28,00	8,00	7,00	15,00	29,00	29,00	18,00	192,00
	P2 (1990–2014)	15,65	11,97	25,50	18,67	14,52	8,31	4,72	10,68	21,51	35,79	22,75	9,52	199,64

Table 2.
Distribution of average monthly precipitation [37].

Station	Period	J	F	M	A	M	J	Jt	A	S	O	X	D
Méchéria	P1 (1913–1938)	6,25	7,60	10,70	14,30	17,60	23,20	27,70	27,45	22,60	16,40	10,10	6,65
	P2 (1990–2014)	6,95	8,06	11,31	14,12	22,91	23,87	27,90	27,12	22,03	17,01	11,01	7,54
Naâma	P1 (1913–1938)	—	—	—	—	—	—	—	—	—	—	—	—
	P2 (1990–2014)	6,07	7,49	11,13	14,35	18,97	24,56	28,61	27,79	22,64	17,20	10,66	7,13
Ain sefra	P1 (1913–1938)	6,05	8,35	10,70	15,65	19,40	24,05	28,40	27,20	23,70	17,05	10,60	6,90
	P2 (1990–2014)	7,44	9,12	12,69	15,95	20,60	25,84	29,53	28,53	23,85	18,24	11,90	8,29

Table 3.
Monthly mean temperatures (°C) (from 1913 to 1938 and 1990–2014) [37].

Station	Period	M (°C)	m (°C)	P (mm)
Mécheria	P1 (1913–1938)	35,1	1,5	293,0
	P2 (1990–2014)	36,78	1,49	243,1
Naâma	P1 (1913–1938)	/	/	/
	P2 (1990–2014)	36,8	0,32	218,7
Aïn Sefra	P1 (1913–1938)	37,6	−0,3	192,0
	P2 (1990–2014)	38,34	0,57	199,6

Table 4.
Average values of temperatures and rainfall in the study stations.

Aridity index (I)	Type of climate
I < 5	Hyper-arid climate
5 < I < 7,5	Desert climate
7,5 < I < 10	Steppe climate
10 < I < 20	Semi-arid climate
20 < I < 30	Mild climate

Table 5.
De Martonne aridity index class [38].

- The semi-continental climate: 25°C < M-m < 35°C;

- Continental climate: M-m > 35°C.

Where: M: average temperatures of the maximums of the hottest month. m: average minimum temperatures of the coldest month.

Seasonal regime: The seasonal regime presents the seasonal variation: the sum of the seasonal rainfall of Winter, Spring, Summer and Autumn. According to Despois [40], the study of the rainfall regime is more instructive than comparing annual averages or totals.

For this purpose, we calculated the amount of rainfall for all the study stations, during the four seasons.

- Autumn (A): September, October, November

- Winter (W): December, January, February.

- Spring (Sp): March, April, May.

- Summer (S): June, July, August.

Climate summary: Climate synthesis is based on the search for formulas that allow the action of several ecological factors to be reduced to a single variable. For this, several climatic indices, taking into account variables such as rainfall and temperatures, have been formulated for a synthetic expression of the regional climate.

We will retain the pluviometric quotient of Emberger [41, 42], which remains the most effective index in the description of the Mediterranean climate, the xero-thermic index of Bagnouls and Gaussen [43] and thermal continentality and rain.

Several methods and indices have been used in the climatic classification of the Mediterranean region, including the method of Bagnouls and Gaussen [43, 44] and that of Emberger [42].

Pluviothermal quotient: Emberger [42] proposed a pluviothermal quotient, which tells us about the xeric character of the vegetation and which takes into account temperatures and rainfall. The latter exercises a preponderant action for the definition of the global drought of the climate. Emberger's quotient is specific to the Mediterranean climate. The quotient Q_2 was calculated by the following formula [42]:

$$Q_2 = [2000R/M^2 - m^2]$$ (2)

Where: Q_2: the pluvio-thermal quotient, R: Average annual rainfall in (mm), M: the average of the thermal maxima of the hottest month in Kelvin, m: the average of the thermal maxima of the coldest month in Kelvin. The Q_2 allowed us to locate our weather stations on the Emberger climagram.

Bagnouls and Gaussen temperature diagram: The ombrothermal diagrams of Bagnouls and Gaussen make it possible to compare the evolution of the values of temperatures and precipitations. On this subject, Emberger specifies: "a climate can be meteorologically Mediterranean, possessing the characteristic Mediterranean pluviometric curve, without being so ecologically or biologically, if the summer drought is not accentuated".

The study region is characterized by minimum temperatures between: - 0.3 and 2.12° C. Le Houérou et al. [45] consider that the Algerian steppes are surrounded by isotherms "m" - 2 and 6° C, and that M-m varies little and remains approximately equal to 32.6–37.9° C. These temperatures explain the absence of certain species whose life is linked to temperate winters.

2.1.3.2 Methods for estimating evapotranspiration

As part of this work, evapotranspiration is calculated for the three meteorological stations in the Naâma region using the Thornthwaite method.

The Thornthwaite Method is an empirical formula for estimating potential evapotranspiration, relatively simple to implement, since it requires few data (average air temperatures, in particular). One of the drawbacks of the Thornthwaite method is its monthly time step for calculating potential evapotranspiration.

Calculation of PET is done by applying Thornthwaite's formula; it is a simple expression suitable for arid climate. This Thornthwaite Method is one of the most widely used formulas for the calculation of evapotranspiration is that of Thornthwaite [46].

It has been tested in several regions of Algeria and in the Mediterranean because it gives acceptable results.

By statistically fitting the results of experimental measurements of the PET to climatological data, Thornthwaite established a non-linear relationship between the mean monthly PET and the monthly mean temperature (Tm) expressed like this:

Potential evapotranspiration: it is the consumption of water, under the combined action of the evaporation of water from the soil and the transpiration of the plant. For its estimation, methods based on climatic variables are used. However, the choice depends mainly on the type of climate data available and the type of climate in the region. However, ETP data for the three stations (Mécheria; Naâma, Ain Sefra) are estimated using the Thornthwaite method [46].

$$ETP = 16(10 \times T/I)^a \, K \qquad (3)$$

Where PET: is the monthly potential evapotranspiration, expressed in mm T: the monthly average temperature of the month considered in degrees Celcius.

a: Coefficient given by the expression: $a = 1.6 \, (I/100)^{+0.5}$.

where the annual thermal index I is equal to the sum of the twelve values of the monthly thermal index: $i = (T/5) \, 1.514$ K: Correction coefficient, which depends on the latitude i: monthly thermal index I: Annual thermal index.

$$I = \sum_{m=1}^{12} i(m) \quad i(m) = \left[\frac{\overline{T}(m)}{5}\right]^{1.514} \qquad (4)$$

Real evapotranspiration (RET) or flow deficit: This is the quality of water that is actually evaporated or transpired by the soil, plants and free surfaces. The RET can be estimated by several methods; for our case we have chosen the Turkish formula [47]:

$$ETR = \frac{P}{\sqrt{0.9 + \left(\frac{P}{L}\right)^2}} \qquad (5)$$

Where: P: designates precipitation in mm.

L: designates a constant dependent on the temperature with $L = 300 + 25\,T + 0.05\,T^3$ andT: is the annual average temperature in °C.

The water balance by the Thornthwaite method

Its purpose is to quantify the water transfers resulting from precipitation, and to characterize a soil from a dryness or humidity point of view.

According to Thornthwaite, the water quality needed for a soil to be saturated is equivalent to a 100 mm depth of water, (this is the generally accepted useful reserve). Still according to Thornthwaite, one can establish a monthly hydrological balance during the period (1990–2014), which makes it possible to estimate for each month: the real evapotranspiration (RET).

3. Results and discussion

3.1 Climate analysis

3.1.1 Rainfall

The climate of the Naâma region is Mediterranean; is characterized by a rainy winter and dry summer. The average annual rainfall for the period from 1990 to 2014 is 243.11 mm in Mécheria. It is 218.75 mm in Naâma, and 199.64 mm in Ain Sefra. The months of July are the driest (5.28 mm for Mécheria 5.64 mm in Naâma and 4.72 mm for Ain Sefra); October is the wettest month (35.79 mm for Ain Sefra, 31.32 mm Naâma and 35.52 mm for Mécheria). The monthly breakdown shows that July and August are the two driest months (5.28 mm for Mécheria, 4.72 mm for Aïn Sefra). On the other hand, the same findings (5 mm for Mécheria, 7 mm for Aïn Sefra, 2 mm Naâma) were recorded for the recent period. On the other hand, the months of October and November are the wettest in the two old and recent periods (43 mm for Mécheria, from 29 to 35.79 mm for Aïn Sefra). The comparison between the rainfall series (1913–1938 and 1990–2014) highlights the nature of the decrease or increase in significant rainfall which is a phenomenon of almost general climatic

evolution and which has affected all of the study region or national territory both
north and south of the country. Analysis of rainfall data (1990–2014) highlights the
nature of the significant decrease or increase in rainfall, which is a phenomenon of
climate change. Analysis of the data shows a decreasing rainfall gradient from north
to south. In the northern part of the high steppe plains (Naâma, Mécheria), the
annual rainfall varies between 200 to 300 mm and in the south (Ain Séfra) the
average annual rainfall is equal to 200 mm /year. In fact, precipitation is generally
concentrated in the autumn season, especially in October in the form of downpours
or thunderstorms. The variability of mean precipitation shows that for the 24-year
series, 4 wet years recorded values below the annual average and 5 dry years. This
latest drought was manifested by rainfall either too low or too irregular during the
year. This period of drought has adverse effects on the steppe environment due to
its long duration. The author Rognon [47] considers that a dry year has a different
effect depending on whether it follows another dry year or a wet year. We know
from the start that the rain regime is irregular in these steppe regions [48]. Several
authors (Despois, [40] and Seltzer [29]), confirm this in their studies. The series of
pluviometric observations is subdivided into two main periods, namely a rainy
period from October to April with a maximum rainfall in October of around 25 mm,
and a second dry period from June to August who's rainfall represents only 11% of
the annual total (0 to 5 mm). On the other hand, the stations which are in the steppe
domain present a structure of precipitation quite different from the Saharan
domain. Indeed, if we consider the long series of Seltzer (1913–1938), precipitation
is mainly concentrated in the winter season. Precipitation of the recent series
(1990–2014), is generally concentrated in the autumn season, especially in October
in the form of downpours or thunderstorms. The variability of the mean precipitation shows that for the series of 25 years, 8 years are considered as wet for the first
period (1913–1938) and 4 years for the recent series (1990–2014), and 17 years are
considered as years dry for the first period and 5 for the second period. In general,
the rainfall remains low, irregular with strong inter-annual variations, it is heterogeneous in time and space, this irregularity of frequencies confirms the appearance
of dry periods which raged in the region during the years 1992, 1995, 1998, 1999,
2001, 2002, 2004 and 2013. All the indicators converge towards a persistent
drought, even if significant rainfall episodes occur they do not manage to fill the
deficit to reverse the trend.

3.1.2 Seasonal regime

In general, precipitation is unevenly distributed during the seasons, as shown in
Table 6. The most important precipitations are those which fall in autumn and
spring, compared to that of winter, although that the latter constitute a significant
contribution (**Table 6**). The table below shows the calculated seasonal regime of
stations in the study region for the two periods.

Period	P1 (1913–1938)					P2 (1990–2014)				
stations	Sp	S	A	W	Regime	Sp	S	A	W	regime
Mécheria	81,6	27,2	84	85.3	WSpAS	75,25	26,02	88,84	52,97	ASpWS
Naâma	—	—	—	—	/	61,68	34,45	81,52	41,09	ASpWS
Aïn Sefra	38	43	73	38	ASWSp	58,69	23,72	80,06	37,16	ASpWS

*Autumn (A); Winter (W); Spring (Sp) and Summer (S).

Table 6.
Seasonal rainfall patterns of the old period.

The analysis of the climatic variability of rainfall totals is due to the spatio-temporal seasonal and annual variability of rainfall; this indicates a change in the climate of the study region (**Table 6**). In general, the rainfall is slightly different, where the autumn maximum is constant; some variations show transformations in the seasonal distribution of rainfall. Dominant autumn rains prevail over most of the study area, but the seasonal pattern may be locally modified slightly. The most remarkable fact is that the raising of altitudes characteristically resuscitates the arid climate. The dominant autumn rains prevail over most of the study region, but the seasonal pattern may locally undergo slight modifications between the 2 periods. This variation in seasonal regimes is explained by its essentially orographic character. The most remarkable fact is that the raising of altitudes characteristically resuscitates the arid climate. From south to north/east the formula becomes: AHAE / AHPE /HPAE/AEHP for the old period (1913–1938).

On the other hand, in the recent period (1990–2014), the most remarkable is the consistency of the APHE-type regime for the majority of the study stations. This transition to the dominant autumn rains is indicative of an accentuation of the oceanic character of the climate. This indicates that the rainfall has therefore increased during the cold season, and summer tends to become the dry period. Consequently, the current seasonal regimes (P2) are markedly changed this is explained by their "degree of continentality".

The distribution of precipitation appears in the study region as an essentially orographic phenomenon: the isohyets reflect the relief. The Tellian and Saharan Atlas plays a much clearer role as a barrier between maritime and continental influences. A succinct explanation of the rains is needed to understand the seasonal variations. In this area the rainy season lasts from 4 to 6 months with some rare local variations, the orientation of the winds appears essential.

Thus, over the past 25 years, the entire study region has been subject to the autumn or winter maximum. The autumn rainy season is prolonged there until December and even January. Here the influence of the relief regenerating the oceanic rain regime is evident. The rainfall figure rises with greater intensity during the winter season: it is therefore a question of relief precipitation. Overall, the evolution of annual precipitation and rainy seasons shows a very moderate decreasing trend between the 2 periods and the following ones, in agreement with observations made at the regional scale. These changes have had repercussions on the vegetation that occurs during the seasonal course of precipitation. The low rainfall is a characteristic of the Saharan climate. However, this region is poorly watered; rainfall is scarce and irregular, often brief (showers), but of high intensity, causing violent floods. The study of seasonal variability is essential, to see if the decrease or increase in rainfall is specific to a particular season or to several seasons, it allows to better visualize the chronology of the seasonal rainfall totals over time. The analysis of monthly average rainfall data makes it possible to better visualize the distribution of the quantities of water recorded at each station and for each month of the year.

3.1.3 Temperatures

Temperatures are an important element for plant life, especially the two extremes: the average of the coldest month's lows and the hottest month's average lows.

Temperatures represent an important element for plant life, especially the two extremes: the average of the minimums of the coldest month and the average of the maximums of the hottest month. We notice a significant increase in maximum temperatures between the two periods; therefore the series of maxima experiences a clear increase which affects all the months of the year, this situation is reflected at the monthly level where the rise in temperatures fluctuates between 0.3°C to 1.5°C inducing to the annual scale an average increase of 0.5° C. This indicates a more marked global warming of the study region. This change in temperature is manifested by consequences on the metabolism and development of fauna and flora, growth, respiration, the composition of plant tissues and the mechanisms of photosynthesis (**Table 7**).

The highest temperatures are generally recorded in July for the three reference stations. The analysis of the maxima highlights the notion of climatic aridity which tends to strengthen from north to south of the region (**Table 7**). The period of high temperatures, lasting from June to October, can cause scalding due to increased sweating. Therefore, the hottest month of the year for the two thermal series (1913–1938 and 1990–2014) is that of July and August with an average temperature of 29.9° C. (Mécheria) at 36.48° C (Naâma). The analysis of the maxima emerges the notion of climatic aridity which tends to strengthen from north to south of the region, so the average thermal amplitude between the southern and northern zones of the region reaches approximately 0.84° C. This value relative to the spatial extent (in the North–South direction) of the region is relatively high. For the period of low temperatures, from November to February, are at the origin of the intensity of winter frosts which can result in vegetative damage such as necrosis. So the coldest and most severe month is that of January for all the stations during the two thermal study series. On the other hand, the minimum series is experiencing a sharp increase affecting all months of the year with the exception of August. This situation is reflected at the monthly level where the rise in temperatures fluctuates between 0.3° C to 1.5° C inducing on an annual scale an average increase of 0.5° C.

3.1.4 Wind

In the arid region, winds have played and still play a major role in the degradation of vegetation and soil destruction and the building of constrained dune systems; they constitute a permanent threat to biodiversity and infrastructure. Therefore, the wind can reach considerable speeds allowing it to exert erosive actions on the ground by the drying out of the superficial parts of the ground.

3.2 Calculation of the different climatic parameters

3.2.1 Bioclimatic indices

3.2.1.1 De Martonne's aridity index

The **Table 8** below shows the average annual temperature, the average annual precipitation and the aridity index calculated for the stations during the two periods.

Stations	m °C		Thermal gap	M °C		Thermal gap
	P1 (1913–1938)	P2 (1990–2014)		P1 (1913–1938)	P2 (1990–2014)	
Mécheria	1,5	1,49	−0,01	35,1	36,78	0,74
Naâma	—	0,32	0,32	—	36,8	36,8
AïnSefra	−0,3	0,57	0,87	37,6	38,34	1,68

Table 7.
Thermal differences between P1 (1913–1938) and P2 (1990–2014).

Stations	Period	R (mm)	T° average	De Martone index	Type of climate
Mecheria	1913–1938	278,1	15.9	10,7	Semi-arid
	1990–2014	243,10	16,65	9,12	Steppic
Naâma	1913–1938	/	/	/	/
	1990–2014	218,75	16,38	8,29	Steppic
AïnSefra	1913–1938	192	15.50	7,53	Desert
	1990–2014	199,63	17,66	7,22	Desert

Table 8.
De Martonne's aridity index.

The comparative analysis of the De-Martone aridity index between the two periods allows us to advance that the study region is strongly marked by increasing aridity which is accentuated from North to South. This is due to the drought induced by the decrease in rainfall and the increase in minimum and maximum temperatures (case of 2001 when the recorded rainfall was 60 mm in Aïn Sefra ...). The values of the aridity index obtained are respectively 8 and 11 depending on the geographical position of the study stations. In the steppe space, for the stations of Mécheria and Naâma are characterized by a semi-arid to steppe climate. In the stations which are in the central part of the region, the Saharan Atlas (Aïn Sefra) the index is 7.53 and reflects a desert-like climate.

3.2.1.2 Thermal continentality

A comparative analysis of the two series (1913–1938) and (1990–2014) recorded at station level (**Table 9**), shows us that the region experiences a contrasting thermal regime, of a continental type. Indeed, the annual thermal amplitude of average temperatures is 30° C to 40° C depending on the North–South orographic gradient. The average seasonal difference can reach more than 30°C, thus promoting soil degradation by the relaxation of friable rocks in terms of erosion in the forms of wind and water erosion.

3.2.2 Climate synthesis

3.2.2.1 The pluviothermal quotient

The mean annual temperatures, the mean annual precipitation, and the calculated rainfall quotients (Q_2) are presented in the following **Table 10**.

Stations	Period	M (°C)	m (°C)	M – m (°C)	Thermal continentality
Mecheria	1913–1938	35,1	1,5	33,6	Semi-continental climate
	1990–2014	36,78	1,49	35,29	Semi-continental climate
Naâma	1913–1938	—	—	—	—
	1990–2014	36,8	0,32	36,48	Continental climate
AïnSefra	1913–1938	37,6	−0,3	37,9	Continental climate
	1990–2014	38,34	0,57	37,77	Continental climate

Table 9.
The thermal continentality of the study stations.

Stations	Period	R(mm)	M(°C)	m (°C)	Q_2	Bioclimatic stage
Mécheria	1913–1938	278,1	35,1	1,5	28,4	Arid Greater Than Cool Winter
	1990–2014	243,10	36,78	1,49	23,6	Arid Greater Than Cool Winter
Naâma	1913–1938	—			—	
	1990–2014	218,75	36,8	0,32	20,6	Medium arid to Cool winter
Aïn Sefra	1913–1938	192	37,6	−0,3	17,4	Saharan Superior to Cold Winter
	1990–2014	199,63	38,34	0,57	18,1	Upper Saharan in Cool Winter

Table 10.
Values of the rainfall quotient.

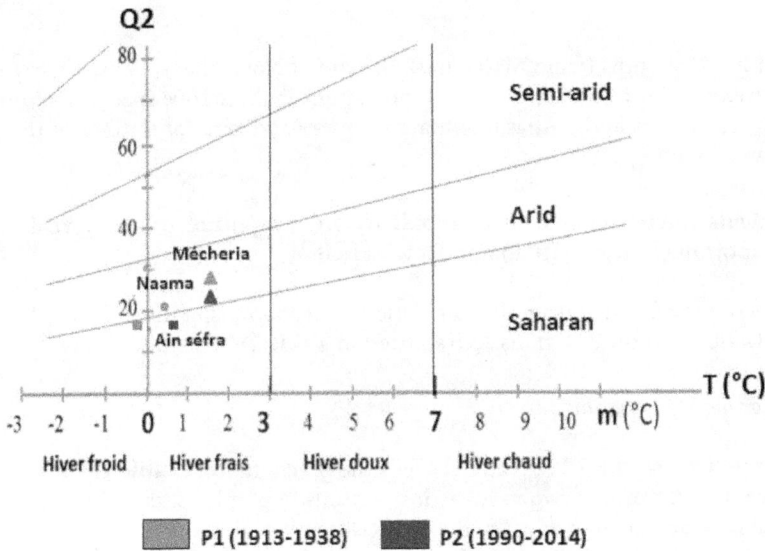

Figure 4.
Variation of the Emberger climagram, of study stations.

The quotients are inversely proportional to the aridity; this Emberger climagram allows us to determine the bioclimatic stages and the thermal variants (**Figure 4**).

The comparative reading of the pluviothermal climagram (**Table 10** and **Figure 4**) shows a slight change in pluviothermal quotients between the old period and the new period. This type of climate change probably also causes a change in plant formation. For example, the Aïn Sefra station moves from the arid lower level with cold winter in the old period to the upper Saharan level with cool winter in recent times. The Mécheria station is moved from the middle arid stage with cool winter in the old period to the lower arid stage with cool winter in recent times.

3.2.2.2 Ombrothermal diagram of Bagnouls and Gaussen

The analysis of the various ombrothermal curves (**Figure 5**) of the stations compared between the two periods (1913–1938) and (1990–2014), allows us to observe a period of drought varies from 5 to 7 months or more (from the month from June to September) in the resorts of the northern part of the region (steppe plains: Mécheria and Naâma). On the other hand, in the central part of the region (Saharan Atlas: Aïn Sefra), it has a fairly prolonged period of drought that varies

T°: Temperature (°C)   R: Rainfall (mm)

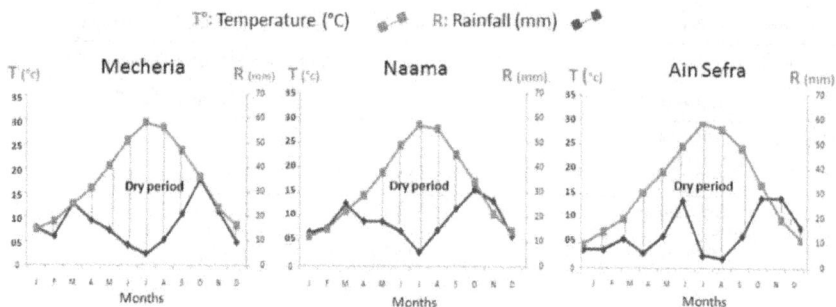

Figure 5.
Ombrothermal diagrams of the study stations.

from 10 to 11 months (from March until the end of November). Thus, a fairly short wet period; varies from 4 to 6 months for stations in the steppe space, from one month for stations in the Atlas mountainous space and zero for stations in the Saharan domain.

3.3 Calculation of the hydrological balance and evapotranspiration, real evapotranspiration (RET) and flow deficit

The results obtained from the calculation of evapotranspiration (PET, RET, EUR, Deficit) for the 3 stations are reported in **Table 11**.

3.3.1 For the Mécheria station

We notice that the PET greatly exceeds the precipitation (**Table 11**); and we observe the existence of two very distinct seasons. A surplus season during which rainfall is greater than or equal to the PET (December–February) and the deficit season from March to November. During the cold season, precipitation covers the needs of potential evapotranspiration and allows the formation of Easily Usable Reserve (EUR). From the month of March we have an exhaustion of the EUR which results in an agricultural deficit.

3.3.2 For the Naâma station

We note that from November the precipitation is greater than the evapotranspiration (R > PET) (**Table 12**). The Easily Usable Reserve (EUR) reaches its maximum in January, February and March. From the month of March, we record an agricultural deficit of 4.72 mm and which reaches its maximum in July with 181.05 mm. The annual deficit is estimated at 677.89 mm, and the actual evapotranspiration (RET) is equal to 207.93 mm or 95.05% of precipitation.

3.3.3 For the Aïn Sefra station

We see that precipitation is less than evapotranspiration throughout the year (R > PET) except in January when 15.65 mm of precipitation is recorded (**Table 12, Figure 6**). The Easily Usable Reserve (EUR) is zero throughout the year. We record an agricultural deficit throughout the year with a minimum of 3.07 mm in February and a maximum of 190.3 mm in July. The annual deficit is of the order of 746.43 mm. The actual evapotranspiration (RET) is equal to 194.53 or 97% of precipitation.

Station of Mécheria (latitude 33°N, I = 80,03, a = 1,78)

Months	J	F	M	A	M	J	Jt	A	S	O	N	D	Total
T(°C)	6,95	8,06	11,31	14,12	22,91	23,87	27,9	27,12	22,03	17,01	11,01	7,54	16,65
R (mm)	18,49	17,88	27,95	26,24	21,05	10,6	5,28	10,14	26,34	35,52	26,98	16,62	243,1
I	1,64	2,06	3,44	4,81	10,01	10,66	13,5	12,93	9,44	6,38	3,3	1,86	80,03
K	0,88	0,86	1,03	1,09	1,19	1,2	1,22	1,15	1,03	0,97	0,88	0,86	
PET	12,44	16,2	29,61	43,95	104,03	111,91	147,74	140,47	97,02	61,23	28,23	14,38	807,21
ETPc (mm)	13,32	17,06	30,64	45,04	105,22	113,11	148,96	141,62	98,05	62,2	29,11	15,24	819,58
R-PETc (mm)	5,17	0,82	-2,69	-18,8	-84,17	-102,5	-143,7	-131,5	-71,71	-26,68	-2,13	1,38	-576,75
RET (mm)	13,32	17,06	27,95	26,24	21,05	10,6	5,28	10,14	26,34	35,52	26,98	15,24	235,72
EUR (mm)	5,17	0,82	0	0	0	0	0	0	0	0	0	1,38	7,37
Deficit (mm)	0	0	2,69	18,8	84,17	102,51	143,68	131,48	71,71	26,68	2,13	0	583,85

Station of Naâma (Latitude 33°N, I = 78,73, a = 1,75)

Months	J	F	M	A	m	J	Jt	A	S	O	N	D	Total
T(°C)	6,07	7,49	11,13	14,35	18,97	24,56	28,61	27,79	22,64	17,2	10,66	7,13	16,33
R (mm)	13,57	15,37	25,47	18,15	18,05	14,18	5,64	14,61	23,51	31,32	26,68	12,14	218,75
I	1,34	1,84	3,35	4,93	7,52	11,13	14,02	13,42	9,84	6,49	3,14	1,71	78,73
K	0,88	0,86	1,03	1,09	1,19	1,2	1,22	1,15	1,03	0,97	0,88	0,86	
PET	10,14	14,66	29,32	45,74	74,55	117,15	153,03	145,43	101,6	62,81	27,19	13,45	795,07
ETPc (mm)	8,92	12,6	30,19	49,85	88,71	140,58	186,69	167,24	104,64	60,92	23,92	11,56	885,82
R-PETc (mm)	4,65	2,77	-4,72	-31,7	-70,66	-126,4	-181,1	-152,6	-81,13	-29,6	2,76	0,58	-667,13
RET (mm)	8,92	12,6	25,47	18,15	18,05	14,18	5,64	14,61	23,51	31,32	23,92	11,56	207,93
EUR (mm)	4,65	2,77	0	0	0	0	0	0	0	0	2,76	0,58	10,76

Station of Naâma (Latitude 33°N, I = 78,73, a = 1,75)

Months	J	F	M	A	M	J	Jt	A	S	O	N	D	Total
Deficit (mm)	0	0	4,72	31,7	70,66	126,4	181,05	152,63	81,13	29,6	0	0	677,89

Station of Ain Sefra (Latitude 32°N, I = 86,99, a = 1,89)

Months	J	F	M	A	M	J	Jt	A	S	O	N	D	Total
T (°C)	7,44	9,12	12,69	15,95	20,6	25,84	29,53	28,53	23,85	18,24	11,9	8,29	17,66
R (mm)	15,65	11,57	25,5	18,67	14,52	8,31	4,72	10,68	21,51	35,79	22,75	9,52	199,6
I	1,82	2,48	4,09	5,79	8,53	12,02	14,71	13,96	10,64	7,09	3,71	2,15	86,99
K	0,89	0,86	1,03	1,08	1,19	1,19	1,21	1,15	1,03	0,98	0,88	0,87	
PET	11,9	17,49	32,66	50,32	81,6	125,24	161,18	151,02	107,64	64,84	28,92	14,6	847,41
ETPc (mm)	10,59	15,04	33,63	54,34	97,1	149,03	195,02	173,67	110,86	63,54	25,44	12,7	940,96
R-PETc (mm)	5,06	-3,07	-8,13	-35,67	-82,58	-140,7	-190,3	-163	-89,35	-27,75	-2,69	-3,18	-741,37
RET (mm)	10,59	11,97	25,5	18,67	14,52	8,31	4,72	10,68	21,51	35,79	22,75	9,52	194,53
EUR (mm)	5,06	0	0	0	0	0	0	0	0	0	0	0	5,06
Deficit (mm)	0	3,07	8,13	35,67	82,58	140,72	190,3	162,99	89,35	27,75	2,69	3,18	746,43

T: Temperatures(°C); R: Rainfall (mm); PET: Potential evapotranspiration (mm); RET: Real evapotranspiration (mm); EUR: Easily Usable Reserve (mm).

Table 11.
Calculation of PET by the Thornthwaite method for the study stations (1990–2014).

Station	R (mm)	T (°C)	L	RET (mm)	Conclusion
Mécheria	243,108	16,65	947,03	248,06	RET > R (Rainfall) but noticeably close
Naâma	218,75	16,38	929,24	225,51	
Aïn Sefra	199,63	17,66	1016,88	207,94	

Table 12.
Results of real evapotranspiration (RET) according to the Turkish method.

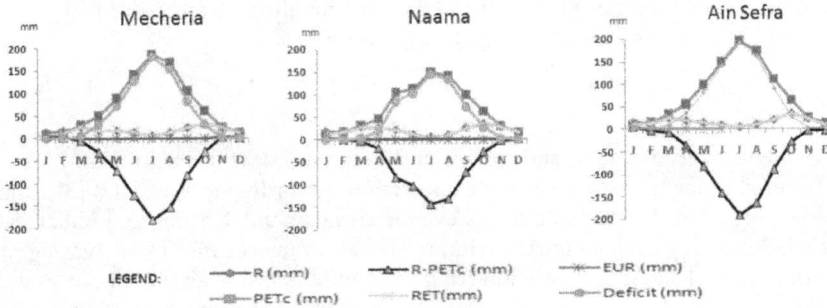

Figure 6.
Graphical representation of the water balance, mean monthly evapotranspiration according to Thornthwaite.

Examination of the graphs (**Figure 6**) shows that on an annual scale, PET greatly exceeds precipitation and on a monthly scale, there are two very distinct seasons. a surplus season during which precipitation is greater than or equal to the ET from November to March and a deficit season from April to October. Potential (PET) and actual (RET) evaporation vary considerably between ecosystems and sometimes according to seasons. During the cold season, the precipitation covers the needs of the potential evapotranspiration and allows the formation of the RFU from where the vegetation appears. The two curves follow the same trend (**Figure 6**). The period from May to September correlates with the deficit period shown in the ombrothermal diagrams. The actual evapotranspiration (REE) is very low, as the lack of water available for the soil and plants due to drought is a limiting factor.

3.4 Effect of climatic factors on evapotranspiration

According to Emberger [42], the climate in the Mediterranean region is based on "the climatic characteristics which most strongly influence plant life". In the steppe region where water is a limiting factor, evaporation is very high and reaches its maximum in summer. The slice of water evaporated annually is almost always greater than the total amount of rain that has fallen [29].

In the study area, we found that the PET is significantly higher than the rainfall received. Thus, we consider that this period is a sequence of water deficit (drought) for spontaneous vegetation. To this end, the dominance of PET generates and/or promotes the process of soil degradation and more particularly the silting up of croplands and steppe rangelands [25].

3.4.1 Effect of aridity

The Naâma region corresponds to an arid area, more or less nuanced according to the orography, and the level of the relief and the capacity of the substrates to retain water from precipitation are low. According to Mjejra [17], in any region marked by aridity, the potential evapotranspiration loss represents 60–80% of the rainfall input.

3.4.2 Temperature effect

Temperature variation between stations shows a period of high temperatures, spanning from June to October, which can cause scalding due to increased transpiration. Periods of low temperatures, from November to February, are the cause of the intensity of winter frosts which can result in vegetative damage such as necrosis. This indicates by Floret & Pontanier [49, 50], the highly contrasted thermal regime is affected by a strong potential evapotranspiration.

3.4.3 Wind effect

Winds are very frequent and violent in the study area, which significantly contributed to the increase in evapotranspiration. According to Escadafal [51], the often strong winds further exacerbate evaporative demand. Khader [52] indicates that wind is a very drying climatic parameter that influences PET by increasing the temperature and simultaneously lowering the humidity of the air which causes it. It accelerates the desiccation of plants, and the increase of evapotranspiration. In the case of the hot and dry southerly "Sirocco" wind blows especially in summer, on average 200 times a year, and lasts more than 45 days a year, accentuating the dry season and bringing back appreciable quantities of sand. This wind causes the soil to dry out by causing a strong evapotranspiration of the plants [53].

3.4.4 Effect of rainfall

According to Derouiche [54, 55], the decrease in rainfall and the increase in temperature represent unfavorable factors for both the soil and the plant. Precipitation cannot compensate for the intense evapotranspiration to which vegetation is subjected during the summer season. The deficit is only made up by the soil's water reserves according to its capacity to store the precipitation it receives.

3.5 Effect of evapotranspiration on plants of steppe

In the arid region of Naâma, the steppe plants are characterized by a low quotient of potential Precipitation/Evapotranspiration. Thus, the potential evapotranspiration is very high due to heat and sunshine, so rain is especially needed when evapotranspiration is high and precipitation is not sufficient for the normal development of the plant.

The decrease of evapotranspiration leads to the change of the surface energy balance, to an increase of temperatures and to a decrease of the soils capacity to store water for vegetation. Evapotranspiration cools the air through the evaporation of water present in the soil and plants as well as transpiration in the leaves. The climatic aridity has a considerable influence on the growth of steppe plants, because vegetation modifies the water balance of the substrate where it grows, by taking water that is lost through transpiration [25].

Maximum temperatures accentuate water stress; in fact excessive heat causes dehydration resulting from accelerated perspiration. If the soil cannot provide sufficient water supply, there is a loss of turgor. In vegetation, potential transpiration increases with temperature and climatic drought. Therefore, vegetation can act as a brake on the diffusion of water vapor [55]. It helps reduce soil evaporation, reducing net radiation and reducing surface temperature [56]. Vegetation can also decrease the amount of solar radiation reaching the soil and the temperature of the soil, which can significantly reduce evaporation compared to bare soil [57].

Stomatal regulation is influenced, degree of opening of the stomata depending on climatic factors of evapotranspiration. As soon as a water deficit occurs, the plant adjusts, quickly and reversibly by the process of transpiration, that is to say the water inputs, are carried out at a rate lower than the thermal needs of the plant, the flows of water which cross it by the closing of its stomata (small openings of the leaves, which regulate the gas exchanges between plant and atmosphere).

3.6 Effect of steppe vegetation on evapotranspiration

Plants then invest in "survival" by reducing the phenomena of evapotranspiration, photosynthetic leaf surfaces, in times of drought. It takes place at the level of the stomata of the leaves by reducing their exchange surfaces and closing their stomata. It turns out that vegetation can have an effect on different components of the water balance. As soon as the water conditions at the root level evolve towards drought, the leaves react by closing their stomata, at the same time reducing evaporation, which has the effect of increasing their surface temperature [55, 58]. Because water stress at the roots has a repercussions on the evapotranspiration regime of the leaves.

According to Le Houérou and Popov [59], the reduction in the maximum daily temperature (2.5°C) by woody vegetation corresponds to a decrease of about 147 mm/year of PET at ground level.

The woody tree cover (*Retama retam, Pistacia atlantica*) directly reduces solar radiation, air temperature and wind speed on the ground, which can reduced the potential evapotranspiration (PET). Indeed, the natures of the vegetation, woody species consume more water by evapotranspiration than herbaceous species.

Vegetation can increase evapotranspiration through transpiration. This can increase water loss through evapotranspiration. This explains by Carminati *et al.* [60], vegetation can influence evapotranspiration in several ways: vegetation can act on the energy state of soil water through transpiration, which has a linear relationship with the suction exerted by the xylem in dry soils.

According to Yagoub (2016) [54], vegetation regulates surface temperature by absorbing radiant energy and re-emitting it as latent heat via the process of evapotranspiration. Among the regulatory mechanisms, plants are reacted by the reduction of aerial organs to reduce the evaporating surface and the taking of reduced forms (reduction of the leaf system, thorns, hairs, etc.) and the distribution and arrangement of the leaves of a plant structure can act on climatic parameters linked to evapotranspiration (wind speed, solar radiation). In fact, in the underground part, the root system can play a more important role than that of the hydrogeological properties in the useful water reserve.

3.7 Mechanism of adaptation and acclimatization of steppe vegetation

Due to the intensity of evaporative transpiration, the steppe vegetation adapts to withstand the harsh climatic conditions. The difficult climatic conditions, in this steppe area, allow the vegetation to develop an adaptation system for its maintenance and survival. These ecophysiological relationships can largely explain the adaptation of steppe species to the arid Mediterranean climate. Despite the very harsh and very restrictive environmental conditions, there are still geomorphological zones offering more or less favorable conditions for the survival and proliferation of a characteristic spontaneous flora adapted to climatic hazards. These adaptations have shown that steppe vegetation adapted to ecological stress uses one or more mechanisms to compensate for the inadequate water balance and mitigate the effect of water deficit. They cover the physiological and morphological regulations that allow plants to adapt to a deficient water supply occurring at different scales [61].

3.7.1 Biological adaptation

Biological types are considered as an expression of a flora adaptation strategy to environmental conditions [62], which represent a privileged tool for the description of the physiognomy of vegetation. Emberger [63] affirms that the rate of therophytes increases with the aridity of the environment. Therophysation is a characteristic of arid areas; it expresses a strategy of adaptation under unfavorable conditions and a form of resistance to climatic rigors [64–66]. Therophytes are more resistant to summer drought than hemicryptophytes and geophytes, since they pass summer as seeds while the others remain as vegetative organs.

According to Raunkiaer [65]; Floret *et al.* [67], chamaephytes are the best adapted for low temperatures and aridity and the absence of these spacies testifie the anthropization of the environment [68–70].

They partially reduce their organs of perspiration and assimilation in the summer, and can develop some forms of adaptation to drought (reduction of the leaf area) as well as by the development of the root system with the proliferation of thorny species such as *Astragalus armatus, Atractylis serratuloides* characteristic of steppe areas [35].

In hemicryptophytes, the perennial organs located at soil level are protected by leaf sheaths (sometimes reduced to fibrils) or old withered leaves as in *Plantago ovata, Echium trygorrhizum*. The roots, sometimes considerably developed, are often sheathed with a thick tomentum which fixes the grains of sand and thus protects them from desiccation (*Malva aegyptiaca, Paronychia arabica*) [35].

The majority of species of this type are nanophanerophytes or shrub-like pseudo-steppes 1 to 4 m long, including *Nerium oleander, Genista saharae, Retama raetam, Retama sphaerocarpa, Rhus tripartita, Tamarix articulata, Ziziphus lotus*. These types of steppes generally occupy sites with a relatively favorable water balance: terraces of the hydrographic network, dayas, cliffs, deep sandy substrates in topographic position. They often constitute relatively favorable environments [35].

3.7.2 Morphological adaptation

For steppe plant formation, evapotranspiration depends on the leaf area and the stage of development of the plants. The perennial steppe vegetation is therefore adapted to morphological modification during the plant's development stage: Among these forms of adaptation we can cite the decrease in leaf area (plants can have small very thick leaves or reduced to thorns, which allows them to limit their water losses (*Fagonia glutinosa, Fagonia latifolia, Zilla spinosa, Launaea arborescens*). One of the strategies is leaf modification (leaf drop, or leaf area reduction), by microphyllia, most steppe species have very small leaves *Salsola vermiculata* or small *Rhanterium suaveolens* [71].

The flattening of vegetative system is fixed on the ground (*Neurada procumbens*).

Cushion formation (pincushion habit) is also a form of adaptation to the xeric environment with a morphological modification, for example the species (*Anabasis aretioides, Teucrium polium, Astragalus armatus, Atractylus humilis*), can take a ball appearance or pincushion and prickly padded xerophyte includes the species Oxyhedron (*Juniperus oxycedrus*). Some plants may have considerably developed underground organs (Rhizomatous) (*Scorzonera undulata*). The leaves can also take shape in needles or scales; it is the case of the following species *Hammada scoparia, Hammada schmittiana, Thymelaea microphylla, Ephedra alata, Genista saharae, Retama retam* [35].

According to Ozenda [72], for the case of amaranthaceae (*Zygophyllum album, Gymnocarpos decander*), are thus carriers of tiny leaves or even are completely leafless, sometimes the leaves are transformed into thorns to constitute reserves in accumulating water in the tissues (crassulescent leaves or at least semi-succulence and aquiferous tissues). The constitution of water reserves in the tissues and color change where the whitish appearance is the most representative: *Thymelia microphylla, Phoenix dactylifera, ...*

A form of sclerophyllia (extravaginate innovations at upper nodes) remarkable in some of the species of Poaceae (*Panicum turgidum*); these innovations protect against wilting [73]. Some trees adapt to the cold by loss of leaves (*Pistacia atlantica, Ziziphus lotus*), others will opt for leaves in needles (*Juniperus oxycedrus, Juniperus phoenicea, Rosmarinus officinalis*) [35].

Thus, the reduction of the vegetative apparatus constitutes a remarkable adaptation to very difficult environmental conditions. This results in the tolerance of certain ligneous plants which opt for a morphological plasticity which reflects the capacity for resilience in response to disturbances of biotic or abiotic origin. These species bury their woody structures below ground level or spread their root system on supports with greater water availability (*Pistacia atlantica, Ceratonia siliqua, Hammada scoparia*, etc.) [35].

3.7.3 Physiological adaptation

According to Scheromm [74], long water deficits result in progressive changes in the structure of the plant, which aim to reduce its transpiring surface (leaf surface, thickening of the cuticles), but which also induce a decrease in its production. Long water deficits induce more irreversible changes, especially in morphology (reduction of evaporation surfaces).

Thickening of leaf cuticles was reduce the rate of evaporation. The leaf surface is covered with a cuticle formed from cutin embedded in a cuticular wax matrix. It therefore reduces the evaporation of water from the surface of the epidermis. Sometimes the plant spends the dry season as a fleshy bulb or rhizome or as a seed (Therophytes) [75].

The increase in the root system increases biomass and consequently transpiration and reduces evaporation from the soil [76]. The significant growth of the root system compared to the aerial system is drawn from the moisture from the depths [77].

The significant development of the root system, was both on the surface and deeper through taproots (*Hammada scoparia*). This modification is manifested by a horizontal extension of the root system (psammophytes) with a horizontal network of roots and especially of rootlets almost in contact with the soil surface to benefit from the slightest rain or dew. For example, the species *Stipagrostis pungens*, however, has strong vertical roots for anchoring and draws from certain moisture from the depths. Or by a vertical extension of the root system is constituted by a taproot in a fairly large number of perennial species (*Moricandia arvensis, Scorzonera undulata, Astragalus armatus ...*). Some perennial species may have roots capable of exploring horizons with several meters deep and most often up to the water table. As it progresses in depth to reach the moisture, the root system will present a particularly dense network of rootlets that colonize the inter-leaf spaces of the soil, this is the case of: *Helianthemum hirtum, Hammada scoparia, Hammada schmittiana, Anabasis articulate* and *Astragalus armatus* (**Figure 7**) [78, 79].

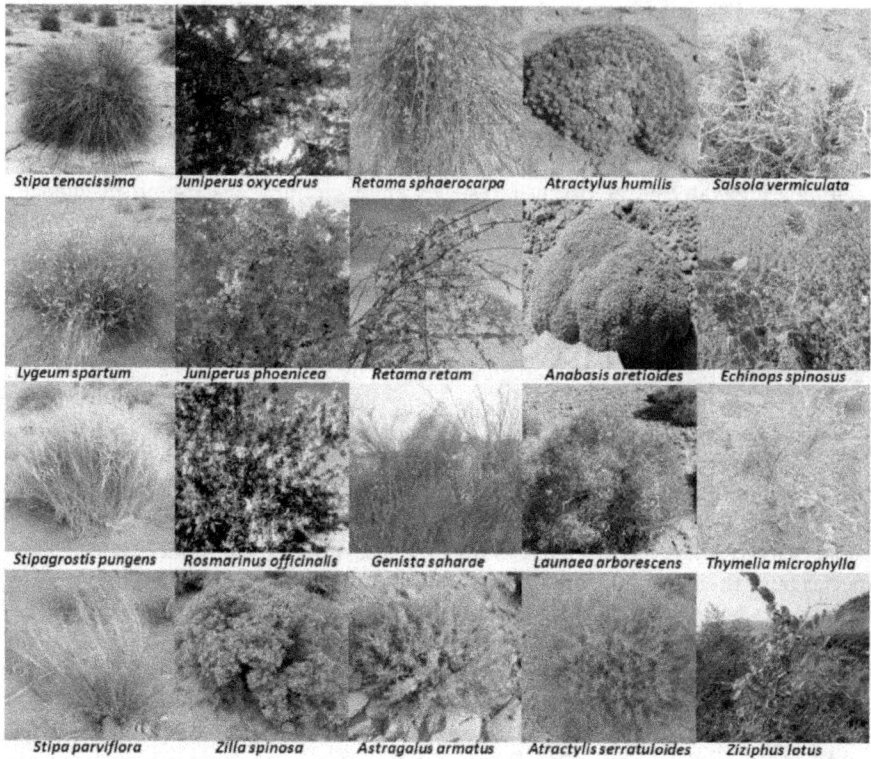

Figure 7.
Some steppe species from the Naâma region (western Algeria).

4. Conclusion

The study region receives an average annual rainfall of less than 300 mm, which explains it's belonging to the arid bioclimatic stage of climate. Precipitation has experienced a very marked interannual irregularity in recent years. The thermal amplitudes were lead to a much faster dieback of annual plants, subjected to intense evapotranspiration.

Climatic data from stations in the study region allowed us to observe the spatio-temporal evolution on a North–South gradient that depends on the irreversible phenomena such as aridity, evaporation (drying out of soils).

This work analyzes the variability of evapotranspiration for steppe vegetation. It is based on climate data measured at three meteorological stations in Naama region. The calculation of evapotranspiration by the water balance method, gave rather satisfactory results insofar as these results oscillate around the normal values of evapotranspiration for the three stations in the study region. Potential evapotranspiration (ETP) data estimated from Thornthwaite's method for the three stations (Mécheria, Naâma and Ain Sefra). The annual average value of potential evapotranspiration is of the order of 807 mm in Mécheria, 795 mm in Naâma de and Ain sefra at 847 mm. It is clearly 3 to 4 times higher than the value of the rainfall received. For this purpose, the PET generates a water deficit (drought) and/or favors a considerable influence on the soil and the growth of vegetation in the steppe ranges. In this steppe area of Naâma, the average annual precipitation is less than two thirds of the potential evapotranspiration (potential evaporation from the ground plus transpiration by plants). The high evapotranspiration confirm the climate aridity of the study area.

The vegetation can have an impact on the water balance by increasing evapotranspiration and reducing runoff, and the vegetation is characterized by various morphological, physiological adaptations such as xerophytes.

It is therefore easy to understand why most steppe species have low woody and herbaceous plants, and they have characteristics of xeropmorphism (*Stipa tenacissima, Lygeum spartum, Stipagrostis pungens*)) and sclerophyllia, especially in phanerophytes (*Quecus ilex, Pinus halepensis, Juniperus phoenicea, Juniperus oxycedrus, Pistacia atlantica*). They have as a result low size of plants and leaves, and low gross production. These plants reduce their exchange surfaces and close their stomata.

In perspective, it is necessary to assess this component of the water balance precisely from field measurements and to establish maps taking into account the particularities of the existing vegetation. This type of study makes it possible to formulate recommendations to better understand the species adapted to these climatic rigors.

Acronyms and abbreviations

HCDS	High Commission for the Development of the Steppe
ET	Evapotranspiration
EUR	Easily Usable Reserve
MET	Maximum evapotranspiration
NMO	National Meteorological Office
PET	Potential evapotranspiration
RET	Real evapotranspiration

Author details

Abdelkrim Benaradj[1*], Hafidha Boucherit[1], Abdelkader Bouderbala[2] and Okkacha Hasnaoui[3]

1 Laboratory of Sustainable Management of Natural Resources in Arid and Semi-Arid Areas, SALHI Ahmed University Center of Naâma, Algeria

2 Department of Earth Sciences, University of Khemis Miliana, Algeria

3 Laboratory of Plant Ecology and Management of Natural Ecosystems –Tlemcen, Dr. Tahar Moulay University of Saida, Algeria

*Address all correspondence to: kbenaradj@yahoo.fr

IntechOpen

References

[1] Boucif H. Contribution à l'étude de la productivité des parcours steppiques de la région de Tlemcen. Mémoire de Master en Foresterie. Département des Sciences de l'Agronomie et des Forêts. Faculté des Sciences de la Nature et de Vie et des Sciences de la Terre et de l'Univers. Université Abou Bekr Belkaid – Tlemcen. 2014. 65p.

[2] Carrière M., Toutain B. Utilisation des terres de parcours par l'élevage et interaction avec l'environnement. Outils d'évaluation et indicateurs. CIRAD/IEMVT, 1995, 103p.

[3] Ben Khouya T., Évaluation de l'impact de la végétation sur le bilan hydrique d'un recouvrement monocouche avec nappe phréatique surélevée. Mémoire de la maîtrise en sciences. Université du Québec en Abitibi-Témiscamingue. 2020. 245p.

[4] Sandu, I., Mateescu, Elena, Vatamanu, V. Schimbări climatice în România şi efectele asupra agriculturii, Editura SITECH Craiova, 2010. 392 p.

[5] Calanca P., Holzkämper A. Conditions agro-météorologiques du Plateau suisse de 1864 à 2050. Recherche Agronomique Suisse 1(9), 2010. 320-325.

[6] Amri R. Estimation régionale de l'évapotranspiration sur la plaine de Kairouan (Tunisie) à partir de données satellites multi-capteurs, Thèse de doctorat de l'Université de Toulouse, 2013. 176p.

[7] Ferchichi A. Etude climatique en Tunisie présaharienne. Medit, vol 7, n.3, 1996. 46-53.

[8] Cosandey Claude, Robinson Mark, 2012. Chapitre 3 : Évaporation et évapotranspiration. Ouvrage d'Hydrologie continentale. Paris, Armand Colin, « U », p. 99-148. DOI : 10.3917/arco.cosan.2012.01.0099.

[9] Cruiziat P. La circulation de l'eau dans la plante en flux non conservatif : quelques faits et problèmes. Non-conservative flux circulation of water in the plant: facts and problems. La houille blanche/n°3/4. 1978: 243-253.

[10] Piel C. Diffusion du CO_2 dans le mésophylle des plantes à métabolisme C_3. Biologie végétale. Université Paris Sud - Paris XI. Thèse de docteur en sciences. Université de Paris XI Orsay, UFR scientifique d'Orsay. 2002. 111p.

[11] Nizinski J-J., Morand D., Loumeto J-J., Luong-Galat A. et Galat G. Bilan hydrique comparé d'une savane et d'une plantation d'eucalyptus dans le bassin du Kouilou (République Populaire du Congo). Climatologie, vol. 5 2008: 99-112.

[12] Gerbier N et Brochet P. L'évapotranspiration. Monographie de la météorologie, 1975. 65p.

[13] Ministère de la coopération,. Évaluation des quantités d'eau nécessaires aux irrigations. Collection Techniques rurales en Afrique. Ministère de l'Agriculture C.T.G.R.E.F. 1979. 204p.

[14] Traore F. Méthodes d'estimation de l'évapotranspiration réelle à l'échelle du bassin versant du Kou au Burkina Faso. Mémoire de fin d'étude de Diplôme d'Etudes Approfondies (DEA) en Sciences et Gestion de l'Environnement. Département des Sciences et Gestion de l'Environnement. Faculté des Sciences. Université de Liège. 2007. 119p.

[15] Annales agronomiques. L'eau et la production végétale. Vol. 14, n°4 et 5. 1963.

[16] Xanthoulis D. Calcul ET_0-Penman. 2010. 54 pp.

[17] Mjejra M. Étude de l'évapotranspiration dans le bassin versant de Mejerda (en Tunisie): apport de la télédétection satellitaire et des Systèmes d'Information Géographique. Géographie. Université Rennes 2. 2015. 273p.

[18] Er-Raki S. Estimation des besoins en eau des cultures dans la région de Tensift AL Haouz : Modélisation, expérimentation et télédétection, Thèse de doctorat en Mécanique des Fluides et Energétique -Télédétection, Faculté des Sciences Marrakech, 2007. 112p.

[19] Piedallu, C. Spatialisation du bilan hydrique des sols pour caractériser la distribution et la croissance des espèces forestières dans un contexte de changement climatique. Thèse, spécialité sciences forestières et du bois. Agroparistech, Nancy, France, 2012. 281p.

[20] Courault, D., B. Seguin and Olioso, A. Review on estimation of evapotranspiration from remote sensing data: From empirical to numerical modeling approaches. Irrig. Drain. Syst. 19, 2005. 223 249.

[21] Pouget M. Les relations sol-végétation dans les steppes Sud-algéroises. Trav. Et. Doc. ORSTOM. Paris. 1980, 555p.

[22] Djebaili S., Achour H., Aidoud F. et Khelifi H. Groupes écologiques édaphiques dans les formations steppiques du Sud-Oranais. Bulletin d'écologie terrestre, Biocenoses. N°1, 1982 : 7-59.

[23] Halitim A. Contribution de l'étude des sols des zones arides (hautes plaines steppiques de l'Algérie). Morphologie, distribution et rôle des sols dans la genèse et le comportement des sols. Thèse. Doct. Univ. Rennes. 1985. 183p.

[24] Haddouche D. Cartographie pédopaysagique de synthèse par télédétection « image Landsat TM » cas de la région de Ghassoul (El Bayadh). Thèse Mag. INA, Alger, 1998. 149p.

[25] Bensaïd A. SIG et télédétection pour l'étude de l'ensablement dans une zone aride : LE CAS DE la wilaya de Naâma (Algérie). Géographie. Université Joseph-Fourier - Grenoble I, 2006

[26] Maire R. Principaux groupements végétaux d'Algérie. Station Centrale des Recherches en Ecologie Forestière C.N. R.E.F.; I.N.R.A. d'Algérie. 1926. 7p.

[27] Quézel P. et Santa S. Nouvelle flore de l'Algérie et des régions désertiques méridionales. Paris : Ed. C.N.R.S. 2 Vol, 1962-1963. 1170p.

[28] HCDS. Carte d'occupation des terres dans la wilaya de Naâma. Identification et cartographie des zones potentielles à l'agriculture en steppe. 2001.

[29] Seltzer p. Le climat de l'Algérie. Inst. Météor. et de Phys. du Globe. Université d'Alger. 1946. 219p.

[30] Bagnoul F. et Gaussen H. Carte des précipitations de l'Algérie et de la Tunisie au 1/500.000°. ING. Paris. 1958.

[31] Djebaili S. Steppe algérienne, phytosociologie et écologie. Alger : OPU. 1984.

[32] Bouazza M. et Benabadji N. Composition floristique et pression anthropozoique au Sud-Ouest de Tlemcen". Rev. Sci. Techn. N°10. Constantine. 1998. 93-97

[33] Aidoud, A., Le Floc'h, E. & Le Houérou, H.N. Les steppes arides du nord de l'Afrique. Sécheresse, 17. 2006: 19-30.

[34] Benaradj A. Étude phyto-écologique des groupements à Pistacia atlantica Desf. Dans le sud Oranais (Sud-Ouest algérien). Thèse Doctorat en Foresterie.

Département des Ressources Forestières. Faculté des Sciences de la nature de la vie et sciences de la terre et de l'univers. Université Abou Bakr Belkaid de Tlemcen. 2017. 269p.

[35] Boucherit H. Etude ethnobotanique et floristique de la steppe à *Hammada scoparia* (Pomel) dans la région de Naâma (Algérie occidentale), Thèse de Doctorat. Département d'Agronomie. Faculté des Sciences de la Nature et de la Vie, des Sciences de la Terre et de l'Univers. Université Abou Bakr Belkaïd Tlemcen. 2018. 175 p +Annexe

[36] Djellouli Y., Louail A., Messner F., Missaoui Kh., Gharzouli R. Les écosystèmes naturels de l'Est algérien face au risque du changement climatique. Geo-Eco-Trop., 2020, 44, 4 : 609-621.

[37] ONM. Bulletin annuelles des données climatique des stations météorologiques d'Ain Séfra, Naâma et Mécheria. 2015

[38] De Martonne E. L'indice d'aridité. Bull. Ass. Geog. France, n°8. 1926.

[39] Debrach J. Notes sur les climats du Maroc occidental. Maroc méridional; 32. 1953: 1122-1134

[40] Despois J. La Tunisie orientale, Sahel et Basse Steppe. Étude géographique. Édit. P.U.F., 2ème édition, Paris. 1955, 554 p.

[41] Emberger L. Nouvelle contribution à l'étude de la classification des groupements végétaux. Rev. Gen. Bot. 45 1933: 473-486.

[42] Emberger L. Une Classification Biogéographique des Climats. Rev. Trav. Lab. Bot. Geol. Zool. Fac. Sc. Montpellier, série bot., n° 7. 1955. pp. 3-43.

[43] Bagnouls F. & Gaussen h. Saison sèche et indice xérothermique. Bull. Soc. Hist. Nat., Toulouse, 88. 1953: 193-239.

[44] Bagnouls F. & Gaussen H. Les climats biologiques et leurs classifications. Ann. Geog., 335. 1957: 193-220.

[45] Le Houerou H. N., Claudin J., Pouget M. Etude bioclimatique des steppes algériennes (Avec une carte bioclimatique à 1j1.000.000ème) Bull. Soc. Hist. Nat. Afr. Nord. Alger, t. 68, fasc. J et 4, 1977.

[46] Thornthwaite, C.W. An approach toward a rational classification of climate. Geographical Review, 38 (1). 1948. 55-94.

[47] Turc L. Evaluation des besoins en eau d'irrigation, évapotranspiration potentielle. Ann. Agron. 12, 1961 : 13-49.

[48] Rognon P. Sécheresse et aridité : leur impact sur la désertification au Maghreb. Sécheresse, 7 (4) 1996: 287-297.

[49] Floret Ch et Pontanier R. Aridité en Tunisie présaharienne, climat, sol, végétation et aménagement. Mémoire de thèse. Travaux et documents de l'ORSTOM. Paris, 1982: 150-544.

[50] Noumi Z., 2010. *Acacia tortilis* (Forssk.) Hayne subsp. *raddiana* (Savi) Brenan en Tunisie pré-saharienne : structure du peuplement, réponses et effets biologiques et environnementaux

[51] Escadafal R. Caractérisation de la surface des sols arides par observations de terrain et par télédétection applications: exemple de la région de Tataouine (Tunisie). Thèse de Doctorat Spécialité: Pédologie (Science des sols). Université Pierre et Marie Curie - Paris - U.F.R des Sciences de la Terre. 1989.

[52] Khader M. Apport de la géomatique à l'analyse spatio-temporelle des parcours steppiques: Cas de la région de Djelfa. Thèse de Doctorat. Département des sciences agronomiques. Faculté des sciences exactes et des sciences de la

nature et de la vie. Université Mohamed Khider de Biskra 2019. 161p.

[53] Djebaili S. Recherches phytosociologiques et phytoécologiques sur la végétation des Hautes plaines steppiques et de l'Atlas saharien ». Thèse Doct. Univ. Montpellier. 1978. 229 p et ann

[54] Derouiche, G. Les risques climatiques et agriculture algérienne. Séminaire sur les risques agricoles assurances et réassurances. B.N.E.D.E.R Algérie Alger-Hôtel El-Aurrassi 10 juin 2007. 12p.

[55] Yagoub H. Cartographie et suivi du couvert végétal des zones semi-arides par l'imagerie satellitaire. Thèse de Doctorat Es-Science. Département de Génie-Physique, Faculté de Physique. Université des sciences et de la technologie d'Oran Mohamed Boudiaf. 2015. 104p.

[56] Monteith, J. L. Evaporation and environment. 19th Symposia of the Society for Experimental Biology. 19, 1965: 205-234.

[57] Chung, S. O., & Horton, R. Soil heat and water flow with a partial surface mulch. Water resources research, 23 (12). 1987. 2175-2186.

[58] Bonn, F. Précis de télédétection. Applications thématiques. Vol. 2. Québec: Presses de l'Université du Quebec/ AUPELEF, 1996. 633 p.

[59] Le Houérou H.-N. et Popov G. F. An ecoclimatic classification of intertropical Africa. 40 p. 3 Cartes. Plant Prad. Paper nO 31, AGP, FAO, Rome, Italy. 1981.

[60] Carminati, A., Passioura, J. B., Zarebanadkouki, M., Ahmed, M. A., Ryan, P. R., Watt, M., & Delhaize, E. Root hairs enable high transpiration rates in drying soils. New Phytologist, 216(3), 2017: 771-781. doi: 10.1111/nph.14715

[61] Frontier S.,Pichod-Vial D., Le Prêtre A., Davoult D., Luczak Ch. Ecosystème, structure, fonctionnement, évolution. 3eme édition. Dunod. Paris, 2004. 549p.

[62] Dahmani M. Diversité biologique et phytogéographique des chênaies vertes d'Algérie. Ecologia méditerranea XXII (3-4), 1996 : 10- 38.

[63] Emberger, L. Aperçu général sur la végétation du Maroc, commentaire de la carte phytogéographique du Maroc au 1/1.500.000. Ed. Hans Huber, Berne. Veroff Geobot. Rubel 1nst. Zurich, 14. 1939: 40-157.

[64] Daget Ph. Sur les types biologiques en tant que stratégie adaptative. (Cas des thérophytes). in : Recherches d'écologie théorique, les stratégies adaptatives. Paris. 1980: 89-114

[65] Raunkiaer, C. The life forms of plants and statistical plant geography. Oxford University Press, London. 1934.

[66] Jauffret S. Validation et comparaison de divers indicateurs des changements a long terme dans les écosystèmes méditerranéens arides : Application au suivi de la désertification dans le Sud tunisien. Thèse de Docteur Ecologie. Université de droit, d'économie et des sciences d'aix-Marseille (Aix-Marseille III). 2001. 334p.

[67] Floret, C, Galan M.J., Le Floc'h, E, Orshan, G. & Romane, F. Growth forms and phenomorphology traits along an environmental gradient : Tools for studying vegetation. J. Veg. Sci. 1. 1990: 71-80.

[68] Le Houerou H.N. La végétation de la Tunisie steppique (avec référence aux végétations analogues d'Algérie de Lybie et du Maroc). Ann. Inst. Nat. Rech. Agron. Tunisie, 42(5), 1969. 1-624 et 1 carte couleur 1/ 500.000.

[69] Aidoud-Lounis, F. (1984). Contribution à la connaissance des groupements à sparte (*Lygeum spartum* L.) des Hauts Plateaux Sud-Oranais. Etude écologique et syntaxonomique. Thèse 3ème cycle. Université des Sciences et Technologies H. Boumediène, Alger.

[70] Aïdoud A. Contribution à l'étude des écosystèmes steppiques pâturés des hautes plaines Algéro-oranaises (Algérie). Thèse Doct. U.ST.H.B. Alger. 1989. pp 43- 210.

[71] Small E. Xeromorphy in plants as a possible basis for migration between arid and nutritionally-deficient environments. Bot. Notiser, 126. 1973: 33-38.

[72] Ozenda P. Flore du Sahara. 2ème Ed. CNRS. Paris, 1977. 622 p.

[73] Poilecot P. La Reserve Naturelle Nationale de l'Aïr et du Ténéré (Niger) MH/E / WWF/ IUCN. 1996. 228p.

[74] Scheromm. La résistance des plantes, la sécheresse. ed. INRA. Centre de mont péllier. 2000.

[75] Dajoz, R. Précis d'écologie. Paris, France, Dunod, 2003. 615p.

[76] Aggarwal, P., Gera, J., Ghosh, S., Mandal, L., Mandal, S. Noncanonical decapentaplegic signaling activates matrix metalloproteinase 1 to restrict hedgehog activity and limit ectopic eye differentiation in drosophila. Genetic 207(1), 2017 : 197-213.

[77] Boudjemaa S., 2010. Cartographie des relations sol-eau-végétation dans un milieu salé (lac Fetzara). Mémoire de Magister. Université Badji Mokhtar Annaba. 115p.

[78] Negre, R.. Recherches phytogéographiques sur l'étage de végétation méditerranéenne aride (sous étage chaud) au Maroc occidental. Trav. I.S.C., sér. bot. Rabat. no 13. 1959. 385p.

[79] Mayouf R. and Arbouche F. Seasonal variations in the chemical composition and nutritional characteristics of three pastoral species from Algerian arid rangelands. Livestock research for rural development. Vol 27 (3). 2015.

Chapter 2

Climate Change Induced Thermal Stress Caused Recurrent Coral Bleaching over Gulf of Kachchh and Malvan Marine Sanctuary, West Coast of India

Mohit Arora, Kalyan De, Nandini Ray Chaudhury,
Mandar Nanajkar, Prakash Chauhan and Brijendra Pateriya

Abstract

Coral reefs are one of the most sensitive, productive, and invaluable biological resources on the earth. However, coral reefs are facing unprecedented stress due to ongoing climate changes and intensified anthropogenic disturbances globally. Elevated Sea Surface Temperature (SST) has emerged as the most imminent threat to the thermos-sensitive reef-building corals. The 2010–2014-2016 El Niño Southern Oscillation (ENSO) caused prolonged marine heat waves (MHWs) that led to the most widespread coral bleaching and mortality in the tropical Indi-Pacific regions. Coral bleaching prediction is vital for the management of the reef biodiversity, ecosystem functioning, and services. Recent decades, satellite remote sensing has emerged as a convenient tool for large-scale coral reef monitoring programs. As thermal stress is a critical physical attribute for coral bleaching hence, the present study examines the effectiveness of the elevated SSTs as a proxy to predict coral bleaching in shallow water marginal reefs. Advanced Very High-Resolution Radiometer (AVHRR) satellite data from the NOAA Coral Reef Watch's (CRW) platform has been used for this study. Coral bleaching indices like Bleaching Threshold (BT), Positive SST Anomaly (PA), and Degree Heating Weeks (DHW) are computed to analyze the thermal stress on the coral reefs. The computed thermal stress from satellite-derived SST data over regions concurrence with the mass coral bleaching (MCB) events. This study concludes that in the last decades (2010 to 2019) the coral cover around these regions has dramatically declined due to higher SST, which indicates that the thermal stress induced recurrent bleaching events attributed to the coral loss.

Keywords: Sea Surface Temperature, Bleaching Threshold, Degree Heating Weeks, El Niño Southern Oscillation, thermal stress

1. Introduction

Coral reefs are one of the most ancient, dynamic, highly sensitive, complex, biologically diverse, highly productive ecosystems, found in the tropical coastal environment between 30° N and 30° S latitudes. Coral reefs provide an conducive environment where one-third of all marine fish species and many thousands of other species are found and offer substantial ecological and economic services to millions of people through fishery and tourism worldwide [1]. Coral reef ecosystems are degrading in rapid pace, and some facing extinction risk due to the synergistic impact of global climate change and chronic human activities including overfishing, pollution, eutrophication, sedimentation, coastal development [2–4]. Rapid decline of coral reef ecosystem health is now the most pressing challenge to the reef managers [1, 5]. Most of the tropical coral reefs are found only within a narrow range of environmental conditions, making them vulnerable to abrupt change in seawater physico-chemical parameters like temperature, and salinity [6–10]. Coral bleaching events are associated with thermal stress are acute disturbances recognized as the primary global challenge to the persistence of coral reefs, which disrupts the mutualistic relationship of corals with the thermo-sensitive endosymbiotic dinoflagellates of the family *Symbiodinaceae* by photoinhibition and their expulsion. Coral bleaching can be divided into two parts: (1) The initial response where corals expel *Symbiodinium*, and (2) The longer-term effect, which may be either coral tissue recovery or mortality [11]. Coral bleaching events occur when sea surface water becomes so warm and remains high for more than 28 days [6, 8, 12]. Coral mortality after bleaching depends on the extent of heat stress, its severity, and duration of bleaching [13, 14]. Coral bleaching prevalence and the extent of subsequent coral mortality patterns are commonly associated with natural and anthropogenic disturbances, which is highly venerable both within and across the region.

The ENSO event is one of the significant climatic events that trigger the rapid warming of the water column of the seas, altering biogeochemical processes and marine life. Thermal stresses associated with the El Niño Southern Oscillation (ENSO) are occurring with increasing frequency and severity [10]. The global SSTs have risen gradually since the 1980s, which have caused mass coral bleaching (MCB) and mortality in more than 90% of reefs since 1997–1998. Many researchers have reported four significant MCB events (i.e., 1982–1983, 1997–1998, 2010, 2015–2016) all over the world over the past four decades due to global warming-induced by the ENSO event [8, 15–17]. The 2015–2016 ENSO event emerged as the most extreme event in terms of ocean warming intensity and extent across the tropical oceans [18–21], which caused one of the most severe and widespread MCB events across the Indo-Pacific [15, 20]. The MHW caused by the 2015–2016 ENSO was unprecedented over the period of two centuries resulting in ecological and economic consequences worldwide [10]. More than 75% of global coral reefs have witnessed MCB and mortality back-to-back from 2014 to 2016 [19, 20]. Corals can re-establish themselves after mass bleaching in some cases; it takes one to two decades for the ecosystem to return to the pre-bleaching state [22]. However, the increasing thermal stress left no window of recovery for corals from the previous bleaching events, leading to mass mortality [19, 20, 23]. Mass coral bleaching events can cause long-term ecological, economic, and social impacts [1, 24]. As increase in the frequency and severity of MCB could overwhelm the ability of coral reefs to recover between events. Consecutive mass coral bleaching episodes and associated coral mortalities could shift coral reefs from coral dominated state to Cyanobacteria and algae dominated state [25, 26].

The objective of this study is to examine the thermal stress that causes coral bleaching over the coral reef regions on the Eastern Arabian Sea in the Indian Ocean using long-term NOAA CRW SST data. We computed the long-term climatologically mean and trend of SST for these coral reef regions and computed coral bleaching thermal indices: BT, PA, and DHW based on the SST analysis. Further, to ground-truth the accuracy of the computed coral bleaching indices, we visit at field sites for coral monitoring and analyzed the bleaching percentage.

2. Study areas

The coral reefs are geographically located at different latitudes on the Indian coast, and are highly important ecosystem for ecosystem service and economy. The climate over Indian coral reef regions is tropical, hot, and humid. Species diversity and reef structure in Indian coral reefs vary significantly between the areas due to differences in the reef extent and environmental conditions. This study was carried out at the Gulf of Kachchh and Malvan Marine Sanctuary, which lies in the Arabian Sea on the Indian coast and harbor some of the most northern reefs in the world [27]. The coral reefs in the Gulf of Kachchh is located between 22° 15' N to 23° 40' N Latitude and 68° 20' E to 70° 40' E Longitude to the north of Saurashtra Peninsula of Gujarat state (**Figure 1**). This region is very rich in terms of biodiversity value and supports varied coastal habitats, including coral reefs, mangroves, creeks, mudflats, islands, rocky shores, sandy beaches, etc. (Arora et al., 2018) and mostly consisting of dead coral boulders and rubbles. The coral species over the Gulf of Kachchh region belonging to the common genera of *Coscinaraea, Favia, Goniastrea, Gonipora, Leptastrea, Porities, Turbinaria*, etc.

It extends over 170 kilometers in length (NNE–SSW) and about 75 kilometers in width (NNW–SSE), covering an area of approximately 7350 km^2 with a mean depth of 30 meters. A total of 76 species of stony coral (Scleractinian) belonging to 30 genera and 12 families [28] and 12 species of soft corals are found in this region [29]. On the other hand, the Malvan Marine Sanctuary, a Marine Protected Area (MPA), is located in the Central West coast of India along the Eastern Arabian Sea, spreads over a 29.122 km^2 area. The marine wildlife sanctuary harbor near shore patch coral reefs mostly dominated by massive and encrusting *Porites* species and

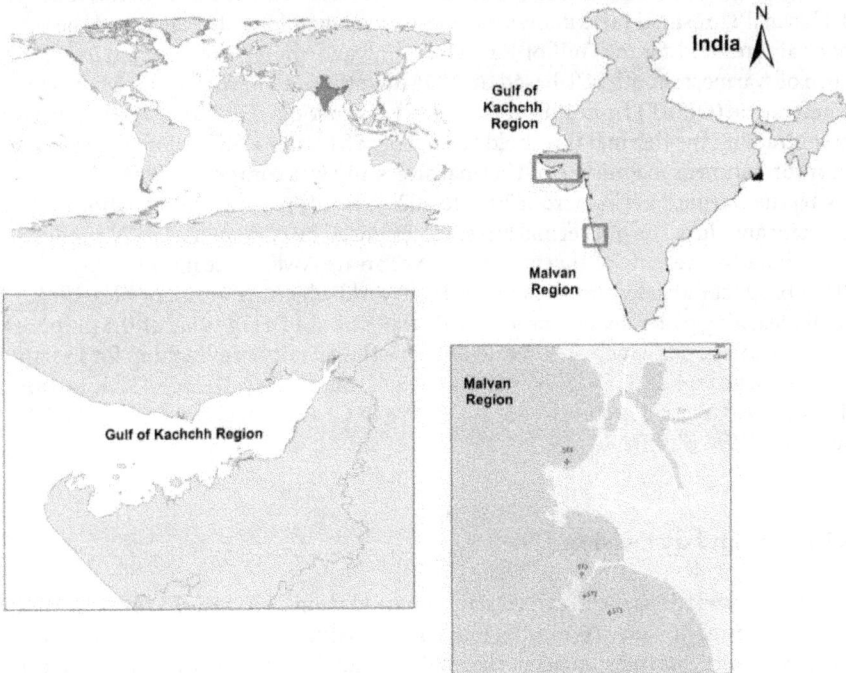

Figure 1.
Study area of Gulf of Kachchh and Malvan Marine Sanctuary, which lies in the EAstern Arabian Sea on the Indian coast.

foliose *Turbinaria mesenterina;* other species includes *Porites lichen, Porites lutea, Porites compressa, Pseudosiderastrea tayami, Siderastrea savignyana, Coscinaraea monile, Favites melicerum, Favites halicora, Cyphastrea serailia, Plesiastrea versipora, Goniopora sp., Tubastraea coccinea* [25, 30]. Corals in the Gulf of Kachchh and Malvan regions are surviving through extreme environmental conditions such as high temperature, high solar radiation, turbidity, salinity changes, and high suspended sediment loads [28, 31]. Both the regions have been designated as a Marine Protected Area (MPA). However, in the Malvan Manrine Sanctuary, the absence of a robust management system and opposition of the MPA from the local population resulted severe local disturbances includes fishing, wastewater drainage, and unregulated recreational activities along with climate change disturbance [30, 32].

3. Materials and methodology

The SST data analysis for the present study was obtained from National Oceanic and Atmospheric Administration (NOAA) Coral Reef Watch's (CRW) (known as 'CoralTemp') high-resolution time-series data product (available from NOAA Coral Reef Watch 2019 https://coralreefwatch.noaa.gov). This data set has a high spatial resolution of 5 km (0.05 × 0.05°C exactly) and a temporal resolution of one day. NOAA CRW global data product provides near real-time SST data from 1985 to the present. This datasets product uses advanced very high-resolution radiometer (AVHRR) satellite data from NOAA Pathfinder SST and has been found in good agreement with *in situ* data from ships and buoys. It also includes a large-scale adjustment of satellite biases with respect to the *in situ*. Bleaching Threshold (BT), Positive SST Anomaly (PA), and Degree Heating Weeks (DHW) are commonly used indices for calculating thermal stress on coral reefs. BT is based on the concept of Thermal Threshold (also known as long-term climatological mean), and the thermal threshold for the Gulf of Kachch & Malvan region was computed using the mean of warmest month SST based on NOAA Optimum Interpolated Sea Surface Temperature (OISST) from 1982 to 2016 (35 years period). SST Anomaly was derived by subtracting the thermal threshold from daily SST values. SST Anomaly provided the information of magnitude of thermal stress and was computed for the 11 years during the warmest period (from 2010 to 2020). DHW provides information on the intensity and duration of thermal stress experienced by coral reefs. DHW product is a cumulative measure of thermal stress over an area over three months [33]. The DHW product indicates the reefs around the world which are at risk of bleaching. Coral bleaching generally begins for corals exposed to a DHW value of 0.5 or more [34]. The categories which are used to describe the severity of bleaching for Indian regions are no stress (0° C<DHW≤2° C), bleach watch (2° C<DHW≤4° C), warning (4° C<DHW≤6° C), alert level-1 (6° C<DHW≤8° C) and alert level-2 (DHW>8° C) based on DHW [8, 15].

4. Results and discussion

The SST variations during the warmest period from 2010 to 2020 for both regions provide information on magnitude, intensity, and duration of thermal stress. The warmest month, warmest quarter, thermal threshold, and bleaching threshold for both regions were computed from NOAA CRW datasets. Based on the maximum frequency of the warmest month recording the maximum monthly mean SST in the year, the climatologically warmest months were identified for both coral reef regions. The warmest month, warmest quarter, Thermal Threshold, and Bleaching

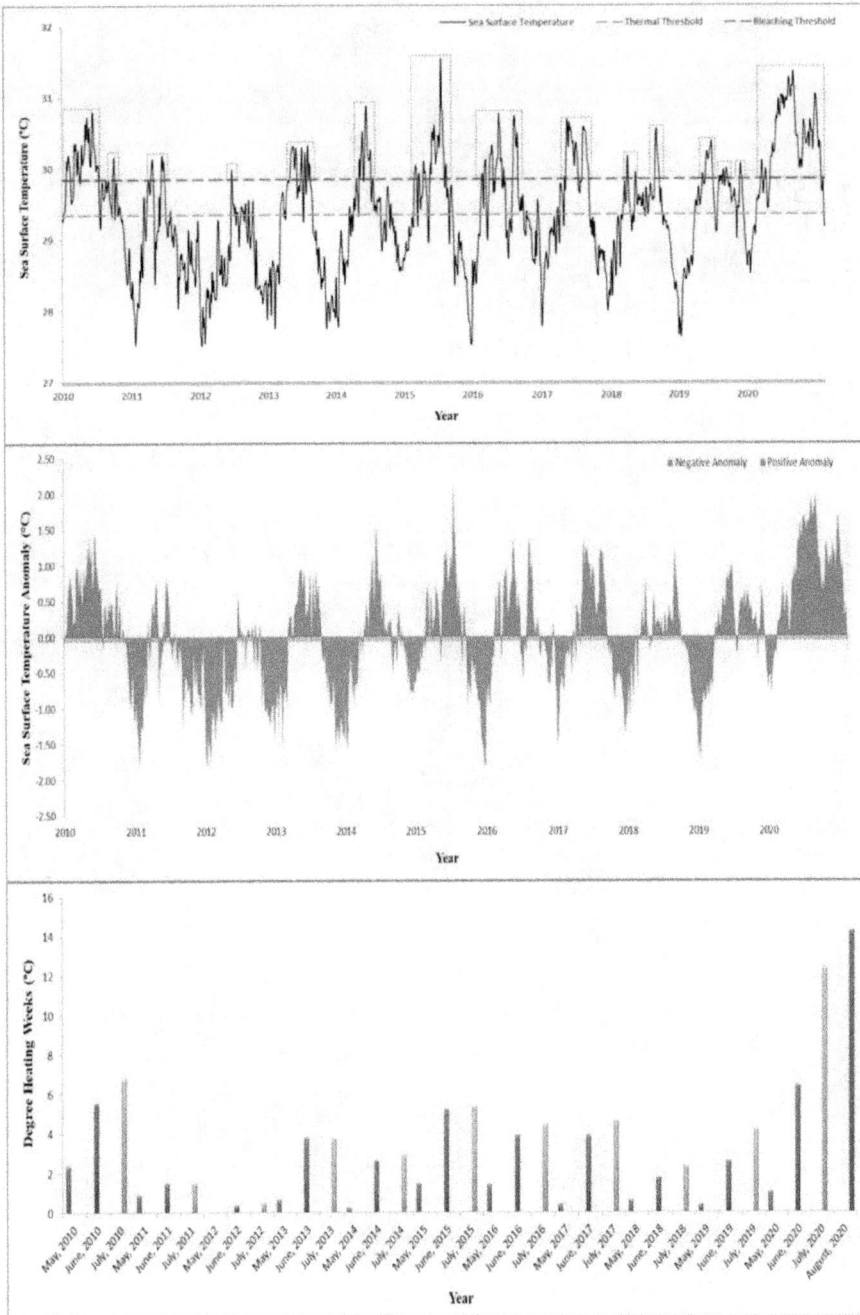

Figure 2.
(Top) Sea surface temperature; (middle) sea surface temperature anomaly; (lower) degree heating weeks variations over Gulf of Kachchh region during warmest quarter (May to July) period from 2010 to 2020.

Threshold for both regions were found to be different. The climatologically warmest month for the Gulf of Kachchh region was June, and the warmest quarter was May to July. Similarly, the climatologically warmest month for the Malvan region was May, and the warmest quarter was April to June. The Gulf of Kachchh region recorded a maximum thermal threshold of 29.35° C(±0.45° C), and the Malvan region recorded

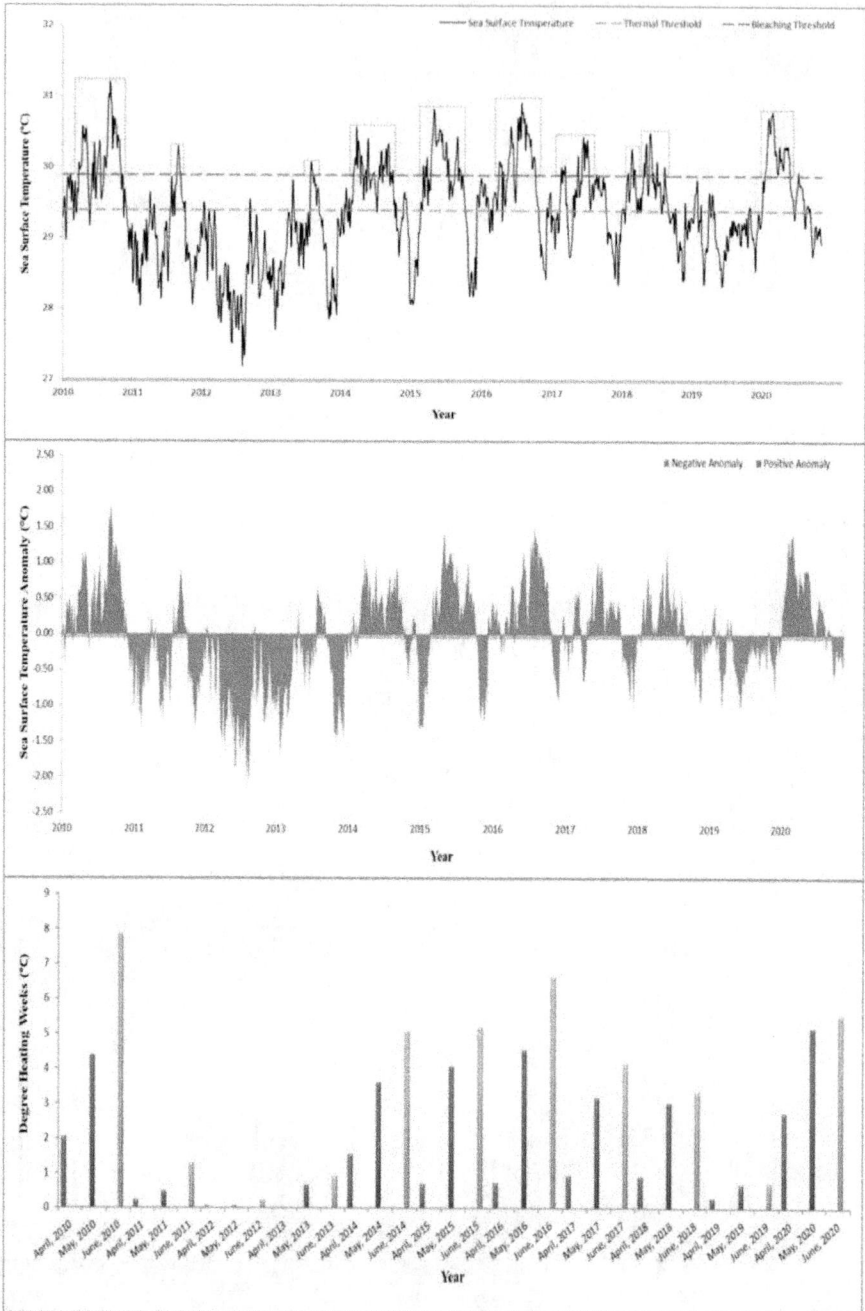

Figure 3.
(Top) Sea surface temperature; (middle) sea surface temperature anomaly; (lower) degree heating weeks variations over Malvan region during warmest quarter (April to June) period from 2010 to 2020.

maximum threshold of 29.39° C(±0.49° C) (**Figures 2** and **3**). In both figures, the orange dotted line shows the thermal threshold, and the red dotted line shows the bleaching threshold for the corals become stressed when SSTs crossed thermal threshold and get bleached if the elevated SST regime is prolonged. BT and daily SST anomalies have been computed on the basis of thermal threshold (sometimes referred as climatologically mean warmest month). Once the climatologically warmest month was identified, the climatologically warmest quarters were identified comparing SST values averaged over three summer months period from a combination of the warmest month and two adjacent months (i.e., the pre and post month). SST anomalies for the warmest quarter were calculated as the absolute difference between the daily SST and the thermal threshold. The warmest quarter anomalies were plotted from 2010 to 2020 for both regions, and positive anomalies were represented in red color while negative anomalies were represented in blue color. It was found that a stark rise in the number of days recording the positive anomalies. Another thermal stress index termed as coral bleaching DHW was calculated in order to assess the accumulative thermal stress. The DHW was generated using the warmest quarter daily SST data. The DHW was also represented in color from blue to green based on their intensity and duration of thermal stress.

5. Gulf of Kachchh region

The diurnal trends of SST in the warmest quarter over a period from 2010 to 2020 provided information on the direction of changes in SST. The BT value for the Gulf of Kachchh region was observed 29.85° C (± 0.45° C) means 0.5° C above the thermal threshold value (**Figure 2**). Over a period from 2010 to 2020, it was observed that the SST value during the warmest quarter had crossed the BT in each year. The maximum SST 31.54° C has been recorded in the year 2015, while the minimum SST 29.97° C has been recorded in the year 2012. During the year 2020, the SST values crossed the thermal as well as BT values and persisted for more than three months. In the same year, the monitoring period was extended up to mid-August. This study also computed and showed variations of daily SST anomaly. The absolute range of positive anomaly for the Gulf of Kachchh region varied between 0.01° C to 2.2° C, and a maximum of 2.2° C was recorded in 2015. The frequency and intensity of positive anomaly have increased continuously over the region. The range of DHW was found to be 0.5° C to 14.26° C, and the maximum was recorded in the year 2020, whereas the minimum was recorded in the year 2012. The "Alert Level-2" status was recorded in 2020, while the "Alert Level-1" status was recorded in 2010. The year 2015, 2016, 2017, and 2019 were recorded "Warning" status.

6. Malvan region

The diurnal trends of SST in the warmest quarter over a period from 2010 to 2020 provided information on the direction of changes in SST. The BT value for Malvan was observed 29.89° C (± 0.49° C) means 0.5° C above the thermal threshold value (**Figure 3**). Over a period from 2010 to 2020, it was observed that the SST value during the warmest quarter had crossed the BT in each year except two years: 2012 and 2019. Year: 2012 and 2019 have experienced significantly less SST as compared with other years. The maximum SST 31.21° C has been recorded in the year 2010. During the year 2020, the SST values crossed the BT values and persisted for three months with two intermittent breaks. This study also computed variations of daily SST anomaly. The absolute range of positive anomaly for the Malvan region varied between 0.01° C to 1.8° C, and a maximum of 1.8° C was recorded in 2010. The frequency and intensity of positive anomaly have increased continuously over the region. The range of DHW was found to be 0.2° C to 7.84° C, and the maximum was recorded in 2010, whereas the minimum was recorded in 2012. The "Alert Level-1" status was recorded in the year 2010 and 2016, while "Warning" status was recorded in the year 2015, 2017, and 2020. The DHW under "Alert Level-2" status was absent over the Malvan region. The year 2019 has shown the least coral bleaching thermal stress indices because of the long period cyclone '*Vayu*' was formed in the Arabian Sea, which triggered heavy to extreme rainfall.

Field observations data were collected over both locations during the period from 2010 to 2020. The recorded satellite-derived SST and field observations data indicated that corals in both regions experienced prolonged heat stress, which is the primary cause of back-to-back coral bleaching events. Field survey at Gulf of Kachchh region revealed bleaching of *Coscinaraea, Favia, Goniastrea, Goniopora, Leptastrea, Porites, and Turbinaria* (**Figure 4**). Field surveys at the Malvan region were revealed partial and whole colony bleaching of *Porites* spp., *Favites spp., Turbinaria mesenterina, Pseudosiderastrea tayami, Cyphastrea serailia, Plesiastrea versipora, Goniopora* spp., *Siderastrea savignyana* (**Figure 5**). Gulf of Kachchh region recorded ~4% and ~10% bleaching during 2016 and 2019 temperature peaks. During field survey at Malvan region in 2014, the bleaching prevalence was observed 14.58%; in the year 2015, the bleaching prevalence was observed 54.20%; in 2016, the bleaching prevalence was observed 46.76%; in 2017, the bleaching prevalence was observed 20.22%; in 2018, the bleaching prevalence was observed 5.07%, and in 2019, the bleaching prevalence was observed 8.37% (**Figure 6**) [25, 30]. The effects of increasing thermal stress on corals were correlated with field observations data. We found that the DHW derived from SST and field observations were positively correlated with a correlation coefficient of 0.71 (**Figure 7**). The significant correlations indicate the SST peaks during the warmest quarter were the predominant cause of mass coral bleaching and mortality.

SST-driven impacts trigger cascading effects at the ecosystem level by reducing coral species heterogeneity, weakening the reef carbonate framework, loss of reef functionality, and negatively impacting the reef-associated biodiversity [9, 24]. This

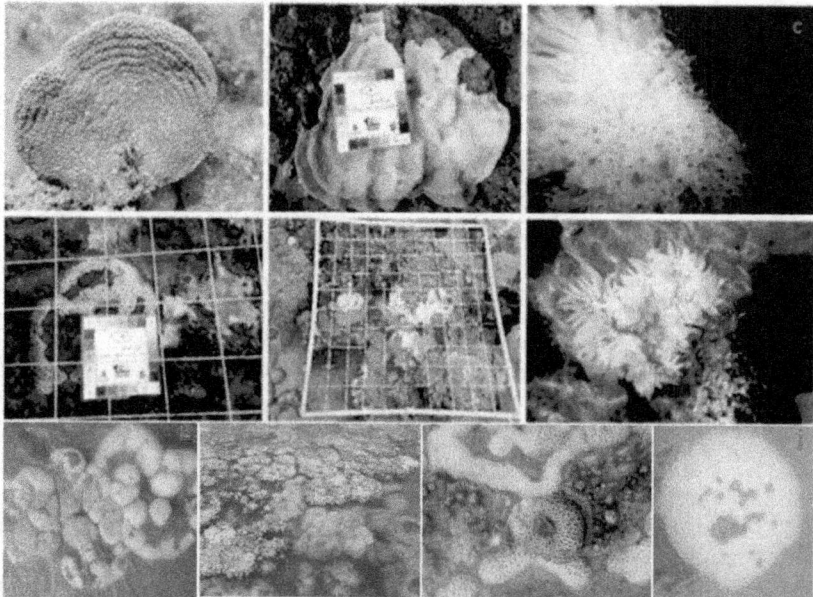

Figure 4.
Coral bleaching observed at Laku Point reef, Gulf of Kachchh region during 2016 and 2019. (a-b) Healthy and bleached Turbinaria sp.; (c) bleached Goniopora sp.; (d) partially bleached Favites sp.; (e) partially bleached Porites sp.; (f) bleached sea anemone Heteractis sp.; (g) bleached colonies of Dipsastraea sp.; (h) bleached colonies of Favites sp. and Porites sp.; (I & j) bleached colonies Dipsastraea sp.

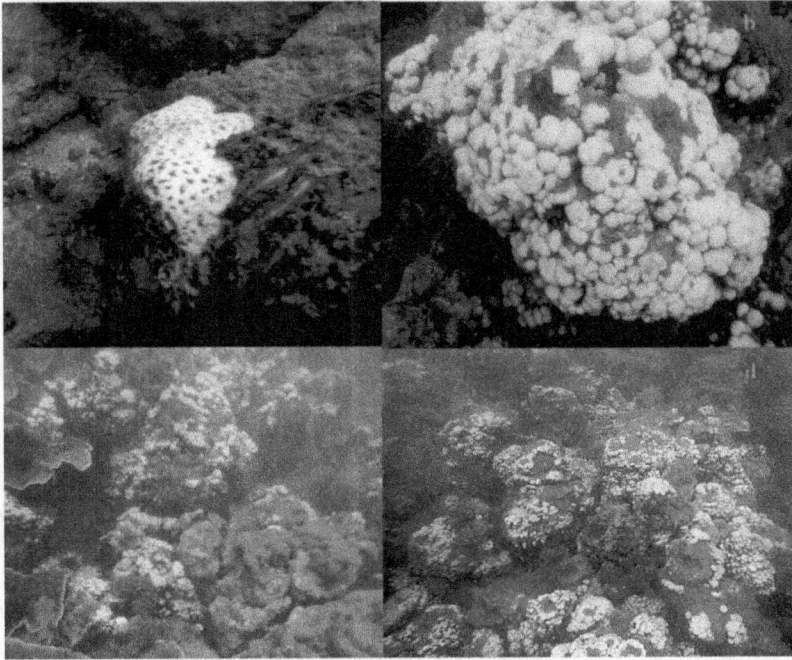

Figure 5.
Coral bleaching observed at Malvan region from 2014 to 2019. (a-b) Bleached Favites sp.*; (c) bleached massive* Porites sp.*; (c & d) bleached and dead coral colonies covered by turf algae and sediments.*

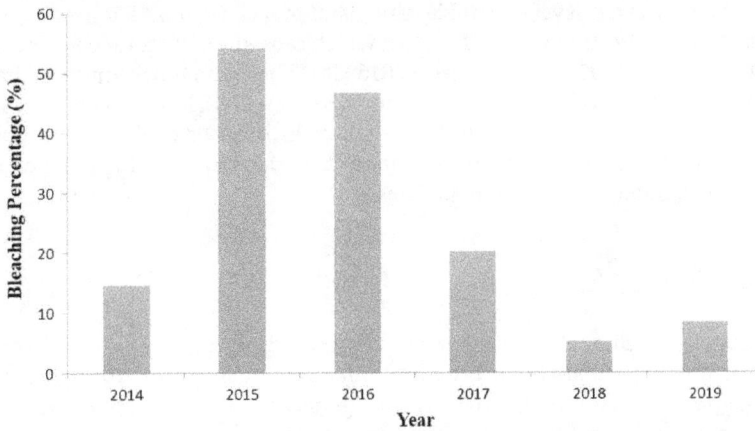

Figure 6.
Coral bleaching prevalence at Malvan Marine Sanctuary, Central West coast of India.

heat shock-mediated bleaching mortality of corals has emerged as the greatest threat to the existence of reefs globally as these habitats fail to recover [10, 16, 19, 35]. The present study highlights that the satellite-derived SST data may be used as a convenient tool for thermal stress-driven coral bleaching events, which will improve reef management practices in the thermal stress-impacted coral reef environment. Climate change poses a threat to the persistence of the coral reefs in tropical seas. The mass coral bleaching estimated during 2010 and 2016 was correlated with a multivariate ENSO index. The year 2010 and 2015–2016 were a strong ENSO years.

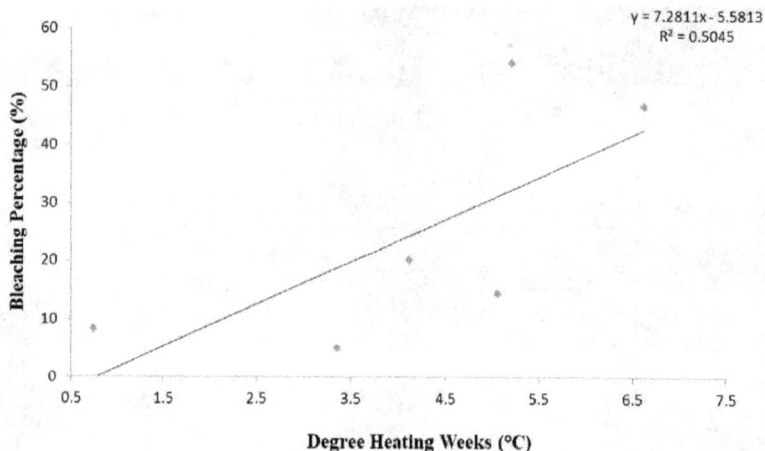

Figure 7.
Correlation of duration of thermal stress (i.e., DHW) and bleaching percentage during the period from 2014 to 2019.

The tropical Indian Ocean is warming rapidly compared to rest of the tropical oceans [36], and warming of the Arabian Sea has increased significantly since the 1990s [37, 38]. In recent years, rapid warming event caused severe negative ecosystem impact in the Indian Ocean. For instances, 2015–2016 ENSO caused a significant decline in oil sardines fishery in South-West India [39], and phytoplankton community shift in the North-eastern Arabian Sea [40], as well as coral bleaching and mortality in Lakshadweep archipelago [41]. Understanding how the Indian Ocean suffered severe coral bleaching and mortality in 2015 following a 7.5 maximum degree heating weeks (DHWs), which caused a 60% coral cover decrease from 30% cover in 2012 to 12% in April 2016 [11]. Therefore, ENSO induced heat stress driven coral bleaching, coral mortality and ecosystem level impact of recurrent mass bleaching events require global scale quantification of the magnitude, intensity, and duration of thermal stress for each reef location for formulation of improved and timely management policies.

7. Conclusion

This study concludes that the increased thermal stress and back-to-back coral bleaching are the particular concern over Indian coral reef regions due to pressure from long-term climate change and anthropogenic activities. This study highlights that the satellite-derived SST data could serve as a useful coral reef monitoring tool along with the field data confirmation. Corals in the Gulf of Kachchh and Malvan regions show distinct regional sensitivity towards BT, SST anomalies, and DHW. NOAA CRW data proves its potential towards a long-term SST. The year 2020 was the warmest in the Gulf of Kachchh region, and 2010 was in the Malvan region during the period from 2010 to 2020, which recorded a high duration of thermal stress over the region. But the highest temperature and highest anomaly was recorded in the year 2015 over the Gulf of Kachchh region, while the Malvan region was received in 2010. In the Gulf of Kachchh region, the year 2020 was recorded high DHW compared to other years, which was under "Alert Level-2" status, and the Malvan region recorded high DHW in 2010 with "Alert Level-1" status. The year 2012 recorded minimum thermal stress over both regions. This study revealed that the high intensity and long duration thermal stress led to bleaching and mortality, which indicates the dire situation of coral reef health degradation. Therefore, the persistence of fragile coral reefs in the Gulf of Kachhh and Malvan Marine Sanctuary are in need of urgent science-informed active conservation, restoration and management intervention.

Acknowledgements

The authors are thankful to National Oceanic and Atmospheric Administration (NOAA) for providing respective high-resolution SST data in the Public domain. Authors are immensely grateful to Dr. Baban Ingole for his support and encouragement for the study. We also thankful to Sambhaji Mote for his help during field observation in Malvan. The authors are thankful to the GEER Foundation, CSIR-NIO, and SAC-ISRO for supporting the field observations. The authors acknowledge the authority of Marine National Park, Jamnagar, for permitting to conduct field observations at different sites.

Disclosure statement

The authors declare no conflict of interest.

Author details

Mohit Arora[1*], Kalyan De[2], Nandini Ray Chaudhury[3], Mandar Nanajkar[2], Prakash Chauhan[4] and Brijendra Pateriya[1]

1 Punjab Remote Sensing Centre, PAU Campus, Ludhiana, India

2 CSIR-National Institute of Oceanography, Dona Paula, Goa, India

3 Space Applications Centre, Indian Space Research Organisation, Ahmedabad, India

4 Indian Institute of Remote Sensing, Indian Space Research Organisation, Dehradun, India

*Address all correspondence to: mohitarorakuk12@gmail.com

IntechOpen

References

[1] Hoegh-Guldberg O, Pendleton L, Kaup A. 2019. People and the changing nature of coral reefs. Reg Stud Mar Sci. 30:100699.

[2] Lam VWY, Chavanich S, Djoundourian S, Dupont S, Gaill F, Holzer G, Isensee K, Katua S, Mars F, Metian M, Hall-Spencer JM. 2019. Dealing with the effects of ocean acidification on coral reefs in the Indian Ocean and Asia. Reg Stud Mar Sci. 28:100560.

[3] Nanajkar M, De K, Ingole B. 2019. Coral reef restoration - A way forward to offset the coastal development impacts on Indian coral reefs. Mar Poll Bull. 149. https://doi.org/10.1016/j.marpolbul.2019.110504

[4] Pendleton L, Hoegh-Guldberg O, Albright R, Kaup A, Marshall P, Marshall N, Fletcher S, Haraldsson G, Hansson L. 2019. The Great Barrier Reef: Vulnerabilities and solutions in the face of ocean acidification. Reg Stud Mar Sci 31:100729. https://www.sciencedirect.com/science/article/pii/S2352485518306959

[5] Madin JS, Madin EMP. 2015. The full extent of the global coral reef crisis. Conserv Biol [Internet]. [accessed 2019 Jan 14] 29(6):1724-1726. http://www.imars.usf.edu/MC

[6] Arora M, Ray Chaudhury N, Gujrati A, Kamboj, RD, Joshi, D, Patel, H and Patel, R. 2019a. Coral bleaching due to increased sea surface temperature in Gulf of Kachchh region, India, during June 2016. Indian J. Mar. Sci. 48(03):327-332.

[7] Arora M, Gujrati A, Chaudhury NR, Patel RC. 2019b. Benthic coverage and bottom topography of coral reef environment over Pirotan reef, Gulf of Kachchh region, India. Geocarto Int. 34(10): 1089-1097.

[8] Arora M, Gujrati A, Chaudhury NR, Chauhan P, Patel RC. 2019c. Assessment of coral reef thermal stress over India based on remotely sensed sea surface temperature. Geocarto Int. https://doi.org/10.1080/10106049.2019.1624983

[9] Hughes TP, Anderson KD, Connolly SR, Heron SF, Kerry JT, Lough JM, Baird AH, Baum JK, Berumen ML, Bridge TC, et al. 2018. Spatial and temporal patterns of mass bleaching of corals in the Anthropocene. Science 359(6371):80-83. http://www.sciencemag.org/lookup/doi/10.1126/science.aan8048

[10] Lough JM, Anderson KD, Hughes TP. 2018. Increasing thermal stress for tropical coral reefs: 1871-2017. Sci Rep 8(1): 6079. http://www.nature.com/articles/s41598-018-24530-9

[11] Head, CEI, Bayley, DTI., Rowlands, G, Roche, RC, Tickler, DM, Rogers, AD, Koldewey, H, Turner, JR and Andradi-Brown, DA. 2019. Coral bleaching impacts from back-to-back 2015-2016 thermal anomalies in the remote central Indian Ocean, Coral Reefs, 38, 605-618.

[12] Vivekanandan E, Hussain Ali M, Jasper B, Rajagopalan M. 2008. Thermal thresholds for coral bleaching in the Indian seas. J. Mar. Bio. Ass. India. 50:209-214.

[13] Ng CSL, Huang D, Toh K Ben, Sam SQ, Kikuzawa YP, Toh TC, Taira D, Chan YKS, Hung LZT, Sim WT, et al. 2020. Responses of urban reef corals during the 2016 mass bleaching event. Mar Pollut Bull. 154.

[14] Xie JY, Yeung YH, Kwok CK, Kei K, Ang P, Chan LL, Cheang CC, Chow W kuen, Qiu JW. 2020. Localized bleaching and quick recovery in Hong Kong's coral communities. Mar Pollut Bull. 153:110950.

[15] Arora M, Ray Chaudhury N, Gujrati A, Patel RC. 2019d. Bleaching stress on Indian coral reef regions during mass coral bleaching years using NOAA OISST data. Curr Sci. 117(2):242-250.

[16] Hughes TP, Kerry JT, Álvarez-Noriega M, Álvarez-Romero JG, Anderson KD, Baird AH, Babcock RC, Beger M, Bellwood DR, Berkelmans R, et al. 2017. Global warming and recurrent mass bleaching of corals. Nature 543 (7645): 373-377. http://www.nature.com/articles/nature21707

[17] McField M. 2017. Impact of climate change on coral in the coastal and marine environments of Caribbean Small Island Development States (SIDS). Caribbean Marine Climate Change Report Card: Science Review 2017. p. 52-59.

[18] Eakin CMM, Liu G, Gomez A. M, Heron SF, Skirving WJ. 2016. Global coral bleaching 2014-2017: Status and an appeal for observations. Reef Encount. 31:20-26.

[19] Eakin CM, Sweatman HPA, Brainard RE. 2019. The 2014-2017 global-scale coral bleaching event: insights and impacts. Coral Reefs 38(4):539-545. https://doi.org/10.1007/s00338-019-01844-2

[20] Skirving WJ, Heron SF, Marsh BL, Liu G, De La Cour JL, Geiger EF, Eakin CM. 2019. The relentless march of mass coral bleaching: a global perspective of changing heat stress. Coral Reefs 38(4):547-557. http://link.springer.com/10.1007/s00338-019-01799-4

[21] Vargas-Ángel B, Huntington B, Brainard RE, Venegas R, Oliver T, Barkley H, Cohen A. 2019. El Niño-associated catastrophic coral mortality at Jarvis Island, central Equatorial Pacific. Coral Reefs 38(4):731-741. https://doi.org/10.1007/s00338-019-01838-0

[22] Baker A, Glynn P, Riegl B. 2008. Climate change and coral reef bleaching: An ecological assessment of long term impacts, recovery trends and future outlook. Estuar. Coast. Shelf Sci. 80(4):435-471.

[23] Thinesh T, Meenatchi R, Jose PA, Kiran GS, Selvin J. 2019. Differential bleaching and recovery pattern of southeast Indian coral reef to 2016 global mass bleaching event: Occurrence of stress-tolerant symbiont Durusdinium (Clade D) in corals of Palk Bay. Mar Pollut Bull. 145:287-294.

[24] Brown B. 1997. Coral bleaching: Causes and consequences. Coral Reefs. 16:129-138.

[25] De K, Nanajkar M, Arora M, Manckam, N, Mote, S, Ingole B. 2021. Application of remotely sensed sea surface temperature for assessment of recurrent coral bleaching (2014-2019) impact on a marginal coral ecosystem. Geocarto Int. 36. https://doi.org/10.1080/10106049.2021.1886345

[26] Frieler K, Meinshausen M, Golly A, Mengel M, Lebek K, Donner S, Hoegh Guldberg O. 2013. Limiting global warming to 2C is unlikely to save most coral reefs. Nat. Clim. Change. 3(2):165-170.

[27] Venkataraman K, Satyanarayana C, Alfred J R B, Wolstenholme J.2002. Handbook on hard corals of India, (Zoological Survey of India, Kolkata, India), pp. 266.

[28] De K, Venkataraman K, Ingole B. 2020a. The hard corals (Scleractinia) of India: a revised checklist. Ind J Geo-Mar Sci. 49(10):1651-1660.

[29] Venkataraman K. 2011. Coral reefs of India. In: Hopley D, editor. Encyclopaedia of modern coral reefs. Springer Netherlands; p. 267-275.

[30] De K, Sautya S, Mote S, Tsering L, Patil V, Nagesh R, Ingole B.

2015. Is climate change triggering coral bleaching in tropical reef? Curr Sci. 109(8):1379-1880.

[31] De K, Venkataraman K, Ingole B. 2017. Current status and scope of coral reef research in India: A bio-ecological perspective. Ind J Geo-Mar Sci, 46: 647-662.

[32] De K, Nanajkar M, Mote S, Ingole B. 2020b. Coral damage by recreational diving activities in a Marine Protected Area of India: Unaccountability leading to 'tragedy of the not so commons'. Mar Poll Bull, 155, https://doi.org/10.1016/j.marpolbul.2020.111190

[33] Strong AE, Arzayus F, Skirving W, Heron SF. 2006. Identifying coral bleaching remotely via Coral Reef Watch: Improved integration and implications for changing climate. In: Phinney JT, Al. E, editors. Coral Reefs Climate Change Science Management (Coastal Estuar Stud vol 61. AGU, Washington, D. C.; p. 163-180.

[34] Done TJ, Turak EI, Wakeford M, Kininmorith S, Wooldridge S, Berkelman R, Oppen MJH van, Mahonly M. 2003. Testing bleaching resistance hypothesis for the 2002 Great Barrier Reef bleaching event. Queensland, Australia.

[35] Sully S, Burkepile DE, Donovan MK, Hodgson G, van Woesik R. 2019. A global analysis of coral bleaching over the past two decades. Nat Commun 10(1):1264. http://www.nature.com/articles/s41467-019-09238-2

[36] Roxy MK, Ritika K, Terray P, Murtugudde R, Ashok K, Goswami BN. 2015. Drying of Indian subcontinent by rapid Indian ocean warming and a weakening land-sea thermal gradient. Nat Commun. 6(1):1-10.

[37] D'Mello JR, Prasanna Kumar S. 2018. Processes controlling the accelerated warming of the Arabian Sea. Int J Climatol. 38(2):1074-1086.

[38] Sun C, Li J, Kucharski F, Kang I, Jin F, Wang K, Wang C, Ding R, Xie F. 2019. Recent Acceleration of Arabian Sea Warming Induced by the Atlantic-Western Pacific Trans-basin Multidecadal Variability. Geophys Res Lett 46(3):1662-1671. https://onlinelibrary.wiley.com/doi/abs/10.1029/2018GL081175

[39] Shetye SS, Kurian S, Gauns M, Vidya PJ. 2019. 2015-16 ENSO contributed reduction in oil sardines along the Kerala coast, south-west India. Mar Ecol 40(6). https://onlinelibrary.wiley.com/doi/abs/10.1111/maec.12568

[40] Vidya PJ, Kurian S. 2018. Impact of 2015-2016 ENSO on the winter bloom and associated phytoplankton community shift in the northeastern Arabian Sea. J Mar Syst. 186:96-104.

[41] Vineetha G, Karati KK, Raveendran T V, Idrees Babu KK, Riyas C, Muhsin MI, Shihab BK, Simson C, Anil P. 2018. Responses of the zooplankton community to peak and waning periods of El Niño 2015-2016 in Kavaratti reef ecosystem, northern Indian Ocean. Environ Monit Assess 190(8):1-22. https://doi.org/10.1007/s10661-018-6842-9

Chapter 3

Mapping and Assessment of Evapotranspiration over Different Land-Use/Land-Cover Types in Arid Ecosystem

Khalid G. Biro Turk, Faisal I. Zeineldin
and Abdulrahman M. Alghannam

Abstract

Evapotranspiration (ET) is an essential process for defining the mass and energy relationship between soil, crop and atmosphere. This study was conducted in the Eastern Region of Saudi Arabia, to estimate the actual daily, monthly and annual evapotranspiration (ETa) for different land-use systems using Landsat-8 satellite data during the year 2017/2018. Initially, six land-use and land-cover (LULC) types were identified, namely: date palm, cropland, bare land, urban land, aquatic vegetation, and open water bodies. The Surface Energy Balance Algorithm for Land (SEBAL) supported by climate data was used to compute the ETa. The SEBAL model outputs were validated using the FAO Penman-Monteith (FAO P-M) method coupled with field observation. The results showed that the annual ETa values varied between 800 and 1400 mm.year^{-1} for date palm, 2000 mm.year^{-1} for open water and 800 mm.year-1 for croplands. The validation measure showed a significant agreement level between the SEBAL model and the FAO P-M method with RMSE of 0.84, 0.98 and 1.38 mm.day^{-1} for date palm, open water and cropland respectively. The study concludes that the ETa produced from the satellite data and the SEBAL model is useful for water resource management under arid ecosystem of the study area.

Keywords: Actual Evapotranspiration (ETa), Landsat-8 data, SEBAL model, Land-use and land-cover (LULC), arid ecosystem

1. Introduction

Evapotranspiration (ET) is an essential component of the water cycle that connects hydrologic and biological processes. It is directly affected by water and land management, land-use change and climate variability [1, 2]. Therefore, estimation of ET in vast areas using efficient tools is important for optimum water resources management over different land-use systems [3].

The knowledge of ETa and its spatial distribution can have a great potential to develop new cost-effective indicators of irrigation performance and increase water use efficiency [4]. The ET can be estimated using many methods and techniques such as lysimeters, sap flow, eddy covariance, Bowen ratio, and scintillometer, which were accurate and efficient at the field scale [5]. However, these techniques cannot be used for large regional scale ET mapping due to prohibitive cost and logistical limitations [6]. Accordingly, remote sensing and biophysical modelling are adequate techniques for evaluating the ET patterns over large scale regions.

SEBAL was selected in this study to estimate the ETa because it can measure the ET without requiring quantifying the other complicated hydrological processes [7]. Also, SEBAL can identify the satellite image's dry and wet pixels by showing a linear relationship between the surface temperature and the temperature gradient difference [8]. The FAO P-M method was adopted to validate the ETa obtained from the SEBAL model, since it used the actual climate data of the study area for quantifying the ETa process.

1.1 Application of LULC systems in hydrology

The study of land use and land cover (LULC) systems directly affects hydrology at different scales. Many studies showed that the LULC changes have clear impacts on the soil surface runoff, evapotranspiration, groundwater recharge, streamflow, and water balance over agricultural areas.

The impacts of LULC changes on hydrology were investigated in the Loess Plateau of China by Lu et al. [9] using hydrological modelling during the period 1995–2010. There was the transformation of farmland into forests, grassland, and built-up land. They showed slight increases in average annual potential evapotranspiration, actual evapotranspiration, and water yield at the basin scale, but soil water decreased between the two intervals. However, in sub-basins, obvious LULC changes did not have clear impacts on hydrology, and the impacts may be affected by precipitation conditions. The streamflow was also affected by the LULC changes in the Dinder and Rahad Rivers basins located in Ethiopia and Sudan [10]. The LULC of the catchment indicated a significant decrease and increase of the woodland and croplands were observed between 1972 and 2011. The effect of LULC change on streamflow was significant during 1986 and 2011, which could be attributed to the severe drought during the mid-1980s and the recent large expansions of cropland.

The hydrological process of a periurban catchment was quantified using high-resolution satellite images (0.50–2.50 m) in the Yzeron district of France [11]. The produced land covers maps of the district categorised into sub-catchments dominated by vegetation and imperviousness areas. According to the image processing and images characteristics, the calculated imperviousness rates were different, and lead to significant differences in the hydrological response. Wolfe et al. [12] compiled climate, geological, topographical, and land-cover data from the Prairie in Canada, and conducted a classification of watersheds using hierarchical clustering of principal components. Their analysis resulted in 7 classes based on the clustering of watersheds. The important defining variables identifying the watersheds clustering were climate, elevation, surficial geology, wetland distribution, and land cover. The authors indicated that developing management strategies is essential to prevent watersheds from future change.

Estimating the actual evapotranspiration and crop coefficients of an almond and pistachio orchard was explored in Central Valley, California, USA during an entire growing season by combining a simple crop evapotranspiration model with remote sensing data [13]. The authors used vegetation index NDVI derived from Landsat-8 to estimate the basal crop coefficient (Kcb), or potential crop water use. Their results showed that the model indicated a difference of 97 mm in transpiration over the season between both crops. However, the soil evaporation accounted for an average of 16% and 13% of the total actual evapotranspiration for almonds and pistachios, respectively. They concluded that the combination of crop evapotranspiration models with remotely-sensed data helps up-scaling irrigation information from plant to field scale and thus may be used by farmers for making day-to-day irrigation management decisions.

Furthermore, the up-scaling of the daily and seasonally ET using multisource remote sensing images was explored by Cha et al. [14] in the agricultural lands of the Kai-Kong River Basin, Xinjiang, China. They proposed a trapezoidal and a sinusoidal method to upscale daily ET values to seasonal ET. Moreover, the actual ET over LULC types in India's Malaprabha River Basin was estimated using Landsat 8 data and surface energy balance models [15]. Their results demonstrate the challenge in actual ET estimation at a fine spatial resolution and highlight the importance of choosing a suitable algorithm.

1.2 Using of SEBAL model for ET estimation

The algorithms that use remote sensing products to estimate the energy balance and ET have become increasingly common [16]. Examples of these models are the two-source energy balance (TSEB), developed by Norman et al. [17]; the surface energy balance algorithm for land (SEBAL), formulated by Bastiaanssen (1995); and the mapping of evapotranspiration at high resolution with internalised calibration (METRIC), developed by Allen et al. [18].

SEBAL model is the most widely used algorithm for estimating ET throughout the world that involves applications for agricultural and water resources management, urbanisation impacts, aquifer recharge, and water balances [16].

The SEBAL model applied to determine the distribution of the ETa for analysing water use patterns over a large basin in Kenya [19]. In Indonesia's Java Province, a method was developed using SEBAL to detect and describe the spatial variability of lowland rice ET [20]. The water used to assess irrigation system performance and management was pointed in the Indus Basin, Pakistan, using the ET estimated by SEBAL and MODIS data [21]. Remote sensing and biophysical modelling were used in recent studies to estimate the ET in different Saudi Arabia regions. Mahmoud and Alazba [22] estimated the ETa in the western and southern regions of Saudi Arabia during 1992–2014 using the SEBAL model, MODIS data and field observations. However, the application of SEBAL in Mara Basin of East Africa indicates that the ETa is measurable over different land-use types in data-scarce regions [23]. Daily ET monitoring using SEBAL model also found to be possible for improving water resources decision support over an oasis in the desert ecosystem [24]. The efficiency of the SEBAL model in estimating ET of pistachio crop in the Semnan province of Iran was investigated. The model shows good efficiency for estimating the actual ET of the pistachio product [25]. Moreover, SEBAL was used to calculate ET during the cultivation and harvesting of wheat crops in the Ilam province, Iran [26]. Thus the evaluation using SEBAL and the FAO-Penman–Monteith method showed that SEBAL has sufficient accuracy for estimating ET.

The Kingdom of Saudi Arabia (KSA) suffers a continuing water scarcity and almost around 90% of the agricultural sector's water budget [27]. Groundwater is the primary source of water in the KSA, considering the limited precipitation and high agricultural demands. Moreover, the increased population of Saudi Arabia resulted in a significant increase in water use [28]. Also, the diversity of the LULC in the arid region is critical to water consumption. Accordingly, for effective water resources management in these regions, the impacts of LULC on hydrology need to be assessed. Hence, precise information of the ETa is crucial for policymakers and water planners to develop and formulate strategies for agricultural water utilisation. Therefore, the objective of this study is to assess the potential of Landsat-8 data and SEBAL model for estimating the daily, monthly and annual ETa under different LULC systems in Al-Ahsa region of Saudi Arabia.

2. Materials and methods

2.1 Study area

The study area is located in the eastern province of Saudi Arabia and lies about 300 km from Riyadh and 70 km west of the Arabian Gulf (**Figure 1**). The study area consists of Al-Ahsa Oasis, one of the leading agricultural centres in Saudi Arabia and dominated by date palm plantation [29]. The climate in the study area is hot and dry during the summer, with a relatively high air temperature that might exceed 45 °C. However, the winter is wet, with a minimum air temperature that might reach 5 °C. The annual rainfall is less than 250 mm and occurs mainly during the winter [30]. The dominated soils of the study area are sandy to sandy loam soil. Groundwater is the primary water source in the oasis and used mainly for irrigation, domestic and industrial purposes [31]. The land-use system in the study area is predominated by date palm plantation as the main agricultural activity. However, the cropping of rice and vegetables is also practiced in the area.

2.2 Data

The data used in this study consist of remote sensing, climate information and field observations. They collected during April 2017–March 2018 to include the summer and the winter season of the study area. The summer season is considered for Aril–September, while the winter is for October–March.

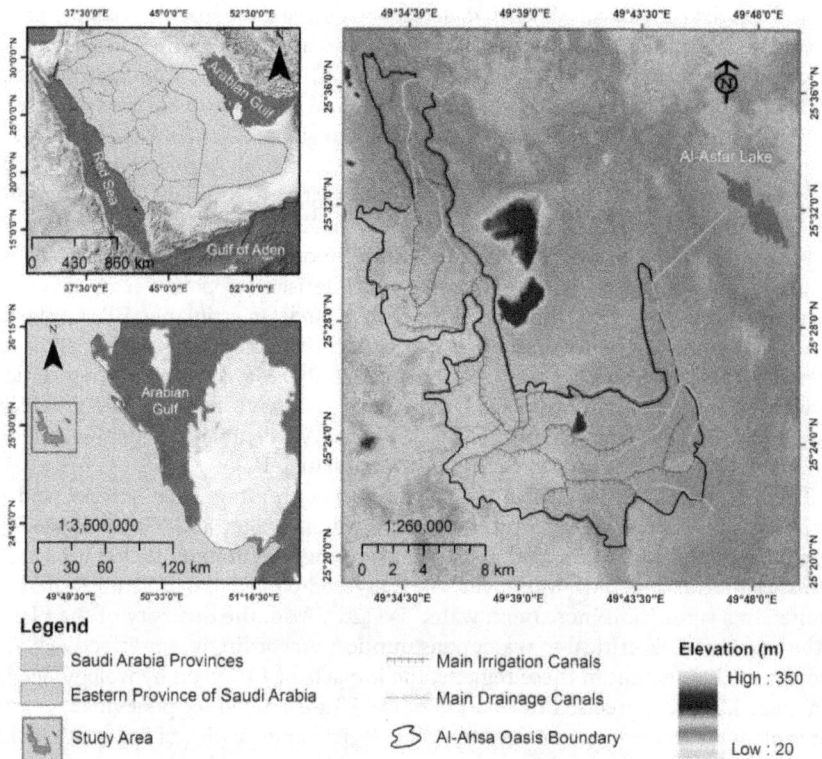

Figure 1.
Location of the study area.

A total number of 22 Landsat–8 images (path/row is 164/042) acquired from the United States Geological Survey (USGS-https://earthexplorer.usgs.gov/) to cover the entire study area, and the characteristics of these data are shown in (**Table 1**). The obtained Landsat-8 images have a cloud cover of less than 10%, and they have been geometrically and radiometrically corrected. All images bands were resampled into a pixel size of 30 m × 30 m using the nearest neighbour method. A global digital elevation model (DEM) is generated from the Advanced Spaceborne Thermal Emission and Reflection Radiometer (ASTER) known as ASTER GDEM obtained from the USGS website. It is a 30 m grid size DEM produced by the National Aeronautics and Space Administration (NASA) and the Ministry of Economy, Trade, and Industry of Japan (METI).

The climate data was collected from two meteorological stations located within the study area. These data include air temperature, relative humidity, wind speed, net radiation, precipitation and vapour pressure, and all data collection on an hourly and daily basis. However, the field observations include the identification of the main land-use system in the study area. Besides field notes, site descriptions, and terrestrial photographs were taken to relate the site location to scene features.

2.3 Analysis of LULC

A field survey was conducted throughout the study area to identify the LULC classes during the study period. Global Positioning System (GPS) instrument was used to obtain accurate location point data for each LULC class included in the classification process. A total number of 115 ground control points (GCPs) were collected. A supervised maximum likelihood classification (MLC) method was employed to classify images. Based on the study objectives, the supervised classification applied in this study does not compare different classifiers. Therefore, the MLC was adopted to be the only classification method for this study. MLC widely used in remote sensing for image classification [33–35]. The accuracy assessment of the classified images was performed using 30% of the collected GCPs. Also, visual interpretation of the unclassified satellite images supported with the field observations was used to validate the LULC maps. However, to reduce bias, the stratified

Sensor	Bands Type	Wavelength (μm)	Spatial Resolution (m) Resolution
Operational Land Imager (OLI) and Thermal Infrared Sensor (TIRS)	Band 1, Coastal aerosol	0.43–0.45	30
	Band 2, Blue	0.45–0.51	30
	Band 3, Green	0.53–0.59	30
	Band 4, Red	0.64–0.67	30
	Band 5, Near Infrared (NIR)	0.85–0.88	30
	Band 6, Short-wave Infrared 1 (SWIR-1)	1.57–1.65	30
	Band 7, Short-wave Infrared 2 (SWIR-2)	2.11–2.29	30
	Band 8, Panchromatic	0.50–0.68	15
	Band 9, Cirrus	1.36–1.38	30
	Band 10, Thermal Infrared 1 (TIRS-1)	10.60–11.19	100
	Band 11, Thermal Infrared 2 (TIRS-2)	11.50–12.51	100

Table 1.
Characteristics of Landsat-8 data used in this study [32].

random sampling method was adopted to classify images [36]. Finally, the overall accuracy, user's and producer's accuracies, and the Kappa statistic were derived from the error matrices [37].

2.4 SEBAL method

The SEBAL model developed by Bastiaanssen et al. [38] was used to calculate the actual evapotranspiration from Landsat-8 satellite images. The model's key inputs were the satellite measurements of surface albedo, normalised difference vegetation index (NDVI) and surface temperature (Ts). Also, the DEM and land-use map were used as additional input data. The DEM was applied for topographic and atmospheric correction [39]. However, the land-use map was used mainly to differentiate between LULC types exist in the study area. In addition to the satellite data, the SEBAL model requires minimum inputs of routine weather data (see the Data section). **Figure 2** shows a flowchart that describes the SEBAL model process. The SEBAL model scripts were formed using the Spatial Modeller Tool of the ERDAS IMAGINE 9.2 software, and the ArcGIS 10.2 software was used for data mapping and visualisation.

The SEBAL algorithm computes the latent heat flux as the residue of the energy balance Equation [38, 40, 41]:

$$\lambda ET = R_n - G - H \tag{1}$$

where R_n is the net radiation over the surface (W/m^2), G is the soil heat flux (W/m^2), H is the sensible heat flux (W/m^2), λET is the latent heat flux (W/m^2) and λ is the latent heat of vaporisation (J/Kg).

2.4.1 Estimation of the Main surface parameters

The main surface parameters of the SEBAL include: the surface albedo, land surface emissivity (ε) and land surface temperature (Ts).

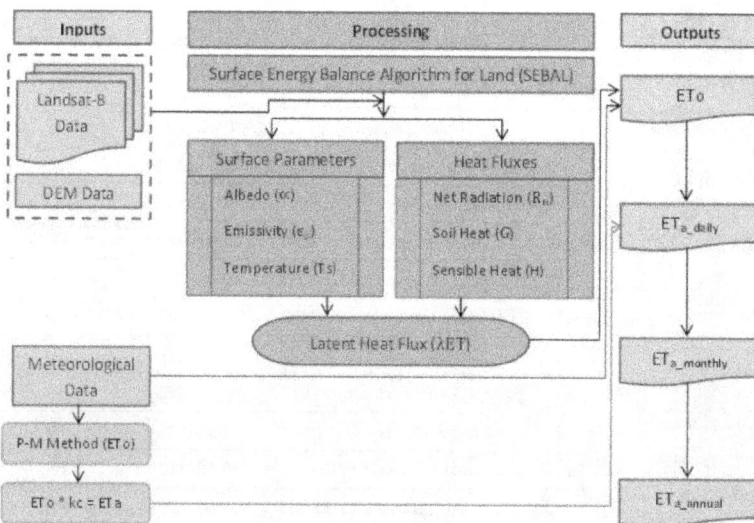

Figure 2.
A flowchart explains the SEBAL model process.

Surface albedo is defined as the ratio of reflected radiation from the surface to the incident shortwave radiation [40]. Thus the albedo is a single value that represents the integrated reflectance across the entire shortwave spectrum as represented by bands 2–7 of Landsat-8 data. Accordingly, the surface albedo (α) was estimated using the following formula [42]:

$$\alpha = 0.2453\alpha_2 + 0.0508\alpha_3 + 0.1804\alpha_4 + 0.3081\alpha_5 + 0.1332\alpha_6 + 0.05221\alpha_7 + 0.0011 \tag{2}$$

where α_2, α_3, α_4, α_5, α_6, α_7 represent the albedo of band2, band3, band4, band5, band6, and band7, respectively.

Surface emissivity is the thermal energy ratio radiated by the surface to the thermal energy radiated by a blackbody at the same temperature [42]. The NDVI is vital for calculating the land surface emissivity since it is used to estimate the vegetation coverage [43]. The calculation of the NDVI and vegetation coverage was as follows [8, 44]:

$$NDVI = \frac{(NIR - RED)}{(NIR + RED)} \tag{3}$$

$$P_v = \frac{(NDVI - NDVI_{min})}{(NDVI_{max} + NDVI_{min})} \tag{4}$$

where NIR is the reflectance in the near-infrared band, which corresponds to band 5 in lansat-8 data, while the RED reflectance corresponds to band 4. P_v is the vegetation coverage.

The surface emissivity was conditionally determined based on the NDVI values using Eq. (5) [45]:

$$\varepsilon_0 = \varepsilon_v P_v + \varepsilon_s (1 - P_v) + C_\varepsilon \tag{5}$$

where ε_0 is the land surface emissivity, ε_v and ε_s are the vegetation and soil emissivity, respectively, and C_ε represents the surface roughness ($C_\varepsilon = 0$ for homogenous and flat surfaces) taken as a constant value of 0.005 [46].

The P_v values are conditioned with the NDVI ones. The land cover is classified as water when the NDVI ‘0. For the NDVI values range between 0 and 0.2, the land is considered covered with soil. The NDVI values of 0.2–0.5 the land cover are considered mixtures of soil and vegetation. However, when the NDVI >0.5, the land is considered covered with vegetation [43].

The land surface temperature was derived from the thermal bands. This step needs the spectral radiance to be converted into a sufficient brightness temperature. That means it has the black body temperature, assuming that the Earth's surface is a black body. Consequently, brightness temperature was determined using the formula [47]:

$$T_b = \frac{K_2}{\ln\left(\frac{K_1}{L_\lambda} + 1\right)} - 273.15 \tag{6}$$

where, T_b is the satellite brightness temperature (C°), $L\lambda$ spectral radiance at top of the atmosphere, K1 and K2 are satellite calibration constants from the image metadata. The absolute zero value of -273.15C° was added to obtain results in Celsius [48].

The land surface temperature (T_s) was computed using Eq. (7) [49]:

$$T_s = \frac{T_b}{1 + \left(\lambda . \frac{T_b}{\rho}\right) . \ln \varepsilon_0} \tag{7}$$

where, λ is the wavelength of emitted radiance (11.5 μm for Landsat 5 &7; 10.9 μm for Landsat 8), ε_0 is the land surface emissivity, $\rho = h.\frac{c}{\sigma} = 1.438 \times 10^{-2}$ mK, σ is the Boltzmann constant $(1.38 \times 10^{-23}$ J/K), h is the Planck's constant $(6.626 \times 10^{-34}$ J s), and c is the velocity of light $(2.998 \times 10^8$ m/s).

2.4.2 Determination of heat fluxes

The net radiation (R_n) was calculated using surface reflectance and surface temperature (Ts):

$$R_n = R_s \downarrow - \alpha R_s \downarrow + R_L \downarrow - R_L \uparrow - (1 - \varepsilon_0)R_L \downarrow \tag{8}$$

where $R_s \downarrow$ is the incoming short wave radiation (W/m^2) (solar radiation), α surface albedo (dimensionless), $R_L \downarrow$ is the incoming long wave radiation (W/m^2), RL↑ is the outgoing long wave radiation (W/m^2), and ε0 is the surface thermal emissivity (dimensionless). The calculation of these radiations was performed in Eqs. (9)–(11) [50]:

$$RS \downarrow = G_{sc} . \cos \theta . r . \tau_{sw} \tag{9}$$

$$RL \uparrow = \varepsilon_0 . \sigma . T_s^4 \tag{10}$$

$$RL \downarrow = \varepsilon_\alpha . \sigma . T_\alpha^4 \tag{11}$$

where G_{sc} is the solar constant, 1367 W/m^2 and $\cos \theta$ is the cosine of the solar incidence angle, r is the Earth-Sun distance (dimensionless), and τ_{sw} is atmospheric transmissivity. RSI↓ values range from 200–1000 W/m^2, depending on the image's time and location and the local weather conditions [42]. σ is the Stefan-Boltzmann constant $(5.67 \times 10^{-8}$ W.m^{-2}.K^{-4}), T_s is the surface temperature (K), ε_α is the atmospheric emissivity, and T_α is the atmospheric temperature (K). The empirical Eq. (6) was used for calculating the ε_α [42].

$$\varepsilon_\alpha = 0.85 \times (-\ln \tau_{sw})^{0.09} \tag{12}$$

The soil heat flux (G) is the rate of the heat flux stored or released into the soil and vegetation due to conduction. The ratio G/R_n was computed using Eq. (13) developed by [51]:

$$G/R_n = \frac{T_s}{\alpha} \left(0.0038 + 0.0074\alpha^2\right) \left(1 - 0.98 NDVI^4\right) \tag{13}$$

where T_s is the surface temperature (C°), α is the surface albedo, and NDVI is the Normalised Difference Vegetation Index (ranged between −1 and + 1). NDVI values between 0 and 0.2 correspond to bare soil or very sparse vegetation, the values greater than 0.2 represent vegetated areas. The typical estimates of G/R_n assumed to be 0.5 for water, 0.05–0.4 for agriculture and 0.2–0.4 for bare soil [42].

The sensible heat flux (H) is the rate of heat loss to the air by convection and conduction, due to a temperature difference. H was determined using the aerodynamic based heat transfer equation as follows [51]:

$$H = \frac{(\rho \times C_p \times dT)}{r_{ah}} \qquad (14)$$

where ρ is air density (kg/m³), C_p is air specific heat (1004 J/kg/K), dT (K) is the temperature difference (T1 – T2) between two heights (z1 and z2), and r_{ah} is the aerodynamic resistance to heat transport (s/m). The r_{ah} is computed for neutral atmospheric stability conditions as:

$$r_{ah} = \frac{ln\left[\frac{Z_2}{Z_1}\right]}{u* \times k} \qquad (15)$$

where Z_1 and Z_2 are heights in meters above the zero plane displacement of the vegetation, $u*$ is the friction velocity (m/s) which quantifies the turbulent velocity fluctuations in the air, and k is von Karman's constant (0.41).

2.4.3 Estimating the evapotranspiration

The instantaneous value of ET in equivalent evaporation depth was computed as:

$$ET_{inst} = 3600 \frac{\lambda ET}{\lambda} \qquad (16)$$

where ET_{inst} is the instantaneous ET (mm/hr), 3600 is the conversion from seconds to hours, λET is the latent heat flux (W/m) consumed by ET, ρw is the density of water (1000 kg/m³), and λ is the latent heat of vaporisation (J/kg) and was computed as:

$$\lambda = \left[2.501 - 0.00236 \, (T_s - 273.15) \times 10^6\right] \qquad (17)$$

The reference ET fraction (ET_0F) or crop coefficient (kc) was calculated based on ET_{inst} for each pixel and ET_0 was obtained from local ground weather stations.

$$ET_0F = ET_{inst}/ET_0 \qquad (18)$$

The daily values of ET (ET24) (mm/day) for each pixel was calculated as follows:

$$ET_a = ET_0F \times ET_0 24 \qquad (19)$$

where ET_0F is the reference ET fraction, $ET_0 24$ is the cumulative alfalfa reference for the day (mm/day), and ET_a is the actual evapotranspiration for the entire 24-hour period (mm/day).

The actual monthly and annual ET was calculated using daily ET data as follows [42]:

$$ET_{a,period} = \sum_{i=m}^{n} ET_0F \times ET_0 24 \qquad (20)$$

$$ET_{a,annual} = \sum ET_{a,period} \qquad (21)$$

Allen et al. [18] showed that one cloud-free satellite image per month is sufficient to develop ET_0F curves for seasonal ET_a estimations.

2.4.4 Validation of the SEBAL model evapotranspiration

The produced ETa from Landsat-8 images and SEBAL model was validated using the FAO P-M method [52]. The FAO P-M was used to calculate the reference crop evaporation (ET_0) from the actual climate data in the study area based on Eq. (22):

$$ET_0 = \frac{0.408\Delta(Rn - G) + \gamma\frac{900}{T+273}U2(es - ea)}{\Delta + \gamma(1 + 0.34U2)} \tag{22}$$

where ET_0 is reference evapotranspiration (mm/day), Δ is slope vapour pressure curve (kPa/°C), γ is psychrometric constant (kPa/°C-1), T is mean daily air temperature at 2 m height (°C), U2 is wind speed at 2 m height (m/s), es is saturation vapour pressure (kPa), ea. is actual vapour pressure (kPa), (es − ea) represents the saturation vapour pressure deficit (kPa).

The crop coefficient (Kc) for the different croplands and the open water determined based on Allen et al. [52]. The ET_0 obtained from the FAO P-M method and the kc were used to calculate the ET_a depending on actual weather data as follows:

$$ET_a = ET_0 \times kc \tag{23}$$

The ET_a resulted from the FAO P-M method was used to validate the ET_a obtained from SEBAL model.

A linear correlation and the root mean square error (RMSE) between the measured (FAO P-M) and the SEBAL daily ETa was computed [13]. The RMSE was calculated as follows [26]:

$$RMSE = \sqrt{\frac{\sum_{i=1}^{N}(O_i - P_i)^2}{N}} \tag{24}$$

where O_i represents the observed values of the FAO P-M method as the standard model; P_i represents the estimated values from the SEBAL algorithm; and $\overline{O_i}$ and $\overline{P_i}$ are the mean values from the FAO P-M method and SEBAL model, respectively.

3. Results and discussions

3.1 LULC mapping

The LULC map of the study area showed that the main identified classes were the date palm, cropland, bare land, urban land, aquatic vegetation, and water (**Figure 3**). The area occupied by each LULC type within the oasis boundaries is shown in **Table 2**. The date palm covers about 131 km² of Al-Ahsa Oasis, and it is the most important land-use class for the local and national economy. Croplands used only 144 km² of the oasis area, and it is dominated by rice and vegetables. The bare land class occupies most of the oasis area with 4759 km². Bare lands dominated by desert and rock outcrops also occurred in this class. Most of the urban land occurs on the oasis periphery, as most oasis land is under agricultural use. The aquatic vegetation and water classes occupy together an area of about 17 km². Al-Dakheel [53] reported that date palm covered about 92% within the oasis boundary.

The overall classification accuracy of the LULC map was 90%, with a kappa index of 88%, while the user's and producer's accuracies differed along with LULC types (**Table 2**). This accuracy level indicates that the classification method adopted in this study effectively produced a compatible LULC map over the study area.

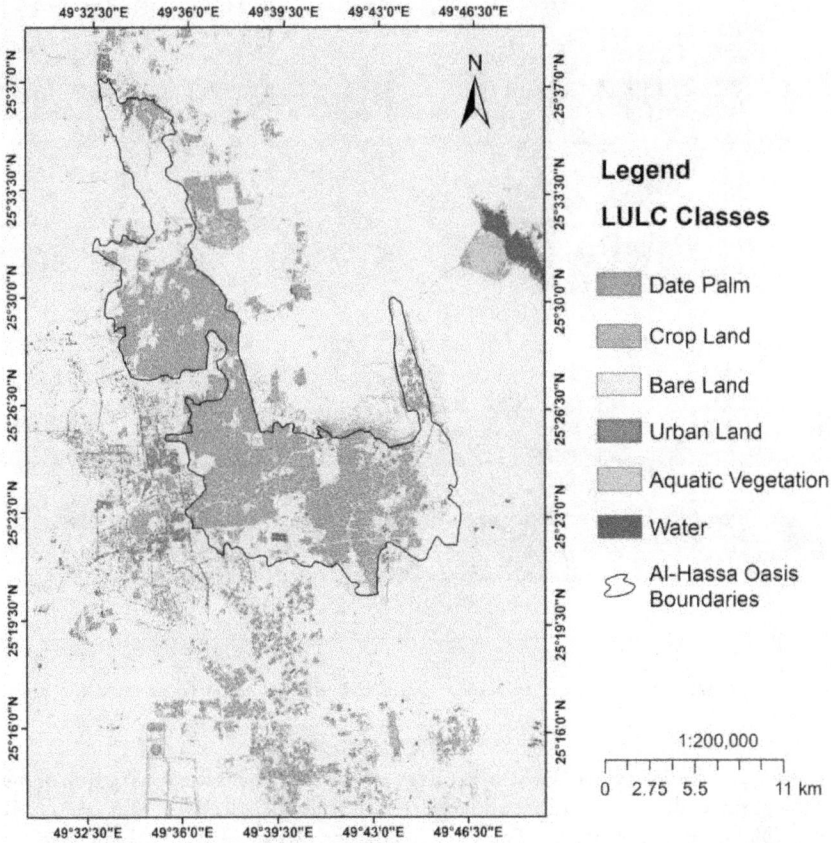

Figure 3.
LULC map of the study area.

LULC	Area		Classification Accuracy (%)	
	Km^2	%	User's	Producer's
Date Palm	131	2.5	95	94
Cropland	144	2.7	83	83
Bare Land	4759	89.9	95	94
Urban Land	241	4.5	82	86
Aquatic Vegetation	7	0.1	86	92
Water	10	0.2	100	100
Overall Accuracy			90%	
Kappa Statistic			88%	

Table 2.
Areas and accuracy assessment of the LULC classes.

3.2 Analysis of land surface parameters

The statistical means values of land surface albedo show that it was raining between 0.46 and 0.51 during Apr. 2017-Mar. 2018 (**Figure 4a**). However, the spatial distribution of land surface albedo indicates higher in the bare lands areas

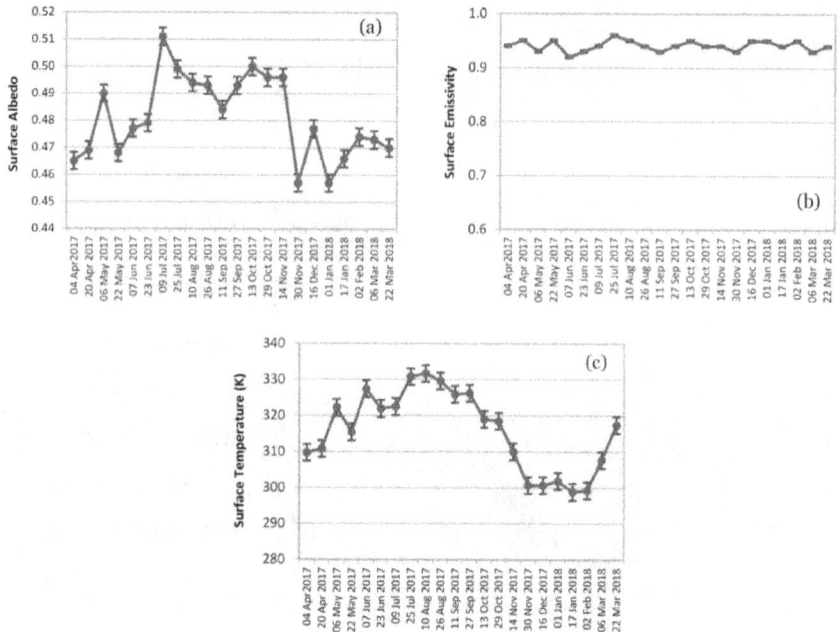

Figure 4.
Mean values of land surface parameters: (a) albedo; (b) emissivity and (c) surface temperature. Bars denoted standard error.

than the other LULC types (**Figure 5a**). Nevertheless, the seasonal variation shows that the surface albedo was higher during the summer (April, July) compared to the winter (November, February). The surface albedos levels were found lower in the vegetation-covered areas than the exposed soil [8].

The land surface emissivity means values clearly show that it does not vary along the study period (**Figure 4b**). However, the land surface emissivity's spatial patterns indicated higher in the open water and lowered in the date palm and croplands classes (**Figure 5b**). The lowest values of the land surface emissivity were observed in bare lands. These results are inconsistent with that reported by Kong et al. [8].

The land surface temperature statistics showed that the average minimum value of 299 K was observed in January and February 2018, while the maximum mean value of 332 K occurred on 10 August 2017 (**Figure 4c**). **Figure 5c** indicated that the highest land surface temperature values were shown in summer, and they associated with the bare lands. Nevertheless, during the winter, the difference in the land surface temperature between the bare lands and the other LULC types was 10–20 K. The land surface temperature is an essential parameter for quantifying the ET process among the different LULC types of the study area. Accordingly, the land surface temperature estimation is essential for land classification, energy budget estimations, and crop production [54].

3.3 Heat fluxes analysis

The surface heat fluxes estimated over the different LULC types can affect the ET amount measured along the study period at varying scales.

The net radiation flux's statistical values show that the values in April, May and June are clearly higher than those in October, November, December and January (**Figure 6a**). As shown in **Figure 7a**, the open water received the highest amount of

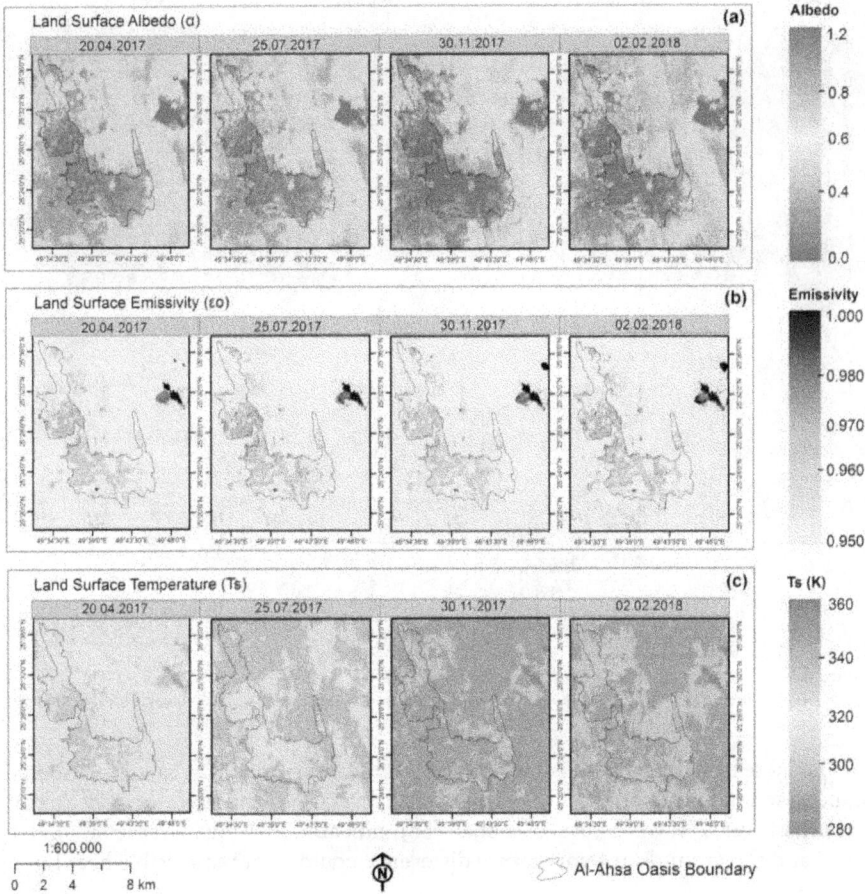

Figure 5.
Spatial distributions of land surface parameters. (a) Albedo; (b) emissivity and (c) surface temperature.

net radiation followed by the aquatic vegetation, date palm, and croplands. However, the lowest net radiation found in the bare lands. Moreover, the net radiation flux increased significantly in April and July compared to November and February.

The increase of the net radiation in April and May might be due to rising ground-air temperature and the gradual death of the sparse vegetation over bare lands. The change in the surface energy budgets due to irrigation results in increased net radiation over agricultural lands [55]. Also, the significant variation of the net radiation flux during the study period could be due to the region's heterogeneous LULC types.

The soil heat flux tendency showed higher in the summer recording average value of 91 W/m² on 23 June 2017, while the lowest mean value was 5 W/m² observed on 16 December 2017 (**Figure 6b**). The spatial distribution of the soil heat flux for the vegetation cover classes was higher than that of the bare lands during the summertime, while the difference was slight in winter (**Figure 7b**).

The sensible heat fluxes mean values do not vary consistently along with the summer and wintertime. The average highest value of 203 W/m² detected on 04 April 2017, while the lowest was on 10 August 2017 (**Figure 6c**). According to the spatial distribution in **Figure 7c**, the bare and urban lands' sensible heat flux show relatively high values compared to the vegetation lands.

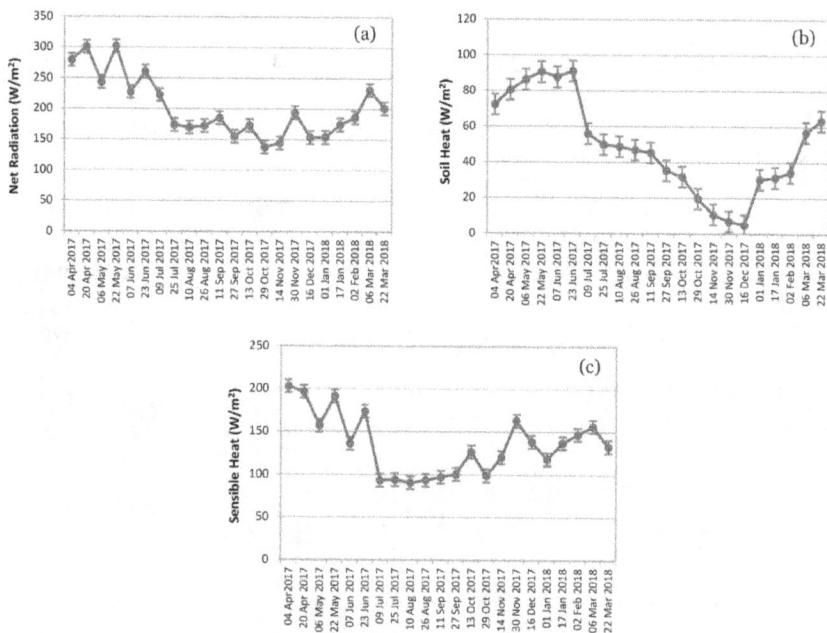

Figure 6.
Mean values of heat fluxes: (a) net radiation; (b) soil heat and (c) sensible heat. Bars denoted standard error.

The increasing trend in sensible heat flux during April and May can be attributed to increased net radiation flux under persistent irrigation land used for agriculture and urban areas. However, Amatya et al. [56] indicated that the increase in wind speed and the ground-air temperature difference could increase sensible heat flux.

3.4 Analysis of the actual evapotranspiration

The spatial distribution of the daily ETa over the study area during 20 April 2017–2022 March 2018 is shown in **Figure 8**. The temporal patterns of the daily ETa showed that the highest values were observed during the peak summertime in July and August. The mean daily ETa values for the different LULC types in the oasis are shown in **Figure 9**. It is clear that the daily ETa for the water bodies and aquatic vegetation was between 5.6–8.7 mm.day^{-1} in summer and about 2.3–5.6 mm.day^{-1} in winter. However, the date palm and croplands showed daily ETa of 3.5–8.0 and 2.0–3.6 mm.day^{-1} for the winter and summer, respectively.

The variability of the daily ETa values for the Date palm and cropland during the summer and winter times mainly attributed to the irrigation water distribution, soil salinity, drainage, and agricultural practices and their impact on moisture and salinity in the root zone. Under Saharan oasis conditions, soil texture, plot size, and farmers' practices in particular irrigation found to have significant effects on the daily ETa [57]. In Saudi Arabia, the daily ETa of date palm was observed to decrease during winter and increased during summer, depending on the crop's growth stage [58]. Also, the daily water consumption of major cropping systems in Saudi Arabia varied spatially depending on cropping practises and climatic conditions [22].

The mean daily ETa for the urban land ranged from 1.3 to 4.5 mm.day^{-1} throughout the study period (**Figure 6**). The urban land is covered with some lanes and parks that make seasonal variations of the daily ETa within the study area.

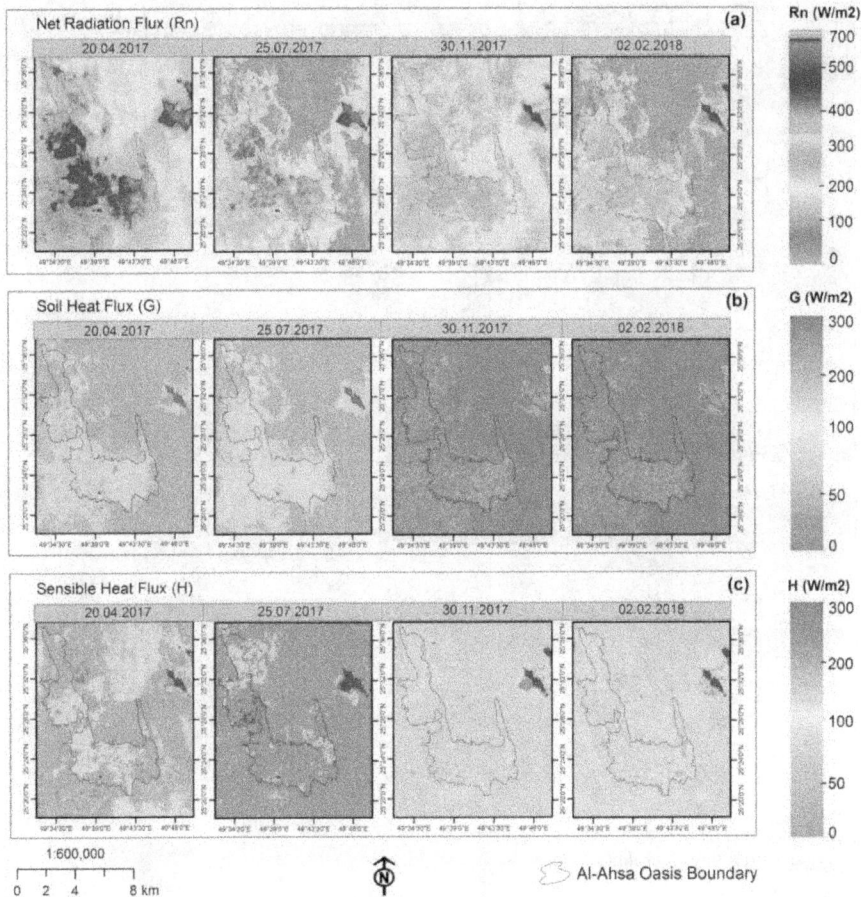

Figure 7.
Spatial distributions of heat fluxes. (a) Net radiation; (b) soil heat and (c) sensible heat.

Nevertheless, the low values of the daily ETa showed in the bare land mainly resulted from the sparse vegetation in this land-use system.

The variation of the daily ETa estimates for the different LULC types under the oasis-arid ecosystem indicates that they change significantly throughout the seasons [24].

Figure 10 shows the spatial pattern of the monthly and annual ETa across the study area. The high rates of the ETa found to be between 80 to 200 mm.month^{-1} during July, August and September for the water, aquatic vegetation, date palm and cropland. However, at the beginning of the summer in April, May and June, the ETa rates were 60–100 mm.month^{-1} for the same LULC types. Nevertheless, in the winter during Oct 2017–Mar 2018 the ETa ranged between 40 to 140 mm.month^{-1} for the water, aquatic vegetation, date palm and cropland land-use systems.

The mean annual ETa produced by SEBAL model for the different LULC types in the study area is shown in **Figure 11**. The ETa rates of date palm trees ranged from 800 to 1400 mm.year^{-1} during Apr. 2017 to Mar. 2018. The annual water consumption for date palm was highly variable. This might be attributed to the type of irrigation system and the age variations of date palm trees along the oasis.

The open water evaporation lost was around 2000 mm.year^{-1}, while an average of 1600 mm.year^{-1} was evaporated from aquatic vegetation. Nevertheless,

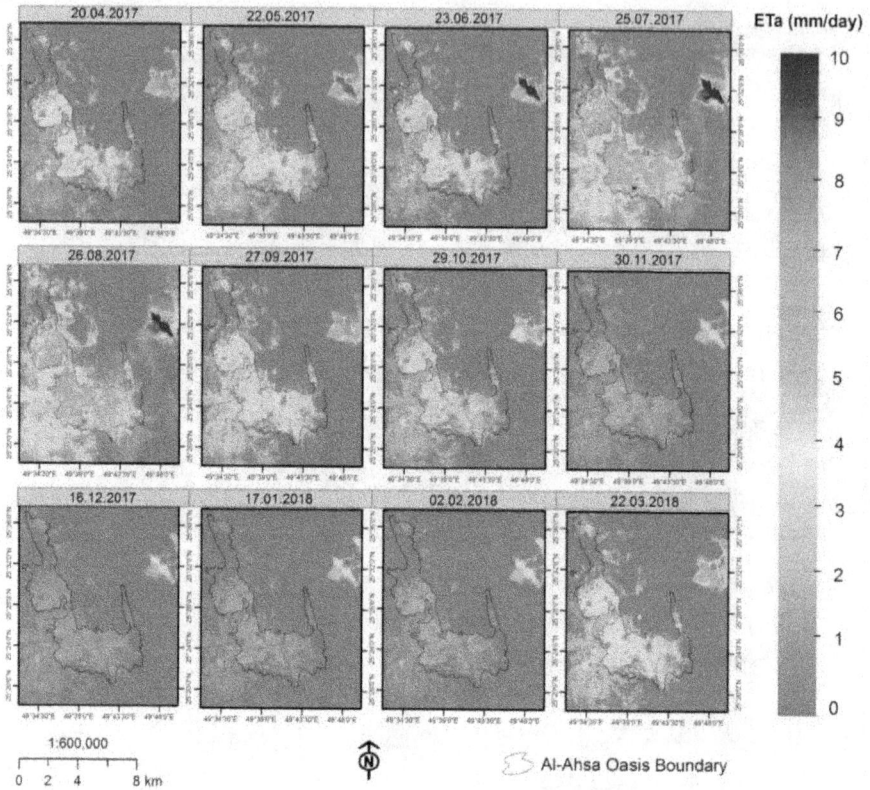

Figure 8.
The spatial distribution of the daily ETa.

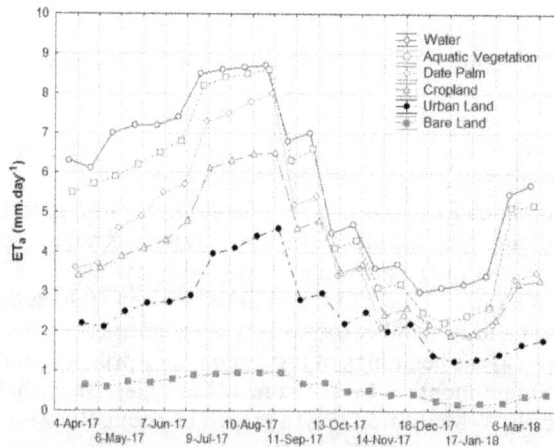

Figure 9.
Temporal variation of mean daily ETa for different LULC systems.

croplands showed lower annual ETa of 800 mm.year^{-1} compared to the date palm. The main crops like rice and vegetables can be cultivated during a particular time of the year in the oasis; therefore, croplands showed relatively low annual ETa compared to date palm areas.

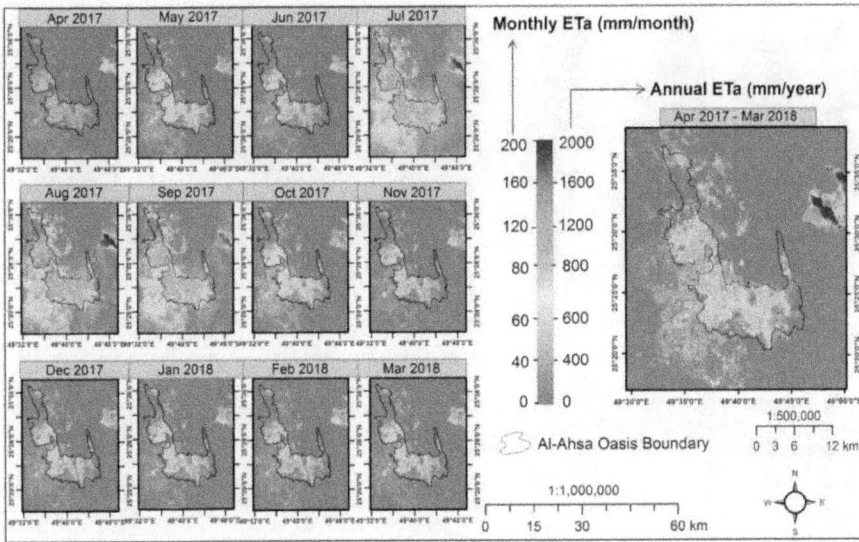

Figure 10.
The spatial patterns of the monthly and annual ETa.

The annual ETa of urban lands was 400 mm.year^{-1}. Urban lands are affected by the irrigation of trees, lanes and parklands, which resulted in consumption of a large amount of the oasis groundwater. The annual evaporation from bare lands was very low (200 mm.year^{-1}) and less than the long-term average rainfall. Bare lands equipped most of the study region areas, and they covered with sand dunes (**Figure 11**). Moreover, the bare lands characterised by low vegetation coverage levels and low water contents in the soil surface [41]. Also, very low rates of ETa from bare soil observed in the western and southern parts of Saudi Arabia [22].

3.5 Validation, limitation and uncertainty of SEBAL model

The FAO P-M was used as a standard method to validate the SEBAL model [8, 59]. The validation measurement for ETa between the SEBAL model and FAO P-M method for the different land-use system is shown in **Figure 12**. A significantly high level of agreement can be observed between the two methods for the selected LULC types. The RMSE for the most validated LULC system in the study area found to be less than 1.0 mm.day−1. However, the RMSE was slightly higher for cropland areas (**Figure 12d**), mainly due to the method used for calculating the crop coefficient (kc) for the different crop types within the croplands system. The SEBAL model does not require kc information because the model biophysical properties estimated kc as part of the SEBAL process. However, the FAO P-M computed kc based on the characteristics and climatic regions for the different crops. Accordingly, the kc of vegetables (tomato and cucumber) ranged between 0.5–1.15, and for the rice, it was 1.0–1.35 [52]. Nevertheless, the date palm's kc was in the range of 0.9–0.95, while it was 1.05 for the open water [52].

Moreover, it seems that SEBAL underestimates the ETa for croplands since they were diverse in terms of crop type and growing season. Also, the kc used to calibrate the ETo was made only for a few experimental plots. Therefore, the variability in the ETa values predicted by SEBAL and measured by the FAO P-M method was slightly higher. However, the low values of ETa in winter do not affect the outputs of seasonal ETa over croplands, as farmers use a little water for irrigation. The kc of

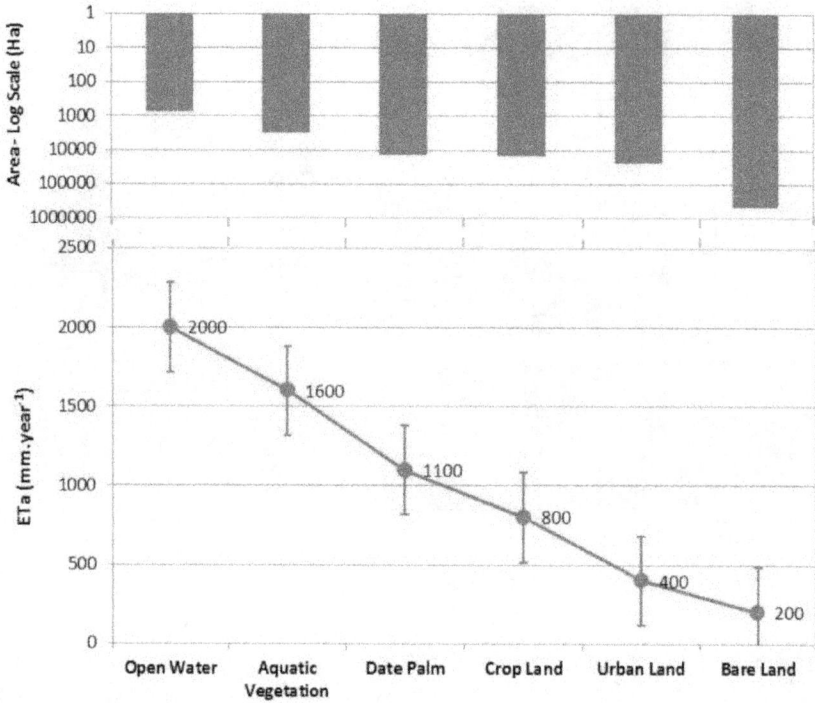

Figure 11.
Mean annual ETa produced by SEBAL for the different LULC types in the study area during April 2017–March 2018. Bars denoted standard error.

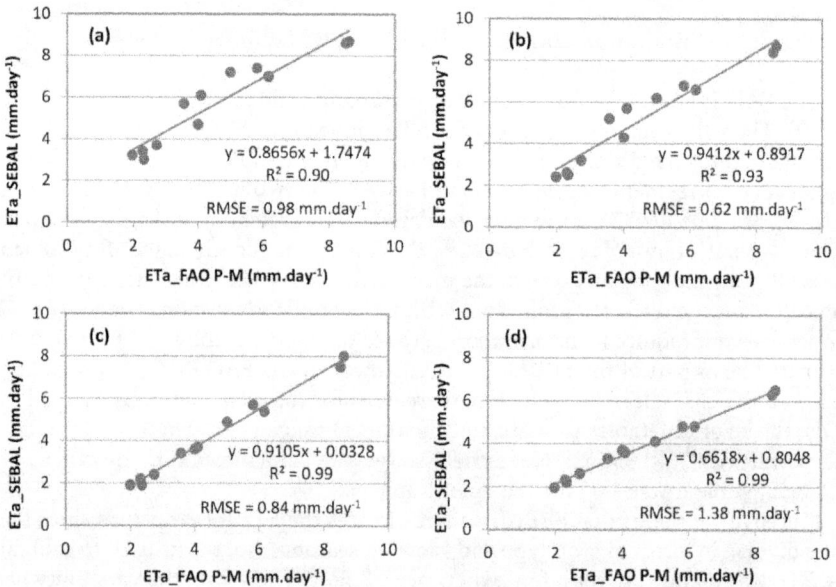

Figure 12.
Linear correlation between the FAO P-M and SEBAL model for the different LULC types. (a) Open water; (b) aquatic vegetation; (c) date palm; (d) cropland.

crops can vary during the growing season, depending on their growth stage [60]. Rahimi et al. [3] reported that the application of SEBAL for estimating the ETa over agricultural land in the Tajan catchment of Iran resulted in RMSE of 1.49 mm.day-1. However, the daily estimated ETa for both cropland and grassland in the Midwestern United States using SEBAL contributed to RMSE ranging between 1.74 and 2.46 mm.day^{-1} [6].

SEBAL model utilises satellite imagery and a small set of surface weather information, including wind speed and air temperature at a reference height of 2 m to solve the energy balance [38]. However, the steps used for the SEBAL process are complicated. The model requires an experienced modeller and a substantial number of working hours for image processing. Also, the large model number of mathematical formulas increases the possibility of human errors [42]. Furthermore, the procedure used by SEBAL for selecting the "hot" and "cold" pixels endmembers involves not only an analysis of the surface temperature but also an understanding of other biophysical properties, such as vegetation indices, surface albedo, and LULC type [16, 61]. Therefore, identifying these endmembers requires modeller intervention and knowledge of the biophysical parameter values and ranges.

4. Conclusions

This study demonstrates the power of remote sensing data and the biophysical modelling for quantifying the ETa process over an arid ecosystem in Saudi Arabia. The estimated mean annual ETa was 2000 mm.year^{-1} for open water and varied between 800 and 1400 mm.year^{-1} for date palm. However, it was 1600 mm.year^{-1} for the aquatic vegetation, while an average of 800 mm.year^{-1} was observed in croplands. The validation measure showed a significant agreement level between the SEBAL model and the FAO P-M method with RMSE of 0.62, 0.84, 0.98 and 1.38 mm.day^{-1} for aquatic vegetation, date palm, open water and cropland respectively.

The obtained ETa information will help Saudi Arabia formulate strategies to reduce the gap between the water supply and demand in the irrigated areas. Furthermore, the ETa patterns mapped over the diverse LULC systems can be used as a baseline framework for sustainable water resources management and agrometeorological services in the different regions of Saudi Arabia. However, conducting long-term ETa studies using remote sensing data coupled with the implementation of different models and field tools may improve the assessment of the ETa dynamic process in arid regions.

Acknowledgements

The author's acknowledge the Deanship of Scientific Research at King Faisal University, for the financial support under the annual research project (Grant No. 186002).

Author details

Khalid G. Biro Turk[1,2*], Faisal I. Zeineldin[1] and Abdulrahman M. Alghannam[1,3]

1 Water Studies Center, King Faisal University, Al-Ahsa, Saudi Arabia

2 Faculty of Agricultural and Environmental Science, University of Gadarif, Gadarif, Sudan

3 College of Agricultural and Food Sciences, King Faisal University, Al-Ahsa, Saudi Arabia

*Address all correspondence to: khalidturk76@yahoo.co.uk; kturk@kfu.edu.sa

IntechOpen

References

[1] Yang Y, Anderson MC, Gao F, Hain CR, Semmens KA, Kustas WP, Noormets A, Wynne RH, Thomas VA, Sun G: Daily Landsat-scale evapotranspiration estimation over a forested landscape in North Carolina, USA, using multi satellite data fusion. Hydrol. Earth Syst. Sci. 2017; 21:1017–1037. DOI: https://doi.org/10.5194/hess-21-1017-2017

[2] Sun SK, Li C, Wang YB, Zhao XN, Wu PT: Evaluation of the mechanisms and performances of major satellite-based evapotranspiration models in Northwest China. Agricultural and Forest Meteorology. 2020;291:108056. DOI: doi.org/10.1016/j.agrformet.2020.108056

[3] Rahimi S, Sefidkouhi MAG, Raeini-Sarjaz M, Valipour M: Estimation of actual evapotranspiration by using MODIS images (a case study: Tajan catchment). Arch. of Agron. 2015;61(5):695–709. DOI: 10.1080/03650340.2014.944904

[4] Awada H, Ciraolo G, Maltese A, Provenzano G, Hidalgo MAM, Còrcoles JI: Assessing the performance of a large-scale irrigation system by estimations of actual evapotranspiration obtained by Landsat satellite images resampled with cubic convolution. Int J Appl Earth Obs Geoinformation. 2019;75:96–105. DOI: https://doi.org/10.1016/j.jag.2018.10.016

[5] Allen RG, Pereira SL, Howell TA, Jensen, ME: Evapotranspiration information reporting: I. Factors governing measurement accuracy. Agr. Water Manag. 2011;98:899–920. DOI: org/10.1016/j.agwat.2010.12.015

[6] Singh RK, Senay G: Comparison of Four Different Energy Balance Models for Estimating Evapotranspiration in the Midwestern United States. Water. 2016; 8, 9. DOI: 10.3390/w8010009

[7] Trezza, R, Allen RG, Tasumi M: Estimation of actual evapotranspiration along the Middle Rio Grande of New Mexico using MODIS and Landsat imagery with the METRIC model. Remote Sens. 2013;5, 5397–5423. DOI: https://doi.org/10.3390/rs5105397

[8] Kong J, Hu Y, Yang L, Shan Z, Wang Y: Estimation of evapotranspiration for the blown-sand region in the Ordos basin based on the SEBAL model. Int. J. Remote Sens. 2019;40:5–6, 1945-1965. DOI: 10.1080/01431161.2018.1508919

[9] Lu Z, Zou S, Qin Z, Yang Y, Xiao H, Wei Y, Zhang K, Xie J: Hydrologic Responses to Land Use Change in the Loess Plateau: Case Study in the Upper Fenhe River Watershed. Hindawi Publishing Corporation, Advances in Meteorology. 2015; Article ID 676030. DOI: http://dx.doi.org/10.1155/2015/676030

[10] Hassaballah K, Mohamed Y., Uhlenbrook S, Biro K: Analysis of streamflow response to land use and land cover changes using satellite data and hydrological modelling: case study of Dinder and Rahad tributaries of the Blue Nile (Ethiopia–Sudan). Hydrol. Earth Syst. Sci. 2017;21:5217–5242. DOI: https://doi.org/10.5194/hess-21-5217-2017

[11] Jacqueminet C, Kermadi S, Michel K, Béal D, Gagnage G, Branger F, Jankowfsky S, Braud I: Land cover mapping using aerial and VHR satellite images for distributed hydrological modelling of periurban catchments: Application to the Yzeron catchment (Lyon, France). Journal of Hydrology. 2013; 485:68–83. DOI: http://dx.doi.org/10.1016/j.jhydrol.2013.01.028

[12] Wolfe JD, Shook KR, Spence C, Whitfield CJ: A watershed classification approach that looks beyond hydrology: application to a semi-arid, agricultural region in Canada. Hydrol. Earth Syst.

Sci. 2019;23:3945–3967. DOI: https://doi.org/10.5194/hess-23-3945-2019

[13] Bellvert J, Adeline K, Baram S, Pierce L, Sanden BL, Smart, DR: Monitoring Crop Evapotranspiration and Crop Coefficients over an Almond and Pistachio Orchard throughout Remote Sensing. Remote Sensing. 2018; 10,2001. DOI: 10.3390/rs10122001

[14] Cha M, Li M, Wang X: Estimation of Seasonal Evapotranspiration for Crops in Arid Regions Using Multisource Remote Sensing Images. Remote Sensing. 2020; 12, 2398. DOI: 10.3390/rs12152398

[15] Paul S, Banerjee C, Kumar DN: Evaluation Framework of Landsat 8–Based Actual Evapotranspiration Estimates in Data-Sparse Catchment. J. Hydrol. Eng., American Society of Civil Engineers (ASCE). 2020;25(9): 04020043. DOI: 10.1061/(ASCE)HE.1943-5584.0001992

[16] Silva AM, da Silva RM, Santos CAG: Automated surface energy balance algorithm for land (ASEBAL) based on automating endmember pixel selection for evapotranspiration calculation in MODIS orbital images. Int J Appl Earth Obs Geoinformation. 2019;79:1–11. DOI: https://doi.org/10.1016/j.jag.2019.02.012

[17] Norman, JM, Kustas, WP, Humes, KS: Source approach for estimating soil and vegetation energy fluxes in observations of directional radiometric surface temperature. Agric. For. Meteorol. 1995;77 (3–4):263–293. DOI: https://doi.org/10.1016/0168-1923(95)02265-Y

[18] Allen RG, Tasumi M, Trezza R: Satellite-based energy balance for mapping evapotranspiration with internalized calibration (METRIC) – Model. American Society of Civil Engineers, J. Irrig. Drain. E.-ASCE. 2007;133380–394.

[19] Mutiga JK, Su Z, Woldai T: Using satellite remote sensing to assess evapotranspiration: Case study of the upper Ewaso Ng'iro North Basin, Kenya. Int. J. Appl. Earth Obs. 2010;12S, S100–S108. DOI: 10.1016/j.jag.2009.09.012

[20] Sari DK, Ismullah IH, Sulasdi WN, Harto AB: Estimation of water consumption of lowland rice in tropical area based on heterogeneous cropping calendar using remote sensing technology. The 3rd International Conference on Sustainable Future for Human Security SUSTAIN 2012. Procedia Environmental Sciences. 2013; 17:298–307.

[21] Usman M, Liedl R, Awan UK: Spatio-temporal estimation of consumptive water use for assessment of irrigation system performance and management of water resources in irrigated Indus Basin, Pakistan. Journal of Hydrology. 2015; 525:26–41. DOI: http://dx.doi.org/10.1016/j.jhydrol.2015.03.031

[22] Mahmoud SH, Alazba AAA: coupled remote sensing and the Surface Energy Balance based algorithms to estimate actual evapotranspiration over the western and southern regions of Saudi Arabia. J. Asian Earth Sci. 2016;124:269–283. DOI: org/10.1016/j.jseaes.2016.05.012

[23] Alemayehu T, van Griensve A, Senay, GB, Bauwens, W: Evapotranspiration Mapping in a Heterogeneous Landscape Using Remote Sensing and Global Weather Datasets: Application to the Mara Basin, East Africa. Remote Sens. 2017;9,390. DOI: 10.3390/rs9040390

[24] Ochege FU, Luo G, Obeta MC, Owusu G, Duulatov E, Cao L, Nsengiyumva JB: Mapping evapotranspiration variability over a complex oasis-desert ecosystem based on automated calibration of Landsat 7 ETM+ data in SEBAL. GISci. Remote

Sens. 2019;56(8):1305–1332. DOI: 10.1080/15481603.2019.1643531

[25] Rahimzadegana M, Jananib A: Estimating evapotranspiration of pistachio crop based on SEBAL algorithm using Landsat 8 satellite imagery. Agricultural Water Management. 2019;217:383–390. DOI: https://doi.org/10.1016/j.agwat.2019.03.018

[26] Ghaderi A, Dasineh M, Shokri M, Abraham J: Estimation of Actual Evapotranspiration Using the Remote Sensing Method and SEBAL Algorithm: A Case Study in Ein Khosh Plain, Iran. Hydrology. 2020, 7, 36. DOI: 10.3390/hydrology7020036

[27] ElNesr M, Alazba A, Abu-Zreig M: Spatio-Temporal Variability of Evapotranspiration over the Kingdom of Saudi Arabia. Appl. Eng. Agric. 2010;26 (5):833–842.

[28] Chowdhury S, Al-Zahrani M: Characterizing water resources and trends of sector wise water consumptions in Saudi Arabia. JKSUES. 2015;27:68–82. DOI: org/10.1016/j.jksues.2013.02.002

[29] Al-Zarah AI: Chemistry of groundwater of AI-Ahisa Oasis Eastern Region Saudi Arabia and its predictive effects on soil. Pak J Biol Sci. 2008;11(3): 332–341.

[30] Al-Taher AA: Estimation of potential evapotranspiration in Al-Hassa oasis, Saudi Arabia. Geo. Journal. 1992; 26 (3):371–379.

[31] Allbed A, Kumar AL, Aldakheel YY: Assessing soil salinity using soil salinity and vegetation indices derived from IKONOS high-spatial resolution imageries: Applications in a date palm dominated region. Geoderma. 2014; 230-231, 1–8. DOI: org/10.1016/j.geoderma.2014.03.025

[32] USGS: Department of the Interior United State Geological Survey (USGS).

Landsat 8 (L8) Data Users Handbook. LSDS-1574 Version 5.0. 2019. Available from: https://www.usgs.gov/media/file s/landsat-8-data-users-handbook [Accessed: 2020-12-24]

[33] Ellis EA, Baerenklau KA, Marcos-Martínez R, Chávez E: Land use/land cover change dynamics and drivers in a low-grade marginal coffee growing region of Veracruz, Mexico. Agroforest Syst., 2010;80:61–84.

[34] Rahman MT, Aldosary AS, Mortoja MdG: Modeling Future Land Cover Changes and Their Effects on the Land Surface Temperatures in the Saudi Arabian Eastern Coastal City of Dammam. Land 2017;6, 36. DOI: 10.3390/land6020036

[35] Abdallah S, Abd elmohemen M, Hemdan S, Ibrahem K: Assessment of land use/land cover changes induced by Jizan Dam, Saudi Arabia, and their effect on soil organic carbon. Arabian J Geosci. 2019;12:350. DOI: org/10.1007/s12517-019-4474-1

[36] Mundia CN, Aniya M: Dynamics of land use/cover changes and degradation of Nairobi city, Kenya. Land Degrad. Dev. 2006;17:97–108. DOI: 10.1002/ldr.702

[37] Congalton RG, Green K. Assessing the accuracy of remotely sensed data, Principles and Practices. 3rd ed. CRC Press, Taylor & Francis Group, Boca Raton London New York: 2019. 328 p.

[38] Bastiaanssen WGM, Menenti M, Feddes RA Holtslag A AM: Remote sensing surface energy balance algorithm for land (SEBAL): 1. Formulation. J. Hydrol. 1998;212–213(1–4):198–212. DOI: org/10.1016/S0022-1694 (98)00253-4

[39] Malbéteau Y, Merlin O, Gascoin S, Gastellu JP, Mattar C: Correcting land surface temperature data for elevation and illumination effects in mountainous

areas: A case study using ASTER data over a steep-sided valley in Morocco. Remote Sens. Environ. 2017;189:25–39. DOI: 10.1016/j.rse.2016.11.010

[40] Bastiaanssen WGM, Noordman EJM, Pelgrum H, Davids G, Allen RG: SEBAL for spatially distributed ET under actual management and growing conditions, ASCE J. Irrig. Drain. Eng. 2005;131(1): 85–93. DOI: org/10.1061/(ASCE) 0733-9437(2005)131:1(85)

[41] Allen RG, Burnett B, Kramber W, Huntington J, Kjaersgaard J, Kilic A, Kelly C, Trezza R: Automated Calibration of the Metric-Landsat Evaporation Process. J Am Water Resour Assoc. 2013;49(3):563–576. DOI: org/10.1111/jawr.12056

[42] Allen RG, Trezza R, Tasumi M, Kjaersgaard J. Metric: Mapping Evapotranspiration at High Resolution using Internalized Calibration. Applications Manual for Landsat Satellite Imagery, Version 2.0.8, March 2012. University of Idaho, Kimberly, Idaho. 2012. P 83.

[43] Avdan U, Jovanovska G: Algorithm for Automated Mapping of Land Surface Temperature Using LANDSAT 8 Satellite Data. Journal of Sensors. 2016; ArticleID 1480307, 8 pages. DOI: https://doi.org/10.1155/2016/1480307

[44] Mohajane M, Essahlaoui A, Fatiha O, El Hafyani M, El Hmaidi A, El Ouali A, Randazzo G, Teodoro, AC: Land Use/Land Cover (LULC) Using Landsat Data Series (MSS, TM, ETM+ and OLI) in Azrou Forest, in the Central Middle Atlas of Morocco. Environments. 2019;5:131. DOI: 10.3390/environments5120131

[45] Sobrino JA, Jim'enez-Mu~noz JC, Paolini L: Land surface temperature retrieval fromLANDSAT TM5. Remote Sensing of Environment. 2004;90(4): 434–440.

[46] Sobrino JA, Raissouni N: Toward remote sensing methods for land cover dynamic monitoring: application to Morocco. International Journal of Remote Sensing. 200;21(2): 353–366

[47] Ahmed B, Kamruzzaman Md, Zhu X, Rahman S, Choi K: Simulating land cover changes and their impacts on land surface temperature in Dhaka, Bangladesh. Remote Sensing. 2013;5 (11): 5969–5998.

[48] Xu H –Q, Chen B –Q: Remote sensing of the urban heat island and its changes in Xiamen City of SE China. Journal of Environmental Sciences. 2004;16(2): 276–281.

[49] Weng Q, Lu D, Schubring J: Estimation of Land Surface Temperature–Vegetation Abundance Relationship for Urban Heat Island Studies. Remote Sensing of Environment. 2004;89:467–483.

[50] Allen RG, Trezza R, Tasumi M: Analytical integrated functions for daily solar radiation on slopes. Agric for Meteorol. 2006;139:55–73.

[51] Bastiaanssen WGM: SEBAL-based sensible and latent heat fluxes in the irrigated Gediz Basin, Turkey. J. Hydrol. 2000; 229:87–100.

[52] Allen RG, Pereira LA, Raes D, Smith M: Crop Evapotranspiration. FAO Irrigation and Drainage Paper 56, Rome. 1998. ISBN: 92-5-104219-5.

[53] Al-Dakheel Y: Assessing NDVI Spatial Pattern as Related to Irrigation and Soil Salinity Management in Al-Hassa Oasis, Saudi Arabia. J Indian Soc. Remote Sens. 2011;39(2):171–180. DOI: 10.1007/s12524-010-0057-z

[54] Ogunode A, Akombelwa M: An algorithm to retrieve Land Surface Temperature using Landsat-8 Dataset. South African Journal of Geomatics.

2017;(6)2, Geomatics Indaba 2017 Special Edition.

[55] Huber DB, Mechem DB, Brunsell NA: The Effects of Great Plains Irrigation on the Surface Energy Balance, Regional Circulation, and Precipitation. Climate. 2014;2, 103–128. DOI:10.3390/cli2020103

[56] Amatya P M, Ma Y, Han C, Wang B, Devkota LP: Recent trends (2003–2013) of land surface heat fluxes on the southern side of the central Himalayas, Nepal. J. Geophys. Res. Atmos. 2015; 120, 11,957–11,970. DOI: 10.1002/ 2015JD023510

[57] Haj-Amor HA, Toth T, Ibrahim MK, Bouri S: Effects of excessive irrigation of date palm on soil salinization, shallow groundwater properties, and water use in a Saharan oasis. Environ. Earth Sci. 2017;76:590. DOI: org/10.1007/ s12665-017-6935-8

[58] Carr MKV: The Water Relations and Irrigation Requirements of the Date Palm (Phoenix dactylifera L.): A Review. Expl. Agric. 2013;49(1):91–113. DOI: 10.1017/S0014479712000993

[59] Sentelhas PC, Gillespie TJ, Santos EA: Evaluation of FAO Penman-Monteith and Alternative Methods for Estimating Reference Evapotranspiration with Missing Data in Southern Ontario, Canada. Agric. Water Manag. 2010;97:635–644. DOI: org/ 10.1016/j.agwat.2009.12.001

[60] Mazahrih NTH, Al-Zubi Y, Ghnaim H, Lababdeh L, Ghananeem M, Abu-Ahmadeh H: Determination of actual crop evapotranspiration and crop coefficient of date palm trees (Phoenix dactylifera) in the Jordan Valley. Am-Euras. J. Agric. & Environ. Sci. 2012;12 (4):434–443.

[61] Bhattarai N, Quackenbush LJ, Im J, Shaw SB: A new optimized algorithm for automating endmember pixel selection in the SEBAL and METRIC models. Remote Sens. Environ. 2017;196:178–192. DOI: https://doi.org/10.1016/j. rse.2017.05.009.

Chapter 4

Evaluation of the Spatial Distribution of the Annual Extreme Precipitation Using Kriging and Co-Kriging Methods in Algeria Country

Hicham Salhi

Abstract

In this chapter, we have conducted a statistical study of the annual extreme precipitation (AMP) for 856 grid cells and during the period of 1979–2012 in Algeria. In the first step, we compared graphically the forecasts of the three parameters of the generalized extreme value (GEV) distribution (location, scale and shape) which are estimated by the Spherical model. We used the Cross validation method to compare the two methods kriging and Co-kriging, based on the based on some statistical indicators such as Mean Errors (ME), Root Mean Square Errors (RMSE) and Squared Deviation Ratio (MSDR). The Kriging forecast error map shows low errors expected near the stations, while co-Kriging gives the lowest errors on average at the national level, which means that the method of co-Kriging is the best. From the results of the return periods, we calculate that after 50 years the estimated of the annual extreme precipitation will exceed the maximum AMP is observed in the 33-year.

Keywords: extreme precipitations, kriging and Co-kriging, cross validation, return levels

1. Introduction

Natural disasters cause loss of human life and damage to infrastructure every year throughout the world. In Algeria, extreme rains are the source of flooding which can cause catastrophic damage both in inhabited areas and in the countryside.

One of the basic problems encountered in meteorology is the need to assess the meteorology risk caused by extreme precipitation in order to avoid human and material losses. Thus, the location and severity of floods can be determined.

In the twenties to the middle of the last century, the theory of extreme values has witnessed a remarkable development [1–3] most studies focused on the monthly or yearly mean values and we find their application in many fields like; rainfall in Algeria [4, 5], extreme precipitations in Argentina [6] Mapping snow depth return levels [7], precipitation and temperature [8, 9].

On the other hand, a lot of studies say there is a great spatial difference in rainfall [10] for this reason it was used interpolation methods, the kriging method is considered as the most used for spatial interpolation of rainfall [11–13], the kriging method has a special feature which is complementing the sparsely sampled primary variable, in the case of secondary variable there is another method called cokriging that outperforms the kriging method [14].

According to [15] In order to examine the spatiotemporal variations of meteorological variables there is a statistical method that allows to apply multiple strategies by cluster analysis to pinpoint the similar places, local and universal meteorology techniques which has been raising lately.

Our goal in this chapter is to compare the two methods kriging and co-kriging using the GEV and determine the location and severity of floods in all regions of Algeria.

2. Methodology and data

2.1 Generalized extreme value distribution

The cumulative distribution function is proposed by [16].

$$F(x) = \begin{cases} exp\left\{-\left[1 + \xi\left(\frac{x-\mu}{\sigma}\right)\right]^{-\frac{1}{\xi}}\right\} & \xi \neq 0 \\ exp\left(-exp\left(-\frac{x-\mu}{\sigma}\right)\right) & \xi = 0 \end{cases} \tag{1}$$

By deriving the Eq. (1) we get the density function

$$f(x) = \begin{cases} \frac{1}{\sigma}\left[1 + \xi\left(\frac{x-\mu}{\sigma}\right)\right]^{-\frac{1-\xi}{\xi}} exp\left\{-\left[1 + \xi\left(\frac{x-\mu}{\sigma}\right)\right]^{-\frac{1}{\xi}}\right\} & \xi \neq 0 \\ \frac{1}{\sigma} exp\left(-\left(\frac{x-\mu}{\sigma} + exp\left(-\frac{x-\mu}{\sigma}\right)\right)\right) & \xi = 0 \end{cases} \tag{2}$$

The logarithm of the likelihood function is given by:
For $\xi \neq 0$

$$l(\xi, \mu, \sigma, Y) = -n \, ln\sigma - \left(1 + \frac{1}{\xi}\right)\sum_{n=1}^{n} ln\left(1 + \xi\left(\frac{x_i - \mu}{\sigma}\right)\right) - \sum_{i=1}^{n}\left[1 + \xi\left(\frac{x_i - \mu}{\sigma}\right)\right]^{\frac{-1}{\xi}} \tag{3}$$

For $\xi = 0$

$$l(\xi, \mu, \sigma, Y) = -n \, ln\sigma - \sum_{n=1}^{n} exp\left(-\frac{x_i - \mu}{\sigma}\right) - \sum_{i=1}^{n}\left(\frac{x_i - \mu}{\sigma}\right) \tag{4}$$

3 and 4 with differentiating the two parameters:

$$\begin{cases} n - \sum_{n=1}^{n} exp\left(-\frac{x_i - \mu}{\sigma}\right) = 0 \\ n + \sum_{n=1}^{n}\frac{x_i - \mu}{\sigma}\left[exp\left(-\frac{x_i - \mu}{\sigma}\right) - 1\right] = 0 \end{cases} \tag{5}$$

2.2 Return period

The return period, also known as a recurrence interval is the estimated average time between events such as earthquakes, floods, landslides, or river floods. From Eq. (1) we can write the return level as following:

$$
Z_T = \begin{cases} \mu + \dfrac{\sigma}{\xi}\left\{1 - ln\left[1 - \dfrac{1}{T}\right]^{\xi}\right\} & \xi \neq 0 \\[2mm] \mu - \sigma\, ln\left(-ln\left(1 - \dfrac{1}{T}\right)\right) & \xi = 0 \end{cases} \tag{6}
$$

2.3 Variogram model

Various parameter variogram models have been used in the literature. Here is some of the most popular content.

Spherical model.
The Spherical model has linear behavior at small separation distances near the origin, but flattens at large distances, which means that it shows a gradual decrease in spatial dependence until a certain distance beyond which the spatial dependence tends to smooth.

$$
\gamma(h) = c_0 + c_1\left(\frac{3}{2}\frac{h}{\alpha} - \frac{1}{2}\left(\frac{h}{\alpha}\right)^3\right) \text{ for } 0 < h \leq \alpha
$$

$$
\gamma(h) = c_0 + c_1 \text{ for } h \geq \alpha
$$

Where c_0 is the nugget effect. The sill is $c_0 + c_1$. The range for the spherical model can be computed by setting $g(h) = 0.95(c_0 + c_1)$.

Gaussian model.
The Gaussian model is used when the data exhibits strong continuity at short lag distances which means the spatial correlation is very high between two neighboring points.

$\gamma(h) = c_0 + c_1\left(1 - e^{-(h/\alpha)^2}\right)$ where c_0 is the nugget effect. $c_0 + c_1$ is the sill. The range is 3α. This model describes a random field that is considered to be too smooth and possesses the peculiar property that $Z(s)$ can be predicted without error for any s on the plane.

2.4 Data description

The precipitation data used in this study are for the National centers for environmental information NOAA of USA, this data used especially in cases where surface data are difficult to obtain or insufficient. Our data represented by the annual daily maximal of rainfall from1979 to 2012 calculated in the 856 Algerian stations (**Figure 1**).

The preliminary analysis of the annual maximum precipitation data during the analysis period (1979–2012) included descriptive statistical calculations (**Table 1** and **Figure 2**). More precisely, we calculated the minimum (Min), the maximum (Max), the mean (Mean), the standard deviation (std.dev) and the coefficient of variation (coef.var). **Table 1** presents the values of the descriptive statistics for the annual time series of maximum precipitation for all stations (from 1979 to 2012). The results show that the maximum values is observed in the years 1982, 1992,

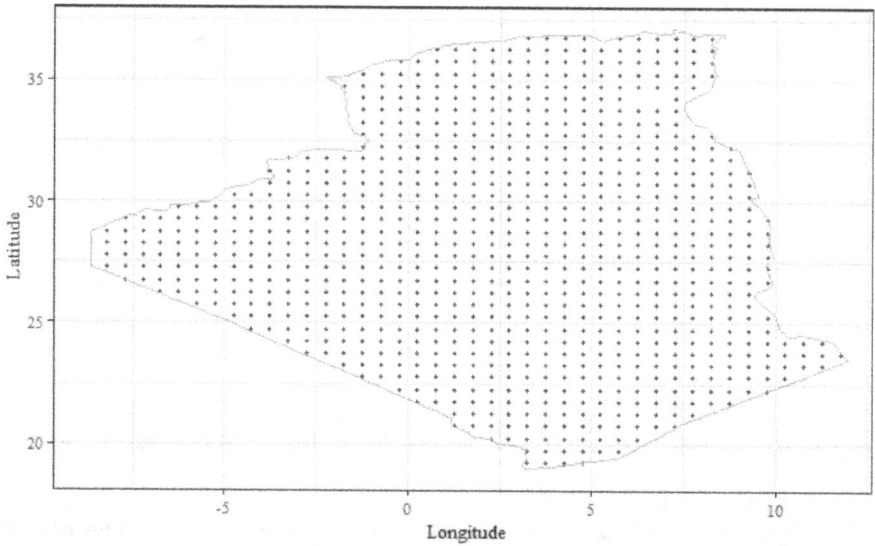

Figure 1.
A map of Algeria showing rainfall stations.

Years	Min	Max	Mean	std.dev	coef.var	Years	Min	Max	Mean	std.dev	coef.var
1979	0.59	88.07	13.46	11.23	0.83	1996	2.93	69.38	13.75	9.39	0.68
1980	1.34	63.85	12.88	11.89	0.92	1997	1.27	63.63	10.84	8.89	0.82
1981	0.84	79.32	9.94	8.88	0.89	1998	0.31	57.40	7.76	8.49	1.09
1982	2.92	123.29	21.36	18.15	0.85	1999	0.19	63.48	14.83	11.01	0.74
1983	4.42	57.48	9.70	7.67	0.79	2000	0.79	81.39	14.14	12.62	0.89
1984	0.21	99.47	7.93	11.09	1.40	2001	0.09	114.35	8.28	10.07	1.22
1985	0.30	83.26	10.09	9.33	0.93	2002	0.66	96.94	12.15	11.07	0.91
1986	2.94	73.87	15.58	11.74	0.75	2003	0.33	84.54	13.69	11.56	0.84
1987	0.82	57.32	8.91	7.43	0.83	2004	0.06	97.24	16.44	16.40	1.00
1988	1.73	52.89	12.16	8.71	0.72	2005	0.90	74.47	18.05	16.24	0.90
1989	0.00	73.35	8.77	9.91	1.13	2006	0.97	115.19	13.49	12.92	0.96
1990	0.95	79.25	14.23	10.69	0.75	2007	0.07	109.16	15.58	15.61	1.00
1991	1.99	57.57	12.51	11.16	0.89	2008	1.92	70.72	15.00	12.09	0.81
1992	1.50	121.05	14.14	16.85	1.19	2009	1.21	97.77	15.28	11.73	0.77
1993	0.39	62.82	11.46	10.77	0.94	2010	1.50	63.11	10.69	9.16	0.86
1994	1.43	123.59	15.24	14.77	0.97	2011	2.93	67.68	14.34	12.03	0.84
1995	0.86	83.47	11.70	9.72	0.83	2012	1.89	91.72	13.50	13.28	0.98

Table 1.
Statistical descriptive of all years.

1994, 2001, 2006, and 2007, while the mean and highest values are observed in 1982 (**Figure 2**). The lowest value of coef.var. is for the year 1996 (68%), and the highest for 1984 (14%). On this basis, the observed data showed that all years had a coef. var. greater than 68%, highlighting the high variability of annual maximum precipitation over Algeria.

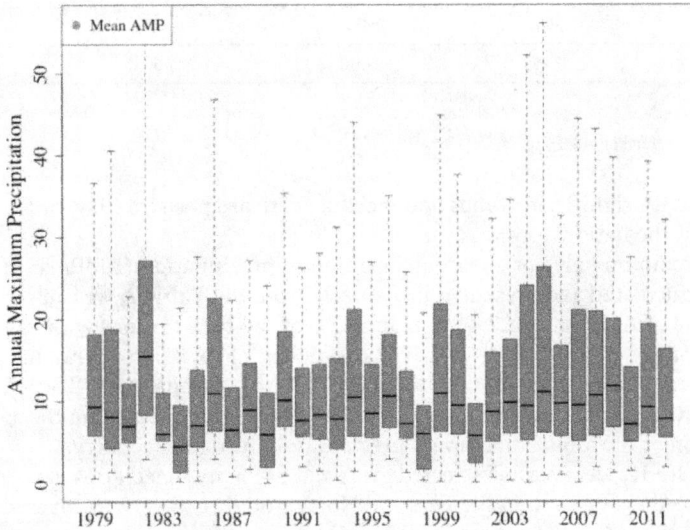

Figure 2.
Boxplots of annual extreme precipitation.

3. Results and discussions

3.1 Trend in annual maximum precipitation

In this study, the Mann-Kendall non parametric test is computed to characterize the time course of annual maximum precipitation at the national scale. The trend is considered significant if the value of the probability (p-value) is greater than 0.05 (95%). The **Figure 3** shows the distribution of the significance trend (red color) and non-significant trend (blue color). From the results, we can observe that 130 (15%)

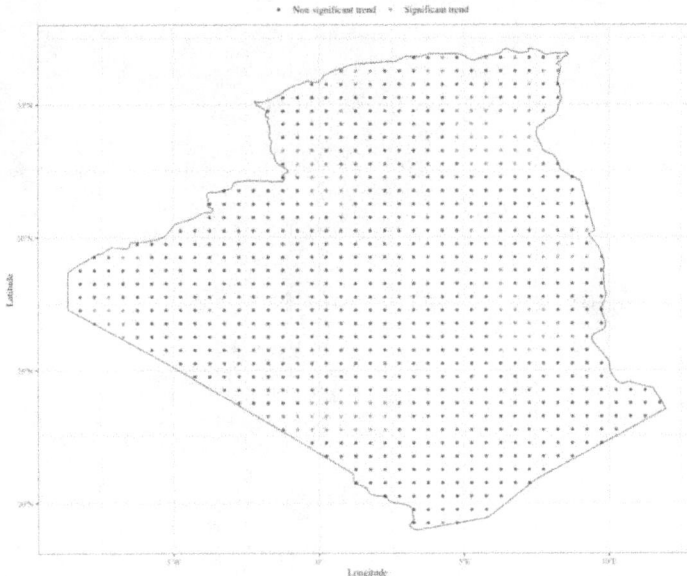

Figure 3.
The spatial distribution of the annual maximum precipitation trend in Algeria country.

Years	Min	Max	Mean	std.dev
5.88	123.59	37.00	22.46	0.61

Table 2.
Statistical descriptive of AMP$_{max}$.

stations have a significant trends and most of them are positive (97% of the total stations) at the national scale.

The maximum value of the annual maximum precipitation (AMPmax) for each station is calculated and presented in **Table 2**. From the **Table 2**, we can see that the mean value of AMPmax is 22.46 mm and 47% of the total stations with values greater than the mean of AMPmax. The coefficient of variation is 61%, indicating the significant spatial distribution of AMPmax at the national scale. Therefore, we applied a Kriging and Co-kriging approaches to better understanding the spatial distribution of the annual maximum precipitation in Algeria country.

In this study, we have 865 selected grids and 9496 predicted grids have locations where spaced every 20 m in the East and North grid directions and covered the irregularly shaped of the country (**Figure 4**). Due to the large numerical range of AMPmax values and to allow easy interpretation of the results, we worked with the logarithmic transformation of the variable. In this application, we chosen a base 10 logarithms (log10) for the data and we randomly selected control and test datasets. In this study, 30% of the total grids were excluded for testing (assessment).

3.2 Choosing the variogram model

We start by plotting the experimental variogram before adjusting the latter with the different models. The sum of the square errors (SSErr) and the regression coefficient (R^2) provided an accurate measure of the fit of the model to the variogram data, with a lower SSErr and a higher R^2 indicating better fit of the model.

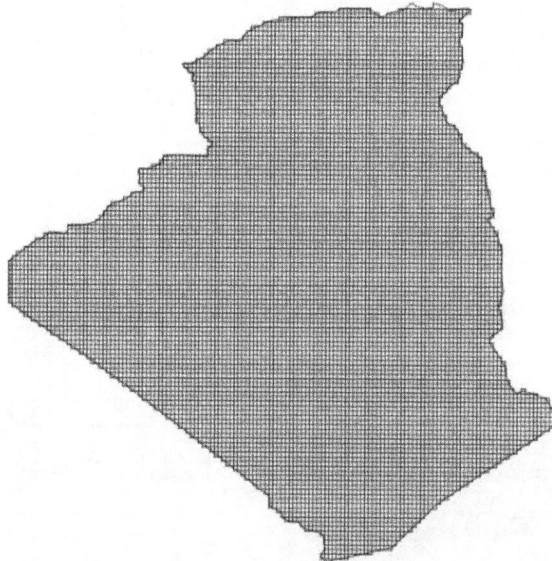

Figure 4.
The spatial distribution of the prediction grid.

Model	Range	Nugget (C0)	Sill (C0 + C)	Nugget/Sill ((C0/C0 + C)*100)	SSErr	R^2
Spherical	1075.984	0.02806	0.11425	24.56	2.26E-05	0.9890
Gaussian	412.298	0.03665	0.10436	35.12	2.28E-05	0.9888

Table 3.
The parameters of each model.

The values of the parameters of the different fitted models are presented in
Table 3.

Theoretical and empirical semi-variogram were prepared for the AMPmax as
shown in **Figure 5**. From the results, we can see that the spherical model has been
found to be the most accurate model for annual maximum precipitation.

The spatial dependence is generally accessible in terms of the ratio between the
nugget (C0) and the sill (C0 + C) expressed as a percentage. The AMPmax is
considered to be a strong spatial dependence when the ratio value is less than 25%,
moderate spatial dependence when this value is between 25% and 75%, and low
spatial dependence when the value is greater than 75%. From **Table 3**, we can
clearly see that the spatial dependence of AMPmax for the best-fitting semi-
variogram model is strong and with a ratio of 24.56%.

3.3 AMPmax interpolation

The Spherical model is used to interpolate the AMPmax for both Kriging and co-
Kriging methods at the national scale. In the first step, we compared graphically the
forecasted and estimated GEV parameters (μ, σ and ξ). From **Figure 6**, we can see
that a very clear spatial pattern for the estimates of the location and scale parameters however an absence of the spatial pattern for the shape parameter. The northern region is very marked compared to the rest of the regions with a significantly

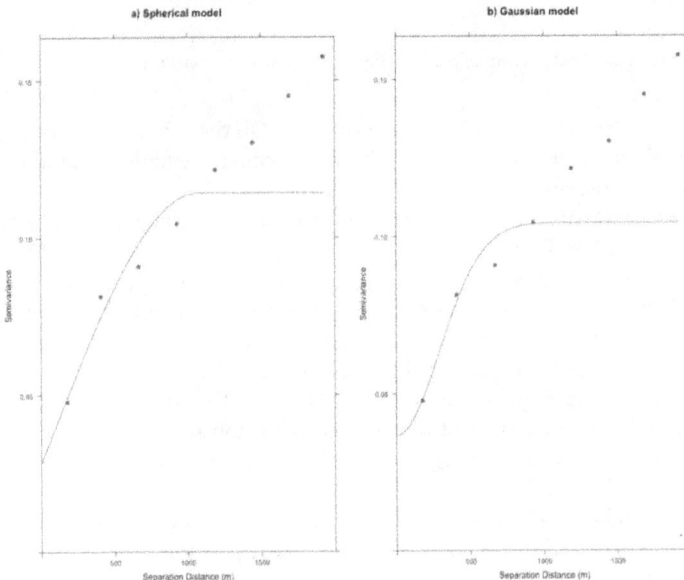

Figure 5.
Empirical semivariance and its fitted model.

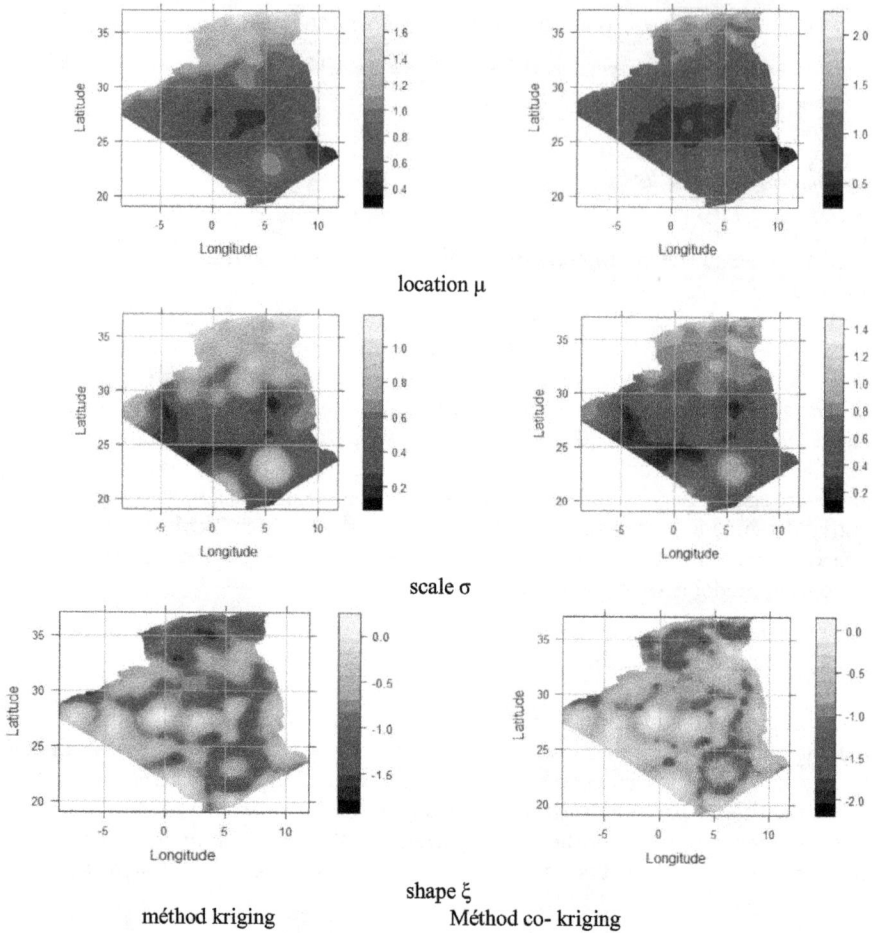

Figure 6.
The spatial distribution of the forecasts of the GEV parameters for the two methods.

higher value of the location and scale parameters. On the other hand, the co-Kriging method clearly provided new regions where the values are high. Generally, the high values could be observed in northern Algeria.

In order to compare the two methods Kriging and Co-Kriging, we used Cross Validation method and some statistical indicators such as Mean Errors (ME), Root Mean Square Errors (RMSE) and Squared Deviation Ratio (MSDR) **Table 4**.

Figure 7 display a bubble plots of the cross-evaluation error of the two methods, where positive values are drawn in green and negative values are drawn in red, and the size of the bubble is proportional to the distance from zero.

From **Table 4** and **Figure 7**, we can clearly notice that the Kriging forecast error map with the three parameters of the GEV distribution shows low errors expected near the stations, while co-Kriging gives the lowest errors on average at the national scale, especially for the shape parameter.

After the validation of the two methods, Co-kriging method used to estimate the return levels (RLs) of the annual maximum of daily precipitations for the different stations using Eq. (6).

The return periods are shown in **Table 5** for the 20, 50, and 100-year. The results show that the maximum annual maximum precipitation observed in 1982,

	M	**Σ**	**ξ**	**M**	**σ**	**ξ**
Min	−0.11691	−0.23323	−2.71689	−0.41537	−0.37368	−2.52461
Max	0.15085	0.19637	0.97323	0.70063	0.72695	0.98770
Mean	0.00013	−0.00011	0.00098	0.00105	0.00082	0.00224
ME	0.00013	−0.00011	0.00098	0.00105	0.00082	0.00224
RMSE	0.02421	0.04819	0.37981	0.07523	0.08761	0.39080
MSDR	0.08987	0.38275	1.91038	0.92698	1.08895	1.70011
	Kriging			Co-Kriging		

Table 4.
Cross-evaluation errors for the two methods.

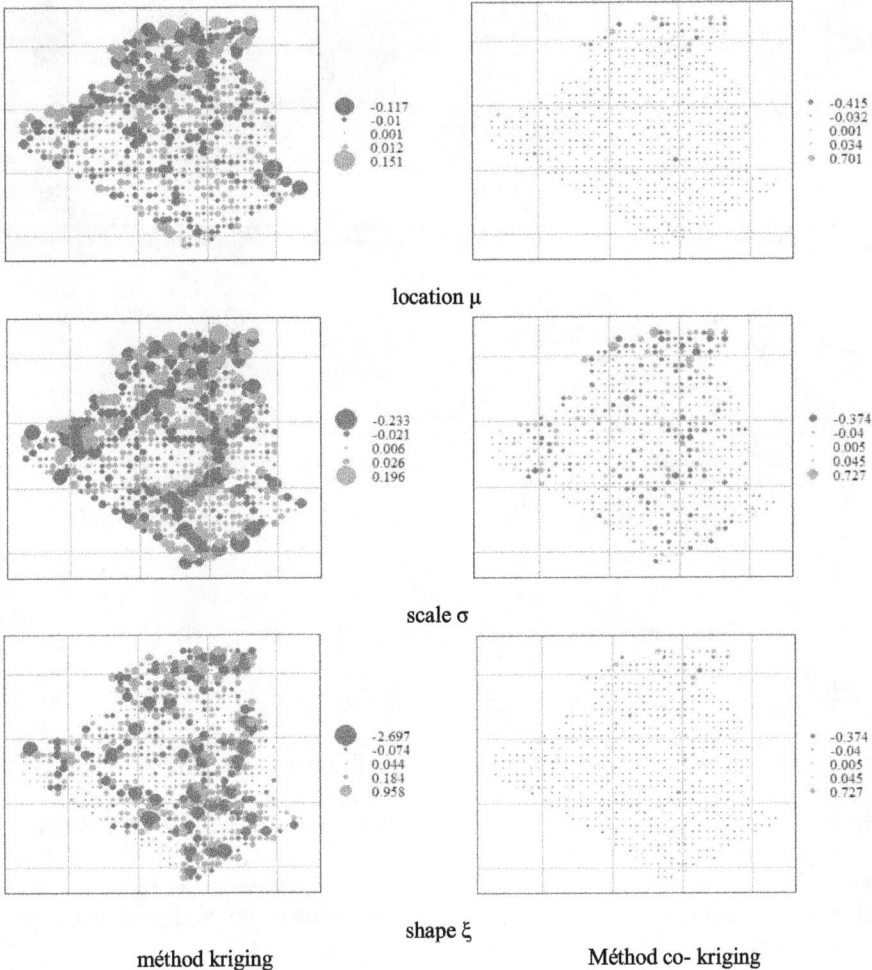

location μ

scale σ

shape ξ

méthod kriging   Méthod co- kriging

Figure 7.
The spatial distribution of cross-evaluation errors for the three parameters of the GEV law by the two methods.

1992, 1994, 2001, 2006, and 2007 exceeds the 20-year regression level. AMP exceeding the maximum AMP during the observation period (123.59) begins to appear in the confidence interval of 50 year.

	Min	Max	Mean	std.dev	coef.var
20 years	5.6697	106.8175	28.7420	17.8361	0.6205
50 years	5.7920	252.1116	40.0470	30.8470	0.7703
100 years	5.8421	493.1154	52.2700	52.0110	0.9950

Table 5.
Statistical descriptive of the return periods.

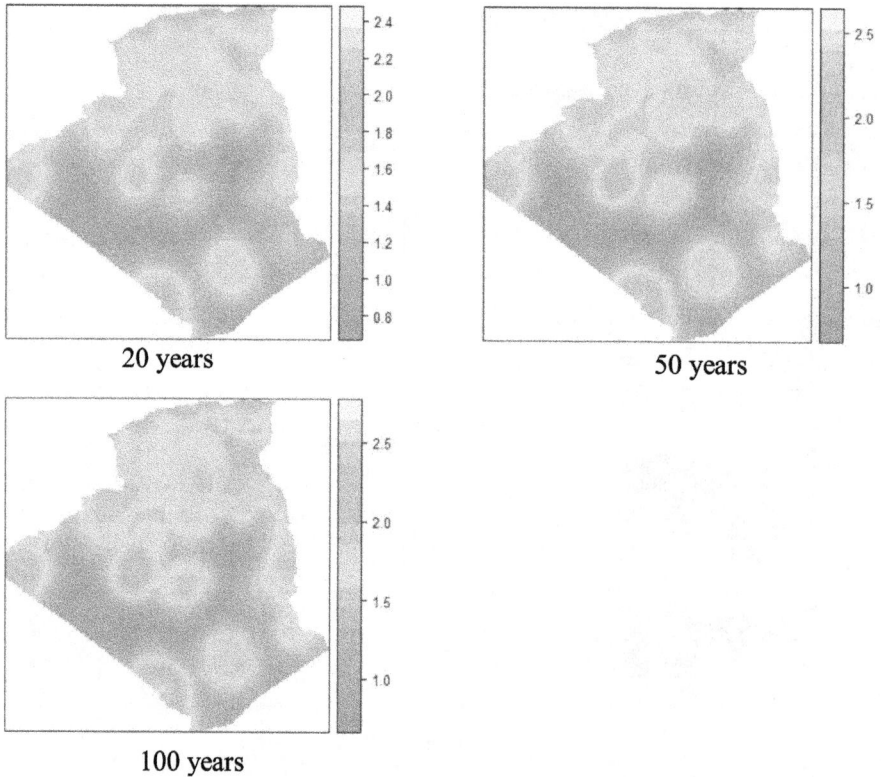

20 years

50 years

100 years

Figure 8.
Spatial representation of return periods (co-kriging method).

Figure 8 shows the results for the three cases RLs considered, in the first and second cases (20 years and 50 years), we notice roughly the same results, although there are some remarks that need to be made. In 50-year RLs, there is an increase in the eastern part, far south and in the center of Algeria. Otherwise in 100-year RLs, we noticed a great difference, especially in the eastern region, in the far southern the state of Tamanrasset, the western region in the state of Tindouf, and in The Middle of the desert the Adrar region.

4. Conclusion

In the current research, we studied the spatial analysis of rainfall data in 856 grid cells during the analysis period (1979–2012). The main conclusions form this study can be summarized as follows:

- We marked that the spherical model was found to be the most accurate model for annual maximum precipitation at the national scale.

- We can clearly see that the Kriging forecast error map shows low errors expected near the stations, while co-Kriging gives the lowest errors on average at the national level, which means that the method of co-Kriging is the best.

- Return levels were estimated for several return time periods. For the return level estimated from the GEV distribution, the point estimate that exceeds the return level of all previous maximum AMP records begins to appear in the 50-year regression period.

Acknowledgements

The author grateful to professor Lazhar Belkhiri for helpful.

Author details

Hicham Salhi
Laboratory of Applied Research in Hydraulics, University of Mustapha Ben
Boulaid-Batna 2, Batna, Algeria

*Address all correspondence to: salhiheat@gmail.com

IntechOpen

References

[1] Fisher RA, Tippet LHC. Limiting forms of the frequency distribution of the largest or smallest member of a sample. Proceedings of the Cambridge Philosophical Society. 1928;**24**:180-190

[2] Gnedenko BV. Sur la distribution limite du terme maximum d'une série aléatoire. Annals of Mathematics. 1943; **44**:423-453

[3] Gumbel EJ. Statistics of Extremes. New York, NY: Columbia University Press; 1958

[4] Naima B, Hassen C, Lotfi H. Modelling maximum daily yearly rainfall in northern Algeria using generalized extreme value distributions from 1936 to 2009. Meteorological Applications. 2017;**24**:114-119

[5] Meddi M, Toumi S. Spatial variability and cartography of maximum annual daily rainfall under different return periods in Northern Algeria. Journal of Mountain Science. 2015;**12**(6): 1403-1421. DOI: 10.1007/s11629-014-3084-3

[6] Ferrelli F, Brendel AS, Aliaga VS, Piccolo MC, Perillo GME. Climate regionalization and trends based on daily temperature and precipitation extremes in the south of the Pampas (Argentina). Geographical Research Letters: Cuadernos de Investigación Geográfica; 2019. DOI: 10.18172/ cig.3707

[7] Blanchet J, Lehning M. Mapping snow depth return levels: smooth spatial modeling versus station interpolation. Hydrology and Earth System Sciences. 2010;**14**:2527-2544

[8] Shrestha AB, Bajracharya SR, Sharma AR, Duo C, Kulkarni A. Observed trends and changes in daily temperature and precipitation extremes over the Koshi river basin 1975-2010.

International Journal of Climatology. 2016;**37**(2):1066-1083. DOI: 10.1002/ joc.4761

[9] Ren Y-Y, Ren G-Y, Sun X-B, Shrestha AB, You Q-L, Zhan Y-J, et al. Observed changes in surface air temperature and precipitation in the Hindu Kush Himalayan region over the last 100-plus years. Advances in Climate Change Research. 2017;**8**:148-156. DOI: 10.1016/j.accre.2017.08.001

[10] Lloyd CD. Assessing the effect of integrating elevation data into the estimation of monthly precipitation in Great Britain. Journal of Hydrology. 2005;**308**(1–4):128-150. DOI: 10.1016/j. jhydrol.2004.10.026

[11] Adhikary SK, Muttil N, Yilmaz AG. Ordinary kriging and genetic programming for spatial estimation of rainfall in the Middle Yarra River catchment. Australia. Hydrology Research. 2016;**47**(6):1182-1197. DOI: 10.2166/nh.2016.196

[12] Moral FJ. Comparison of different geostatistical approaches to map climate variables: Application to precipitation. International Journal of Climatology. 2010;**30**(4):620-631. DOI: 10.1002/ joc.1913

[13] Yang X, Xie X, Liu DL, Ji F, Wang L. Spatial interpolation of daily rainfall data for local climate impact assessment over greater Sydney region. Advances in Meteorology. 205:12. DOI: 10.1155/2015/ 563629

[14] Goovaerts P. Geostatistical approaches for incorporating elevation into the spatial interpolation of rainfall. Journal of Hydrology. 2000;**228**(1–2): 113-129. DOI: 10.1016/S0022-1694(00) 00144-X

[15] Rad AM, Khalili D. Appropriateness of clustered raingauge stations for

spatio-temporal meteorological drought
applications. Water Resources
Management. 2015;**29**:4157-4171

[16] Jenkinson AF. The frequency
distribution of the annual maximum
(or minimum) of meteorological
elements. Quarterly Journal of the Royal
Meteorological Society. 1955;**81**:158-171

Social Consequences of Climate Change

Flooding and Flood Modeling in a Typhoon Belt Environment: The Case of the Philippines

Fibor J. Tan

Abstract

Flooding is a perennial world-wide problem and is a serious hazard in areas where the amount of precipitable water has potential to dump excessive amount of water. The warming of the Earth's climate due to the increase in greenhouse gases (GHGs) increases the availability of water vapor and hence, of extreme precipitation as observed and forecasted by researchers. With rainfall intensity too high, the torrential rains coupled with weather systems that enhances its effects, flooding not only submerges anything low-lying, it also washes away living and non-living things along the course of the river and the floodplain. The flooding is even worsened by the increase in velocity of flow caused by unsustainable urbanization and denudation of the watershed at the headwaters. Nature's strength is an order of a magnitude that is way beyond that of the strength of men but human ingenuity enables us to transform our living environment into models that could help us better understand it. Flood modeling provides us decision support tools to deal better with nature. It also enables us to simulate the future especially nowadays that changes in our climate is imminent and even happening already in many parts of the world. Therefore, strategies on how to cope with our ever changing environment is very important particularly to countries that are at more risk to climate change such as the archipelagic Philippines.

Keywords: Flooding, flood modeling, climate change, hydrology, LiDAR

1. Introduction

An expression of nature when excess rainfall flows through the floodplain, flooding is a problem our ancestors, who started civilization had already dealt with. And now, even in our modern times, it is still a world-wide problem that seems to be a worsening problem due to increasing population, and hence, of urbanization that is usually leading to unsustainable development. Moreover, climate change makes the problem more intense as extreme precipitation, rainfall, in particular, near the equator is observed and projected to increase by researchers from the Intergovernmental Panel on Climate Change (IPCC) [1]. To be able to manage our watersheds, we have to understand it. Knowing how our rivers behave will give mankind a harmonious existence with the rest of the environment.

2. Rainfall potential

In the tropics, the amount of solar energy is the most radiated here since it is perpendicular, to near perpendicular to the sun's rays. This results to more energy converting the liquid water into water vapor, resulting to an increase in humidity in the area. The amount of precipitable water that could potentially become rainfall is a function of specific humidity which is proportional to the amount of vapor pressure which is a function of the temperature in the area [2]. This makes the wet tropics vulnerable to effects of climate change. Moreover, in this area in the tropics, since the sun's rays is concentrated, differential heating of the oceans and the lands will cause the air masses of different energy to move relative each other. Aside from the relative movement of the air masses, their differential densities will tend warm air mass to be lifted. Lifting of an air mass cools down the temperature adiabatically and results to condensation in the form of precipitation, or in the tropics, rainfall.

The effect of the changing sun's available energy which is caused by the varying tilt of the Earth's axis as it revolves around the sun results to the changes in seasons that the planet experiences yearly. Near the equator, it creates two distinct seasons while beyond the Tropic of Cancer in the northern hemisphere and Tropic of Capricorn in the southern hemisphere, it results to four different seasons in a year. In the Philippines, which is near the equator, the seasons is composed of the wet and the dry seasons that happen around May and November, respectively. The seasons is dictated by the presence of the monsoon winds – the Southwest Monsoon (SW Monsoon), locally known as Habagat season and the Northeast Monsoon (NE Monsoon), locally known as the Amihan season. During the SW Monsoon season, the wind comes from the southwest of the Philippines which is a moisture-laden air mass originating from the Indian Ocean which when it weakens, will be replaced by the NE Monsoon coming from the northwest in the cold northern regions such as Siberia.

At the time that SW Monsoon is strong, the same solar energy received at the region in the Pacific Ocean brews low pressure area (LPA) that usually develop into tropical cyclones (TC) or typhoon as we call it in the Philippines and other countries in the Pacific. Forming over oceans where sea surface temperature as well as air temperatures are greater than 26°C, TC accumulates large amounts of sensible and latent heat as it spirals toward the center and further get strength [3]. The counter-clockwise rotation of the Earth that produces the trade winds or easterlies tends the low pressure area that forms at low latitudes but greater than 5degrees, to move westward toward the Philippines. Depending on the strength of the monsoons, the incoming TC moving as the trade winds will have its characteristic bend that determines if it will make landfall or not to the country. If the SW Monsoon is prevailing, it could interact with the TC and intensify the rainfall that the TC could bring to the area. **Figure 1** shows a record compiled by the Philippine Atmospheric, Geophysical, and Astronomical Services Administration or PAGASA showing the TC tracks per month from 1948 to 2016 that at times deflect away the country depending on the relative strength of the monsoons.

On average, 20 Tropical Cyclones enter the Philippine Area of Responsibility (PAR) every year, 8 to 9 of which make landfall, as tallied by PAGASA. Coupled with the SW Monsoon, the problem of flooding in the country is exacerbated by TCs which attracts the monsoon's moisture-laden airmass. Even during the cool months, some parts of the country experience excessive rainfall due to the tail-end of the cold front or TECF which happens when the cold air from the prevailing NE Monsoon meets with the warm Easterlies. The lifting of the warm air mass as it moves past the advancing cold air mass (a cold front) causes much rain that floods the eastern and south eastern sea board of the country.

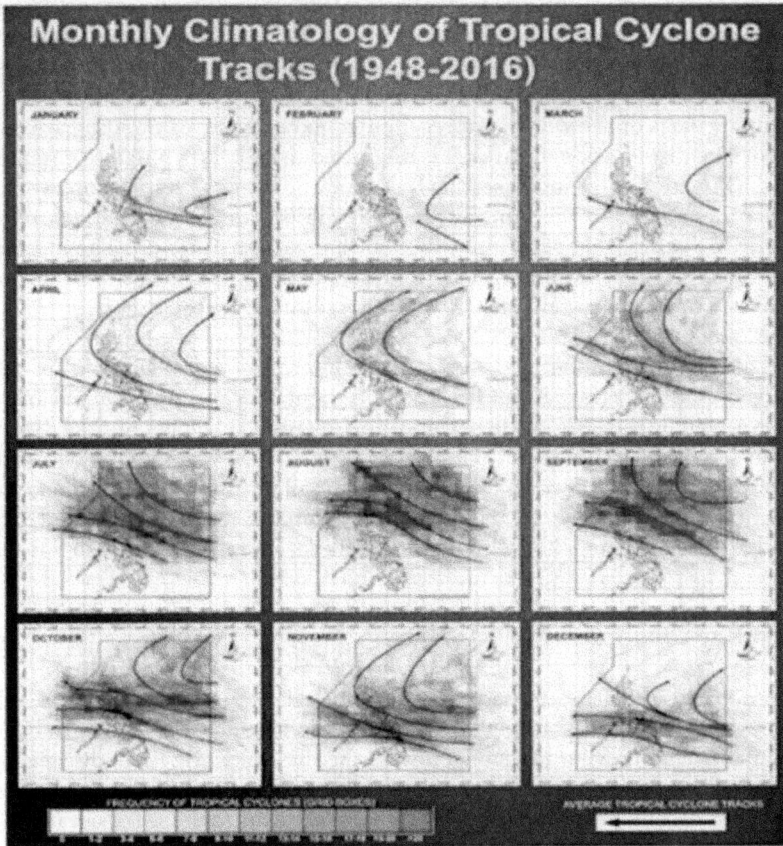

Figure 1.
Tropical cyclone tracks in the Philippines averaged from 1948 to 2016. Source: PAGASA, available at http://bagong.pagasa.dost.gov.ph

The same scenario happens in other countries in the tropics that are near a large body of water and are influenced by prevailing wind patterns that come together with different air mass of different thermal properties. Countries in the Indian Ocean experience cyclones while hurricanes are in the Atlantic, and typhoon in the Pacific – different naming that pertain to Tropical Cyclone that forms in the warm waters of the oceans and affects many countries in the tropics and dissipates as it reaches the colder latitudes.

3. Excess rainfall

Flooding is commonly brought about and intensified by tropical cyclones along with other weather systems that generate storms. The presence of abundant moisture in the air mass carried by the different weather systems, when precipitated, make our cities prone to excessive surface runoff from streams and through floodplains where most of these cities are located. Flooding is a natural reaction of a watershed when the excess rainfall which is the amount of rainfall that is not absorbed by infiltration to the ground, flows as surface runoff. However, urbanization tends to alter the permeability of the soil and reduces the infiltration rates of the surface by replacing the permeable ground with impermeable concrete pavements and building structures. Natural waterways and floodplains that are the natural drainage systems and detention basins of floods are also blocked or replaced with built up environment. This makes flooding a major problem in areas where people and anthropogenic activities are thriving.

In many parts of the world, unprecedented experiences of flooding are documented. The distribution of rainfall is disrupted usually attributed to climate change. Climate change increases the frequency of heavy precipitation events (or the proportion of total rainfall from heavy falls), an observation reported in the IPCC Technical Paper VI on Climate Change and Water [1]. The increase will in turn increase the annual average runoff making previous efforts of flood control and drainage system insufficient to accommodate the runoff.

The amount of excess rainfall is also dictated by the residence time the rain water will remain on the surface such as by the interception of vegetation and the ponding on soil surfaces to be infiltrated. Since rainfall events are observed and even projected by the IPCC [1] to become more intense, i.e. higher rainfall depth on a shorter period of time due to climate change in the wet tropical regions, risk of flooding is projected to rise. Higher intensity rainfall results to lesser residence time for the rain water to be intercepted by the plant's foliage and to be infiltrated to the ground, thereby, causing the hydrograph to peak earlier resulting to flash floods.

4. Flooding consequences

In tropical watersheds, the natural environment is naturally designed to assimilate in its system the excess rainfall that flows as direct runoff through the hillslope, accumulating and flowing as channel flow in the river network. From the steep slopes of the headwater source, it reaches the flatter terrain of the floodplain that functions as natural detention basin before it finally exits to the sea or ocean.

The floodplain serves as a natural barrier to make the flow of water more efficient to in maintaining equilibrium in the watershed system. This equilibrium defines the natural environment of the watershed that hosts different flora and fauna unique to the environment.

Human intervention to promote development to support the demands of mankind and in modern civilization to support the requirements of the economy of a country has put the natural balance of the watershed environment in danger. The modification of our environment leads to the alteration of the hydrological processes. From the watersheds headwaters, the forests are cleared at a rate that leaves the mountains denuded reducing the capacity of the watershed to capture the rainfall for infiltration and recharge of the groundwater. The effect is an increase in surface runoff that endangers the downstream. Moreover, clearing of the forests also reduces the surface roughness of the terrain that tends to increase the velocity flow. Climate change, on the other hand, also modifies the hydrologic cycle and its effects is cascading. An increase in temperature will affect the atmospheric water and hence, the water balance. With increase flowrate and velocity due to anthropogenic alterations coupled with climate change, more floodwaters flow downstream.

Development often favors the flat terrain of the floodplain for accessibility and ease. It is also advantageous for growing crops as it is fertile enriched with nutrients coming from the mountains. Because of this, our floodplains became the usual sites of urbanization. The replacement of the natural environment with the built environment robbed the floodwaters of its own space. Every rainy season or typhoon season people's habitation are inundated with flood. It even encroaches now areas that we thought of not flood prone but are actually are. We just occupied the natural course and detention area of the floodwaters. During typhoons that amplify the

Figure 2.
Effects of flooding in Metro Manila, Philippines in 2012.

effects of the monsoon season, low-lying populated areas in the floodplains are a common scene in countries like the Philippines.

In 2009, the Tropical Cyclone Ketsana (Tropical Depression Ondoy which did not reach the typhoon category) wreak havoc to the capital and its suburbs that dumped a month's rainfall in just more than 3 hours. It was an unprecedented flooding event with the scale of its damage. Just three (3) years after in 2012, the same areas were affected even just by the Habagat or Southwest Monsoon rains enhanced by a tropical storm. **Figure 2** shows a swollen river and flooding in low-lying areas.

In countries frequented by tropical cyclones, riverine flooding is even exacerbated by sea level rise brought about by climate change and amplified by land subsidence, making worse the net sea level rise. Moreover, during typhoons, the effects of storm surge brought about by the gusts of winds, even worsens the flooding problem. In the Philippines, when the Tropical Cyclone Haiyan (Typhoon Yolanda) hit the central islands in 2013, it was estimated that more than 6000 people died and left the coastal towns and cities devastated.

5. Modeling flood inundation

The watershed is a very dynamic environment especially if it is being altered by anthropogenic activities. Human activities modify the hydrologic cycle to meet the demands of the population in order to survive. More often than not, the modifications we do to our watersheds lead to the deterioration and destruction of our natural environment impairing their natural purpose of efficiently conveying excess surface runoff downstream.

Because of the necessity to alter our watersheds to satisfy the needs of the society, it is imperative that we simulate the options that we want to do to minimize the problems that could be brought about by the modification that we will do. To alleviate the already existing flood problems we encounter in the floodplains we propose flood control measures that need to be simulated as well. The simulation of the flow of flood in a watershed or flood modeling is important as a decision support tool or system to achieve better options in solving our flooding problems given our limited resources.

In flood modeling, one of the most important inputs other than the rainfall, is the elevation. Water seeks its own level and therefore, elevation has to be accurately represented.

In the Philippines, extensive flood modeling of many of its riverbasin happened along with the introduction of LIDAR which stands for Light Detection and Ranging to create 3D digital elevation model (DEM). The LIDAR DEM generated has an accuracy of 20 cm in the vertical and 50 cm in the horizontal directions generated from at least 2 data points averaged per square meter. A Digital Surface Model (DSM) kind of model is produced from this scan of the earth's surface. Sample DEM that shows purely terrain elevation called Digital Terrain Model or DTM is shown in **Figure 3**. It has a resolution of 1 m x 1 m as compared to Synthetic Aperture Radar (SAR) and Interferometric Synthetic Aperture Radar (IfSAR) DEMs which has a resolution of 30 m x 30 m and 10 m x 10 m or 5 m x 5 m, respectively.

Starting 2012, all of the 18 major river basins of the Philippines and its 247 principal river basins and other smaller river basins had their floodplains modeled using LiDAR DEM which was a priority project of the Department of Science and Technology (DOST) through its Nationwide Operational Assessment of Hazards or Project NOAH's Disaster Risk and Exposure Assessment for Mitigation or DREAM Component Project and the Phil LiDAR Program.

5.1 LiDAR data processing and validation

For the said projects, floodplains that did not have LiDAR data yet, manned airborne LiDAR data acquisition was conducted. LiDAR data collection and prepro-cessing is divided into three parts: 1) Data acquisition, 2) Point cloud generation, and 3) Data classification. Data acquisition included flight planning, system instal-lation, data collection and data download. After data acquisition, pre-processing was done. Point cloud generation included the accuracy determination that will need the hardware and the software that came with the LiDAR system. For the data classification, Terrasolid was used for classifying infrastructures from the bare earth. The end-products of the preprocessing stage were unedited digital elevation models or DEMs.

Pre-processed LiDAR data were then subjected to processing. The processing of the LiDAR data began with manually editing some features of the unedited DEMs so that water will flow in a realistic manner. In the DEM, the bridges were removed, pits were filled and important features were removed in the last return digital

Figure 3.
A LiDAR DEM processing output for Labo River basin in Camarines Norte Province, Philippines. It features a burned bathymetric data on LiDAR digital terrain model (DTM) processed by the Mapua-Phil LiDAR 1 Project of Mapua University through the Department of Science and Technology.

terrain model (DTM) retrieved from the secondary DTM. DTM, together with digital surface model (DSM) is a type of DEM. When DEMs were already edited, they were mosaicked into a bigger DEM and integrated with the bathymetric data gathered from the field. A minimum of ArcGIS 10.2 (or more recent version) was used in the project for the processing data since the scripts of Phil-LiDAR 1 Project was used. Simultaneously, feature extraction or digitizing features on LiDAR DEM were done. This also used ArcGIS for processing.

Validation was done by getting the difference in elevation of each check point and its corresponding LIDAR data point where the checkpoint locations were based on the guidelines provided by the American Society for Photogrammetry and Remote Sensing (ASPRS). Reference points and benchmarks established by National Mapping and Resource Information Authority (NAMRIA), Land Management Bureau (LMB) and other agencies were researched and checked on the field. The elevation of the geographic location labeled (X,Y) of these bench marks were determined through static observation using dual frequency GNSS (*Global Navigation Satellite System*) receivers. For water surfaces, detailed survey of the cross-sections and bathymetry were done.

5.2 Topographic and hydrographic surveys

Information on river topography and geometry and lake bathymetry is important in watershed and flood modeling. These features, nevertheless, are poorly represented in the LIDAR data due to low penetration of light signals in water. River profiling, cross-section and bathymetric surveying for lakes must be collected and integrated into the LIDAR data. GNSS kinematic surveys and topographic surveys using total station or digital level, and bathymetric mapping using an echo sounder equipped with GNSS positioning were done to obtain the above data needed. The collected data were analyzed and filtered to remove the outliers and then interpolated to come up with a continuous surface with the same spatial resolution as the LIDAR-derived elevation dataset. This surface data was subjected to accuracy assessment using an independent dataset and was ensured to have the same vertical accuracy as the validated LIDAR data or better. This was embedded into the LIDAR-data and used in modeling of the watershed and flood hazards.

Modeling and simulations were done using a combination of software that included GIS software, hydrologic modeling software and river and floodplain hydraulic software to generate flood hazard maps at different rainfall scenarios of 5, 10, 25, 50, and 100-year return periods. **Figure 4** shows sample two (2) dimensional flood hazard maps produced from flood hazard computer models simulated in 5, 25, and 100-yr return periods.

5.3 Hydrological assessment

In the development of the watershed, GIS proved to be useful in the derivation of the the the catchment and subcatchment boundaries and properties. These are then used to create the hydrologic model with additional input of the rainfall data at different return periods. The hydrologic models were calibrated using actual or observed data taken during a rainfall event caused by storms or typhoons to capture the direct runoff that exhibits the response of the watershed to excess rainfall.

The hydrologic model simulations generate river flow or discharge that are input to the river hydraulics model which simulates the depth and velocity of flow in the river channel and the floodplain.

For the projects, hydrological measurements such as for rainfall, water level and discharge were done in the sub-basins for use in the flood model calibration and

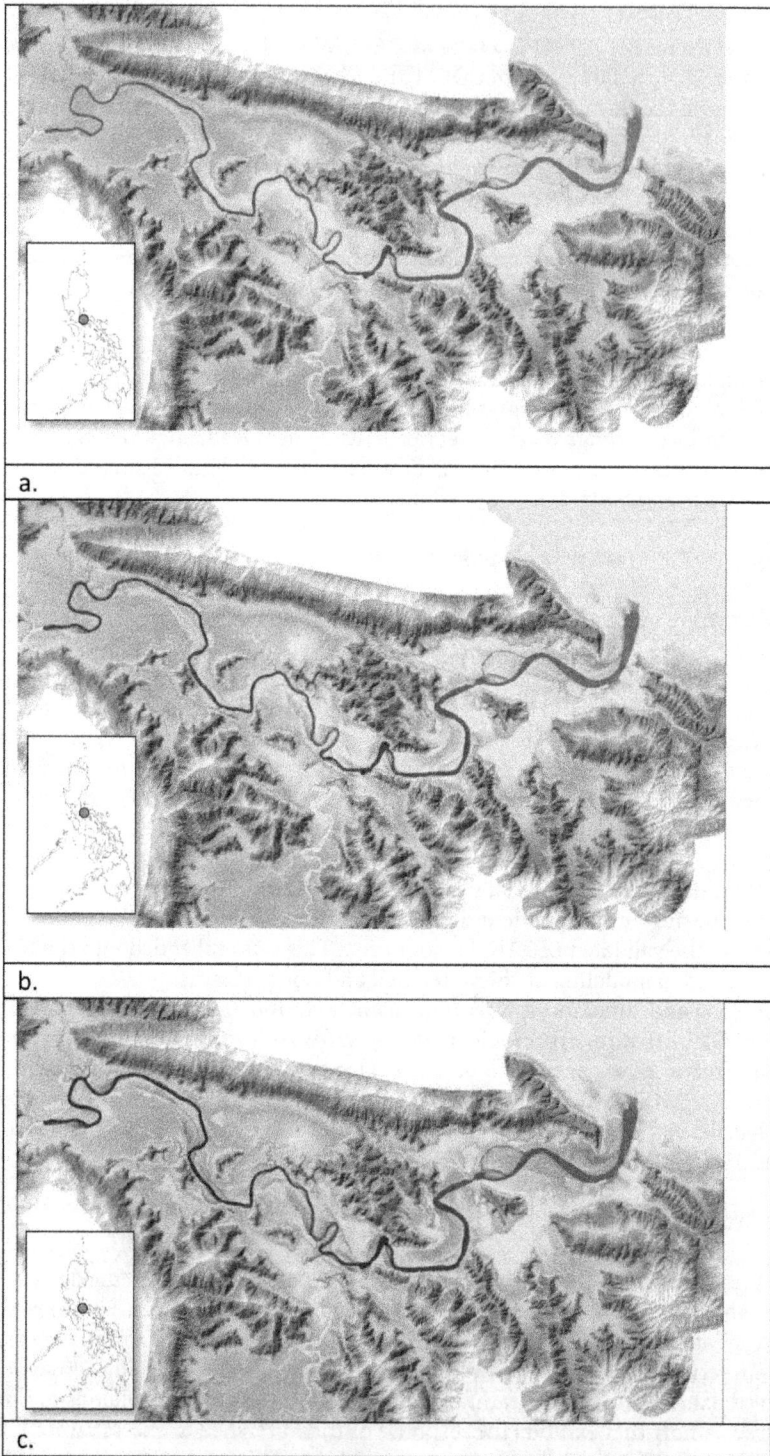

Figure 4.
Flood hazard map generated using LiDAR digital elevation model in Maling River in Atimonan, Quezon Province, Philippines produced by the FRAMER Project of Mapua University and Department of Science and Technology (DOST).

validation. Existing data from different government agencies e.g. PAGASA were also collected and assessed. For ungauged watersheds, rain gauges, water level sensors, and velocity meters were used for at least a year to capture seasonal changes. At least one rain gauge was ensured to be present near the center of each river basin. On the other hand, water level and velocity sensors were used at a selected point in the river in the usually with the presence of a bridge for ease and safety in data gathering. At that location, called project point, cross-section measurements were done so that the rate of discharge can be computed. The water level and the computed discharge were needed in generating rating curves to be used in forecasting water stage.

5.4 Watershed and flood modeling and hazard mapping

It is in the hydraulic model that the LiDAR DEM plays an important part of giving accurate flood depths. The flood depths are represented geographically in a map to generate the flood hazard map of an area.

With the validated LiDAR-DTM, the sub-basins were delineated using watershed delineation algorithm in GIS. HEC-HMS (Hydrologic Modeling System) and HEC-RAS (River Analysis System) programs of the Hydrologic Engineering Center (HEC) of the US Corps of Engineers were utilized [4]. The hydrological model for the upstream watersheds were developed using HEC HMS where soil type and land-cover related parameters needed in program were derived through analysis satellite images, and from the Department of Agriculture's Bureau of Soil and Water Management (DA-BSWM) soil maps. Other parameters of the model were set to initial values and were adjusted during calibration. After calibration, the model were validated with an independent set of rainfall and discharge data [5].

The hydraulic model for the floodplain were developed using HEC-RAS. Cross-section data extracted from the validated LiDAR DTM was used in this program to create the geometry of the river system in each sub-basin. The land-cover related parameter which is the Manning's roughness coefficient were determined through analysis satellite images.

The model generated from the combined HEC-HMS and HEC-RAS programs were used to run the actual flood events for each sub-basin using historical flood data. Flooding due to hypothetical extreme rainfall events of various return periods obtained from PAGASA were simulated to see the effects of such events in the watershed. The flood simulations were then used to generate water surface elevation grids which were overlain into the high resolution LiDAR-DTM to generate flood hazard maps. To enhance the flood hazard maps, a 2D hydrodynamic model was utilized as well [6]. This 2D model in HEC-RAS utilized the discharge hydrographs generated by HEC HMS.

6. Flood modeling results

The use of a more sophisticated equipment to measure elevation at higher resolution is necessary when it comes to modeling flood inundation. Elevation is one of the most important inputs as this will tells us the depth of water which is a serious concern to people and for properties.

In **Figure 4**, sample flood hazard maps for one of our study areas in Quezon Province show the different flood inundation levels at different rainfall return periods, i.e. 5, 25, and 100 years. The flood hazard maps serve as an important tool for use by the Local Government Units (LGUs) planners and responders, and by the public in general. The maps produced show the flood inundation that can be related to the forecast of rainfall event the weather agency releases. Across the Philippines,

PAGASA has synoptic weather stations, each has corresponding rainfall intensity, duration, frequency curves (RIDF). A station nearest to the area of study adopts this specific RIDF.

During a storm or a typhoon, forecast of rainfall depth in time or intensity in an area i.e. with respect to a watershed, is checked with this curves for equivalent return period. The determined return period will tell us which flood hazard map should be used. From the specific inundation map, we can clearly see the extent of the flooding as well as the relative depth of flood. The geographic location of flooding can be easily visualized when it is overlain on the elevation DEM. When data of exposure is included, the number of households and establishments affected can be determined.

As shown in **Figure 4**, flood hazard maps are an important tools that can be used to see spatially the occurrence of flood and the potential impact it may cause to the area. Moreover, a risk assessment of the flood hazard that incorporates the aspects of exposure and vulnerability of elements such as the population and economic activities in the area has to be done in order to identify flood prioritization response for cost efficiency on responses [7]. Flood hazard and flood risk assessments are two different but related procedures whose outputs are an important decision-making tools for identifying flood prone areas needed for the design of flood control and drainage projects, and for the prioritization of responses for saving lives and properties, respectively.

Rainfall data such as in the RIDF should be continuously updated to account for new entries that reflect the effects of extreme climate variabilities such as climate change. Although there are still uncertainties in the projections of climate scenario due to limited knowledge, it has been observed that there are widespread increases in heavy precipitation events which are associated with the increase in water vapor in the atmosphere consistent with the observed warming [1]. Aside from rainfall, behavior of storms and tropical cyclones that brings with them rain are becoming stronger as manifested in the introduction of a higher category of Tropical Cyclones in the Philippines by its weather agency, PAGASA. In May 1, 2015, a new category Super Typhoon with speeds more than 220kph is added in the Tropical Cyclone Classification as a consequence of the strongest typhoon TC Haiyan (Typhoon Yolanda as it is locally known). Changing precipitation volumes and intensities increase the risk of flash flooding and urban flooding in many areas in the region and even the world.

7. Conclusions

Flooding, especially in tropical countries that has the hydrometeorological factors that generate huge quantities of precipitable water, is a yearly event that disrupts people's lives. Flood control structures and proper drainage design can help mitigate this but quantification of the surface runoff should be accounted for correctly. Present technologies such as computer models aid us in doing this. Along with the better input data we have nowadays, these models can give us flood hazard information that enable us to quantify the amount of flood water our mitigating measures should accommodate to reduce or even eradicate flooding problems in certain areas. Continuous update of data parameters, nevertheless, should be done to account for rapid changes happening in the environment including climate change. Specifically, affected by climate change is the rainfall variable that designers of flood control and drainage structures should consider. This is to ensure that our mitigating solutions are not just satisfying future capacity requirements but are also climate-proof for better life and continuous progress.

Flooding and Flood Modeling in a Typhoon Belt Environment: The Case of the Philippines
DOI: http://dx.doi.org/10.5772/intechopen.98738

Acknowledgements

The models in maps were produced through the Mapua-Phil LiDAR 1 Project of Mapua University with our partner University of the Philippines - Diliman and the Flood Risk Assessment for Mitigation and Effective Response FRAMER Project of the Mapua University. Both projects were funded by the Philippine Council for Industry, Energy, Emerging Technology Research and Development (PCIEERD) of the Department of Science and Technology (DOST).

Author details

Fibor J. Tan
Mapua University, Manila, Philippines

*Address all correspondence to: fjtan@mapua.edu.ph

IntechOpen

References

[1] Intergovernmental Panel on Climate Change (IPCC). Climate Change and Water. IPCC Technical Paper VI. 2008

[2] Ven Te Chow, David R. Maidment, and Larry W. Mays. Applied Hydrology. 1988

[3] PAGASA. Learnings. Available at: http://bagong.pagasa.dost.gov.ph/learnings/faqs-and-trivias Accessed: 2021-06-10

[4] Fibor J. Tan, Edgardo Jade R. Rarugal, Francis Aldrine A. Uy. One-dimensional (1D) River Analysis of a River Basin in Southern Luzon Island in the Philippines using LiDAR Digital Elevation Model. International Journal of Engineering and Technology. Vol. 7 No. 3.7. 2018. DOI: 10.14419/ijet.v7i3.7.16200

[5] Fibor J. Tan; Francis Aldrine A. Uy; Cris Edward F. Monjardin; Chennie Carissa A. Caja; Roa Shalemar R. Pornasdoro; Jeffrey Dave R. Sy; Larriz M.Samudio; Marc Julius A. Bunag; Jonel B. Tarun; Czeskian Z. Realo; Jasmin M. Domingo; Brylle C. San Agustin; John Cedrec D. Recalde; Myra Donne T. Chua; Adonis B. Sigua Event Flow Measurements in Remote Tropical Watersheds in the Philippines: The Need for Automated Weather-proof Devices. .2020. In: Proceedings 2020 IEEE Conference on Technologies for Sustainability, SusTech 2020

[6] Veronika Röthlisberger, Andreas P. Zischg, Margreth Keiler. Identifying spatial clusters of flood exposure to support decision making in risk management. 2017. Science of the Total Environment. DOI: 10.1016/j.scitotenv.2017.03.216

[7] Md. Sanaul Haque Mondal, Takehiko Murayama, Shigeo Nishikizawa. Assessing the flood risk of riverine households: A case study from the right bank of the Teesta River, Bangladesh. International Journal of Disaster Risk Reduction. 51 (2020) 101758. DOI: 10.1016/j.ijdrr.2020.101758

Chapter 6

Changing Climatic Hazards in the Coast: Risks and Impacts on Satkhira, One of the Most Vulnerable Districts in Bangladesh

Md. Golam Rabbani, Md. Nasir Uddin and Sirazoom Munira

Abstract

Changes in the climate due to anthropogenic and natural variation are indicated by parameters including temperature and rainfall. Climate change variability with changing trends of the two have been unpredictable and unprecedented globally leading to changing weather patterns, natural disasters, leading to sectoral impacts on food and water security, livelihood, human health among others. This research analyses the changing patterns of these parameters over the last 35/37 years of Satkhira district of Bangladesh to assess the state and trend across spatial and temporal dimensions. Such, the study validates to rationalize the observed seasonal changes that persist in Satkhira of Bangladesh. Both in terms of intensity and frequency of the occurrences of natural disasters, the series of natural events have been triangulated, with impacts and vulnerability being assessed from temperature variations, erratic rainfall, cyclone, flood and water logging etc. The study's prime contribution remains in attribution of climate change in relation contextual circumstances in the region including sea level rise, salinity intrusion. Therefore, the risk and climatic hazards and its resulting impacts over time has been assessed to draw deeper connection between theoretical and practical values. The series of analyses also draw conclusion that assets are at risk from changing climatic condition.

Keywords: temperature, rainfall, climate risk, impact, Satkhira

1. Introduction

Bangladesh has long been recognized as being very vulnerable to both environmental degradation and climate change impacts [1, 2]. Increasing changes in climate variability and extreme events have pushed the country towards greater vulnerability. This vulnerability is compounded by low economic progress of the country, inadequate infrastructure, lack of institutional capacity, improper management practices, and increased dependency on the natural resource base which makes coping strategies difficult to implement.

It has been predicted that "climate change impacts will be differently distributed among different regions, generations, age classes, income groups, occupations and genders" [3]. The IPCC also notes: "the impacts of climate change will fall disproportionately upon developing countries and the poor persons within all countries, and thereby exacerbate inequities in health status and access to adequate food, clean water, and other resources". It became an unkind or terrible reality for the communities of Bangladesh to face extreme climatic events e.g. recent prolonged and devastating floods (1998, 2004, 2007) and cyclonic events in last decades. IPCC Fifth Assessment Report (AR5) shows that the global average mean sea level rose by 1.7 mm per year during 1900–2010 and it further increased to 3.2 mm per year during 1993–2010 [4]. The most recent IPCC (2018) special report states that the global mean sea level rise is expected to be around 0.1 meter less with global temperature of 1.5°C compared to 2°C [5]. It is predicted that the sea level is likely to rise 30 and 50 cm by 2030 and 2050 respectively (World Bank, 2000). A recent report states that the range of sea level rise in the coast of Bangladesh is 6 mm to 21 mm/year during 1980–2010 [6]. This may have adverse impacts on natural resources in different degrees in different ecosystems (terrestrial and aquatic). IPCC Fifth Assessment Report (AR5) indicates that sea level rise and wave action are causing decline of vegetated coastal habitats across the world (Wong, 2014). Some other reports also state that the coastal communities of Bangladesh are exposed to risks of weather and climate related extreme events [7].

Variation in temperature, erratic behavior of rainfall, cyclonic events [Cyclone Sidr (2007), Cyclone Bijli, Cyclone Aila (2009), Cyclone Mahasen (2013), Cyclone Roanu (2016), Cyclone Fani and Cyclone Bulbul (2019) and Cyclone Amphan (2020)], salinity intrusion, droughts, extreme heat waves, cold wave etc. made their life and livelihoods miserable in the coast of Bangladesh in last decade.

Changes of climate may directly and indirectly affect freshwater resources (water availability and deteriorate water quality), fisheries biology and aquatic ecosystem, human health (increase incidences of water borne diseases e.g. diarrhea, cholera, dysentery etc.) and agriculture crops. These would result increase demand and consumption of water due to increase of temperature, increased pressure on groundwater, shortage of food due to decreased agriculture production and finally increase of morbidity and mortality of the communities with low resilience in the coast especially in Satkhira.

2. Materials and methods

The research delved into the state and trend of some key climate related primary and secondary hazards including variation in temperature and rainfall, cyclonic events, salinity intrusion and sea level rise that are affecting the Southwest coast, especially Satkhira district of Bangladesh. Cyclone and storm surge, salinity intrusion and SLR are the key climate induced secondary hazards or disaster that are affecting the coastal communities. It is expected that the vulnerability may be enhanced in the future because of increased frequency and intensity of such events.

The key primary elements including temperature and rainfall of the research district (Satkhira) were collected from the Bangladesh Meteorological Department (BMD) and analyzed to see the trend of temperature over the 35 years long-period (1981–2015) and rainfall over 38 years long period (1981–2018). The research mainly considered at annual and seasonal trend of temperature and rainfall. The secondary literatures collected from different sources were used to analyze the state and trend of other secondary climatic hazards in the southwest coast of Bangladesh. The following sections provide the details of the trend of temperature and rainfall pattern in the study location (Map in **Figure 1**).

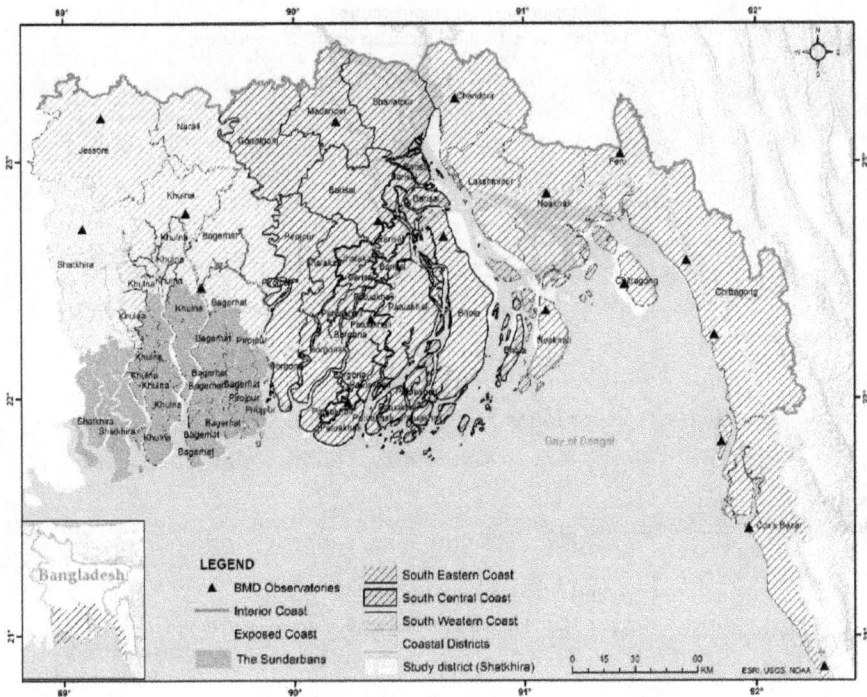

Figure 1.
Study area (Satkhira) map.

3. Results and discussions

3.1 Temperature

Bangladesh Meteorological Department (BMD) provided the climatic data for Satkhira district over the period 1981–2015. The climatic data comprised monthly, seasonal and annual average maximum and minimum temperature for the above-mentioned district. The following figure (**Figure 1**) shows that the 35 years annual mean maximum temperature follows a decreasing trend over the period of 1981–2015 while the annual mean minimum temperature go along with an increasing trend for the same period. Over the period, the highest annual mean minimum temperature (22.44°C) was observed in 2010 while the lowest (20.53°C) was in 1999 (**Figure 2**). The highest annual mean maximum temperature (32.51°C) was in 1987 while the lowest (30.08°C) was in 1981. It was also observed that both maximum and minimum (annual mean) temperature were on increasing trend for last ten years (2006–2015).

As mentioned above that the liner trend analysis shows the decreasing trend (0.008°C/yr) of mean maximum temperature during 1981–2015 in Satkhira, which is statistically non-significant trend. The Mann-Kendal (MK) test also shows insignificant decreasing trend of annual mean maximum temperature in Satkhira during the same period. In terms of five-year moving average, the annual mean maximum temperature shows an increasing trend until mid-1980s and then suddenly drops in 1987. From 1990 onwards, no trend is observed; rather the temperature shows regular fluctuations.

The annual mean minimum temperature shows an increasing trend (0.02°C /yr) during 1981–2015, which is highly statistically significant(p = 0.002). In non-linear trend, the MK test also found an increasing trend with statistical significance.

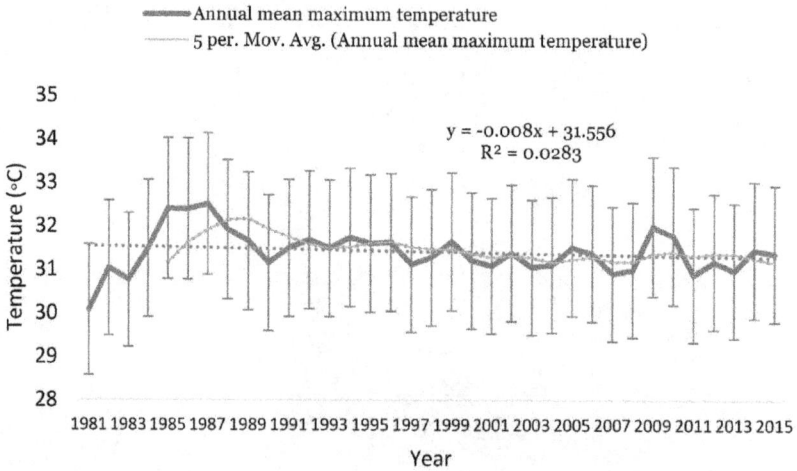

Figure 2.
Annual mean maximum temperature (1981–2015).

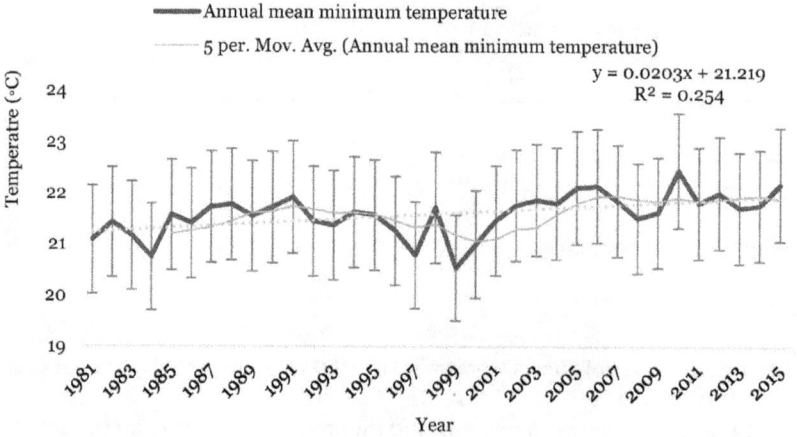

Figure 3.
Annual mean minimum temperature (1981–2015).

On seasonal analysis, average maximum temperature for the period of 1981–2015 was on decreasing trend in pre-monsoon, post-monsoon and winter while it was on increase (**Figure 3**) in monsoon. With the application of MK test, the mean maximum temperature in early winter (December, January) and late post monsoon (November) depicted statistically significant decreasing trend over the period of 1981–2015 in Satkhira. The mean minimum temperature was on increasing trend in all the seasons between 1981 and 2015 (**Figure 4**). The linear trend of annual mean minimum temperature in monsoon (p = 0.009) and post monsoon (p = 0.002) was found statistically highly significant increasing trend. Over the period of 1981–2015, the MK test also shows statistically significant increasing trend of mean minimum temperature in each of the months from June to December, except October. But again, both mean maximum and mean minimum temperature in monsoon was on increasing trend over the last 35 years (1981–2015).

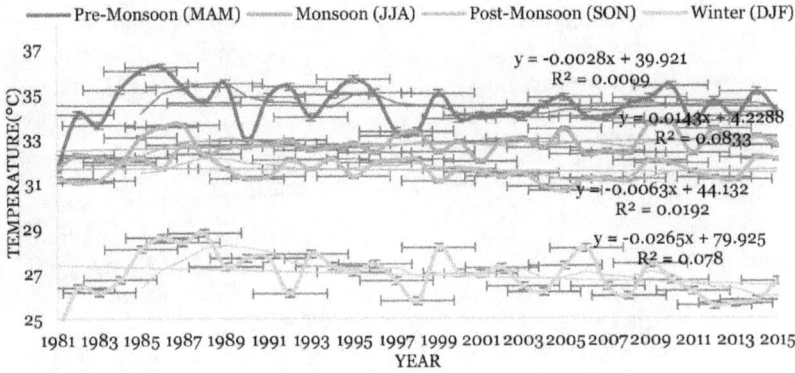

Figure 4.
Seasonal mean maximum temperature (1981–2015).

3.2 Rainfall pattern

Annual and seasonal total rainfall of the study area was observed. During 1981–2015, the trend of total rainfall in Satkhira was on decreasing trend (**Figure 5**) (**Rabbani et al., 2018**). This linear trend of total rainfall over the period of 1981–2015 was found statistically non-significant decreasing trend. MK test also depicts that this rainfall trend is on decreasing trend without any statistical significance. In terms of five-year moving average, fluctuation was observed from mid 1980s to late 1990s. From 1999, it shows an upward trend but again it repeatedly dropped from 2007 to 2014.

On seasonal mean rainfall pattern, Pre-monsoon, monsoon and winter rainfall followed a decreasing pattern from 1981 to 2015 while post monsoon shows an increasing trend for the same period (**Figure 6**). On the other hand, the difference between the trend of total rainfall of monsoon and the rest three seasons is much closer in latter half of last 35 years (1981–2015) period.

In terms of five-year moving average, pre-monsoon mean rainfall did not show any notable trend between mid-1980s and mid-1990s. From 1995 it was on upward trend till 2001 but then it constantly dropped till 2012. Monsoon mean rainfall pattern shows irregular pattern from 1985 to 2005. From 2005 it was on decreasing trend. Similar pattern was observed for post-monsoon rainfall from 1985 to 2005

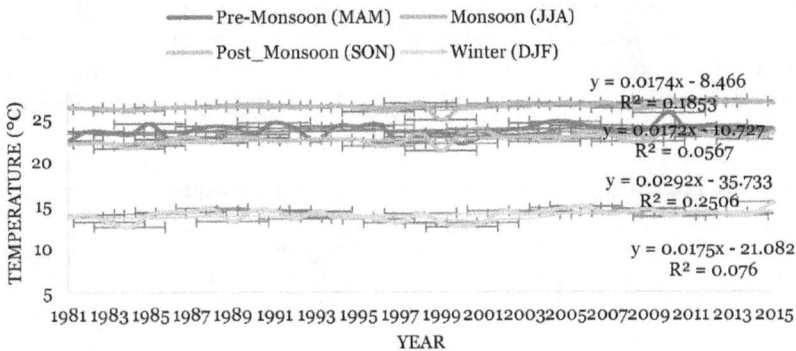

Figure 5.
Seasonal mean minimum temperature in Satkhira (1981–2015).

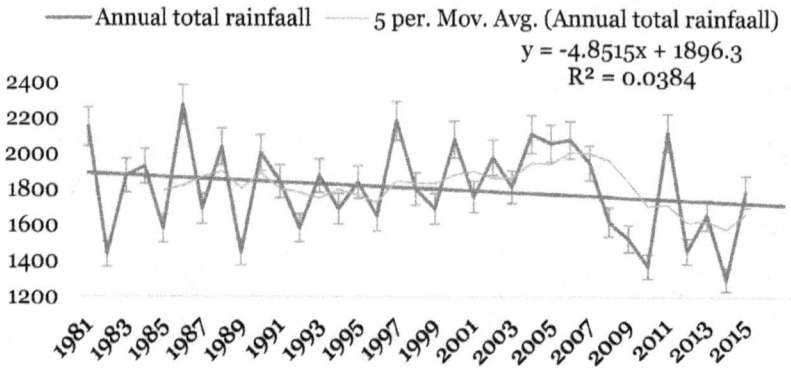

Figure 6.
Annual (total) rainfall in Satkhira during 1981–2015.

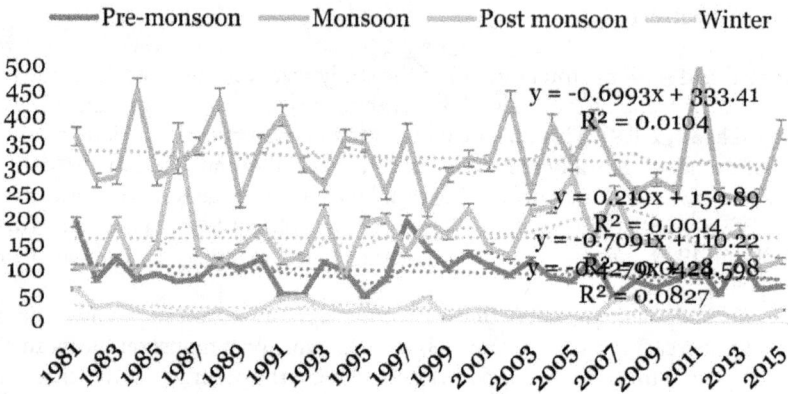

Figure 7.
Trend of seasonal rainfall (total) in Satkhira during 1981–2015, modified from [8].

but from 2007 it showed sharp downward trend till 2015. Five year moving average does not show any remarkable upward or downward trend over the 35 years period.

The study observed that the no of days without rainfall was on the increasing trend over 1981–2017, which is highly statistically significant (p = 0.0001) (**Figure 7**). The trend of days with over 100 mm and 150 mm rainfall were found to be on the decrease, which did not show any statistical significance (Rabbani et al., 2018). Five-year moving average of the trend of no of days without rainfall shows an upward trend from early 2000s. While five year moving average of the trend of days with above 100 and 150 mm rainfall during 1981–2017 in Satkhira shows a downward trend since 2006/2007.

3.3 Cyclone and storm surges in the southwest coast including Satkhira

On the number of cyclonic events that hit Bangladesh varies in different studies. [9] indicates that the coast of Bangladesh experienced 154 cyclonic events of different classes[1] between 1877 and 1995 while [10] refer 117 cyclonic events from

---

[1] The classes of cyclonic events include: Super Cyclonic Storm (greater than 220 km/hour), very severe cyclonic storm (119–220 km /hour), severe cyclonic storms (90–119 km/hour), cyclonic storms (60–90 km/hour), Deep depression (51–59 km/hour), Depression (32–50 km/hour) (Dasgupta et al., 2010).

1877 to 2003. Another study indicates that 149 cyclones hit Bangladesh between 1891 and 1998 [11]. This difference in number of cyclonic reference probably occurred because of the scope of different studies. However, it appears that 38 cyclonic events affected South-West coast (including Satkhira) between 1877 and 2010 [10]. Between 1970 and 2010, seven severe cyclone (>90 km/hour wind speed) devastated south-west coast (including current Satkhira District) and the local communities. Some studies clearly point that the storm surge accompanying with cyclone cause huge damage of the infrastructure and the wetlands resources. When it coincides with high tide, it becomes almost catastrophic with above 5-meter depth of inundation along the coast [12].

The following **Table 1** indicates that 40 cyclonic events affected South-West coastal region (Greater Khulna) including Satkhira during 1877–2010. Four severe cyclonic storms and one super cyclonic storm affected this region since 1973. Cyclone Sidr and Cyclone Aila devastatingly affected the study region. Increased intensity of the cyclonic events may increase risks of life and livelihoods of the local communities. Moreover, the compound effects of different climatic hazards/disasters may alter existing livelihoods opportunities. The poor households will have limited options for water supply, sanitation and farming practices if the ecosystems are more severely affected in future.

On Cyclonic events, IPCC Fourth Assessment Report predicts about the intensification of the extreme weather events such as cyclones and associated storm surges especially along the Bay of Bengal. There are evidences of decreasing frequency of monsoon depression and formation of cyclone but increase of intensity in the Bay of Bengal since 1970 [13, 14]. This prediction of decreasing frequency of cyclone is in line with earlier findings of Ali in 2000 although it partially differs with SMRC findings [15] which reveals that frequency of intense cyclone during post-monsoon (November) has been increasing. But many studies were conducted to understand the role of SST in formation and intensity of cyclonic events [e.g. 17, 18]. It is reported that increase of 2°C and 4.5°C of SST would cause increase of 10% and 25% wind speed of the cyclone respectively [11]. This generally means that the intensity of cyclone will be increased with increase of temperature.

In addition, increase in SLR will bring the water line further inwards. Consequently, the effect of storm surge will penetrate deeper into the landmass. These are going to largely affect agricultural production, health, loss of livelihoods and increase in poverty of this Southwest coastal region including Satkhira. The recent Cyclone Amphan of 2020 killed 31 people and affected over 10 million people altogether in Bangladesh[2]. Total damage from cyclone Amphan worth BDT 1100 crore or Taka 11 Billion (USD 130 Million).

3.4 Salinity intrusion and sea level rise (SLR)

Sea level rise and salinity intrusion are already affecting the coastal communities in Bangladesh. It is projected that the possible SLR may severely affect the coast of the country. It has been reported that a vast and diverse coral reef of South Asia were lost in 1998 due to coral bleaching induced by the 1997/98 El Niño event [16–18]. A study report shows that Bangladesh would face the largest impacts due to SLR [19]. The possible SLR may affect Bangladesh by inundating coastal areas. As mentioned above it has been predicted that by 2030 and 2050 at least 30 and 50 cm sea level will rise respectively. Another report shows that if 25 cm sea level rises then 40 percent of Sundarban will be submerged, and in case of rising sea level by above 60 cm, the whole Sundarban will disappear [20]. A recent report

---

[2] The Dhaka Tribune, 22 May 2020.

SL	Year	Month of occurrence	Type of Cyclonic event
1	1877	6–8 August	Tropical Depression
2	1877	18–20 August	Tropical Depression
3	1881	26–28 May	Tropical Depression
4	1885	16–19 June	Tropical Storm
5	1890	22–24 June	Tropical Depression
6	1895	30-Sep	Not Available
7	1899	13–16 July	Tropical Depression
8	1904	10–12 June	Tropical Depression
9	1906	25–27 July	Tropical Storm
10	1907	23–26 June	Tropical Storm
11	1908	16–22 June	Tropical Storm
12	1916	4–6 June	Tropical Storm
13	1919	5–10 August	Tropical Depression
14	1919	23–25 September	Not Available
15	1928	1–2 August	Tropical Depression
16	1932	7–10 June	Tropical Depression
17	1932	11–15 November	Tropical Depression
18	1937	11–14 October	Tropical Storm
19	1938	25–28 May	Tropical Depression
20	1939	30-Sep	Tropical Depression
21	1944	24–31 August	Tropical Depression
22	1959	9–18 July	Tropical Depression
23	1963	October	Tropical Depression
24	1965	9–12 May	Tropical storm
25	1966	September	Severe Cyclonic Storm
26	1968	June	Tropical Depression
27	1969	October	Tropical Storm
28	1970	October	Tropical Storm
29	1973	December	Very Severe Cyclonic Storm
30	1986	November	Severe Cyclonic Storm
31	1988	October	Tropical Storm
32	1988	November	Very Severe Cyclonic Storm
33	1996	October	Tropical Storm
34	1998	November	Very Severe Cyclonic Storm
35	2000	October	Tropical Storm
36	2002		Tropical Storm
37	2007	15-Nov	Super Cyclonic Storm
38	2009	24–25 May	Very Severe Cyclonic Storm
39	2019	9–11 November	Cyclone Bulbul (Very Severe Cyclonic Storm)
40	2020	16–21 May	Cyclone Amphan (Super Cyclonic Storm)

Source: [10, 11].

Table 1.
Cyclonic events that affected south-west coast (greater Khulna) since 1877.

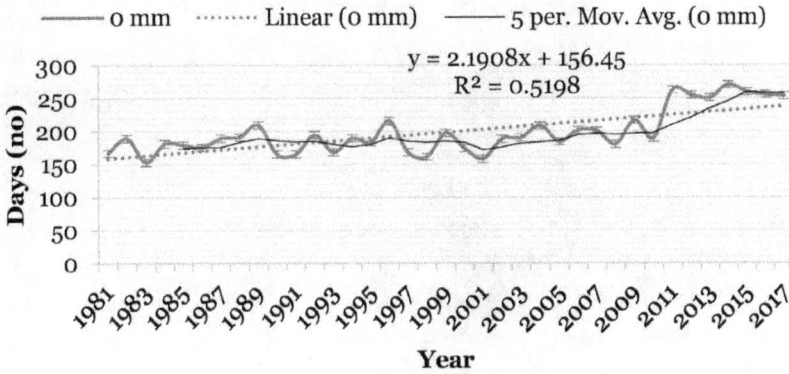

Figure 8.
Trend of "no of days without rainfall" during 1981–2017 in Satkhira, modified from [8].

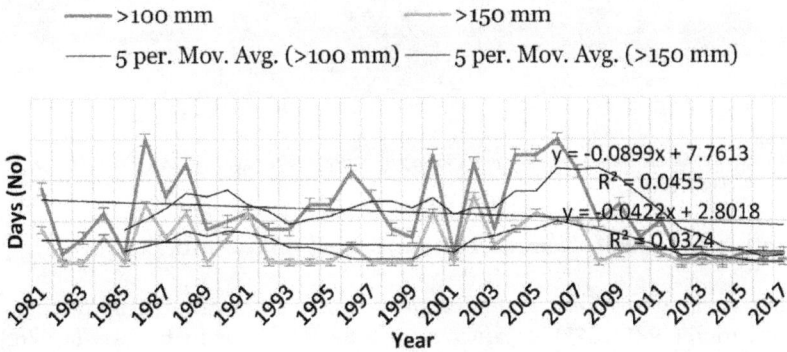

Figure 9.
Trend of days with above 100 and 150 mm rainfall during 1981–2017 in Satkhira, modified from [8].

shows that there is a trend of increasing SLR at different points by 6–20 mm/year during 1983 to 2012 [21] (**Figures 8–10**).

In fact, the SLR is likely to inundate the coastal wetlands, lowlands, accentuate coastal erosion, increase frequent and severe floods, create drainage and irrigation problems and finally dislocate millions of people from their homes and occupation [22]. This may catalyze the increasing rate of rural –urban migration within the country. An estimation based on a coarse digital terrain model and global population distribution data, shows that SLR will directly affect more than 1 million people in 2050 in each of the Ganges- Brahmaputra-Meghna delta in Bangladesh [23]. On the other hand, salinity became one of the major problems for the coastal zones of Bangladesh. This is happening may be due to low flow of fresh water from the Ganges and ingress of salt water from Bay of Bengal. So the compound effect of SLR and salinity may disrupt agriculture (e.g. reduction of rice), mangroves including the Sunderbans and coastal ecosystem including ponds and create additional health problems in the local communities. The recent reports state that the coastal community may suffer more with water borne diseases and other physical problems (e.g. menstruation problems of the women from drinking of saline water) due to SLR and salinity intrusion [24]. However, the poor and marginal groups would be critically affected by the possible SLR and salinity intrusion in coastal zone of Bangladesh.

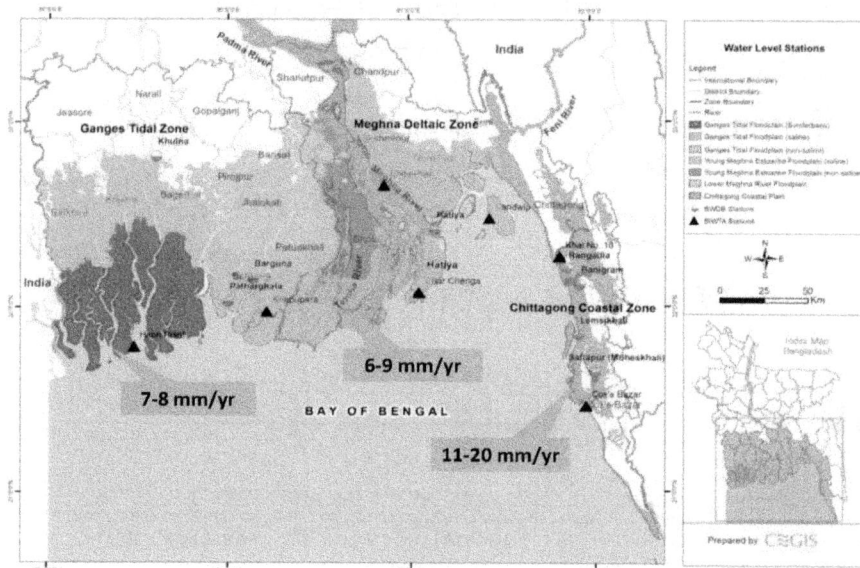

Figure 10.
Water level trends for the Ganges, Meghna and Chittagong coastal sub zone of Bangladesh based on the data of last 30 years. Source: [21].

3.5 Synthesis of climate hazards, risk and impacts in Satkhira

Analysis shows that the trend of mean maximum temperature in monsoon over a 35-year period (1981–2015) is on increasing trend while it is on decrease in all other seasons in the study area of Satkhira. According to Mann Kendall test, the trend of mean maximum temperature of November, December and January is on decrease significantly over the period of 1981–2015. It was also observed that the average minimum temperature was increasing trend for the last 35 years period (1981–2015) with statistical significance. Both maximum and minimum temperature trend is also on increase in monsoon for the same period. It indicates that monsoon was warmer in last 35 years. Early winter days were also on cooling trend. Nights are on warming trend in every season over the 35 years period in Satkhira. Rainfall patterns are changing in the study district. Pre-monsoon rainfall is also decreasing over the above-mentioned period. Likewise, the number of days without rainfall has increased with statistical significance over the same period. It specifies that the study communities are losing the rainy days even in monsoon. It may be argued that a longer span of rainless days, over extraction of water during warmer seasons and high rates of evaporation, especially during pre-monsoon periods, may present possible reasons for depleting water levels. Existing literature demonstrates that changes of temperature and rainfall pattern have an impact on wetlands [25–27]. [28] mention that climate change and water-related infectious diseases are also intertwined.

The literatures on major secondary elements of climate change including cyclone and storm surges, salinity intrusion and sea level rise (SLR) were collected and reviewed to see the changes over the decades. As the study area is close to the Bay of Bengal, the risks related to cyclonic events and storms surges, salinity intrusion in water and soil and potential sea level rise is high for the natural resources and local communities. The following sections provide details on the changes of key secondary events of climate change in the study location. The research summarizes the details on key secondary events of climate change in the study location, Satkhira. Existing literature encompasses key major secondary elements including cyclones, sea level rise and salinity intrusion, all of which are detrimental events in the face of climate change. Since 1970s the frequency of monsoon depression and cyclone formation in the Bay of Bengal has reduced whereas the intensity has increased [29]. Rabbani et al., [30] posits that the acceleration in the wind speed may as well bring unprecedented casualties in the coastal ecosystems, pushing the vulnerable communities towards greater risks of losing lives.

This research also speculates deeper practical connections between theory and practice. It comes to the conclusion that the compound effects of different climatic disasters may alter existing livelihoods opportunities. The poor households will have limited options for water supply, sanitation and farming practices if this area is more severely affected in future. It may be inferred that Cyclone Sidr (2007), Cyclone Aila (2009) and Cyclone Amphan (2020) largely impacted on the study region while it occurred causing increased intensity of the cyclonic events to the risks of many. The research therefore posits connection between life and livelihood of the communities susceptible to the impacts of climate change and overall resilience it is failing to build. Literature shows the various times of the year when cyclonic events have been in effect from 1877 to 2020. It can be indicated that cyclone usually hit the Southwest coast between May and December, whereas

SL	Climate Change Related Key Hazards	Key Impacts and vulnerability
1	Temperature variations	• Challenges in domestic and irrigation water supply • Surface water quality deteriorates • Inadequate safe drinking water • Sanitation problems increase
2	Erratic behavior of rainfall	• Affects agricultural yields • Increases incidences of water borne diseases and associated cost • Loss of working days especially in monsoon
3	Cyclone and storm surge	• Damage of infrastructure (roads, culverts, embankments and so on) • Lack of water for domestic purposes including drinking, sanitation and hygiene practices • Loss and damage of fisheries and dependent livestock • Increased water borne diseases • Forced migration
4	Salinity and potential Sea level Rise	• Salinity affects water and soil/lands • lack of freshwater supply including drinking water • Loss of crops/lands • Forced migration
5	Flood/water logging	• Affects standing crops, fisheries and livestock • WASH challenges • Increases incidences of water borne diseases and associated cost • Increases incidences of water borne diseases and associated cost • Damage housing and infrastructure • Enhance river bank erosion

Table 2.
Climate change related hazards and associated impacts in the study location, Satkhira.

Monsoon and Post-Monsoon are the main seasons for cyclonic events in greater Khulna region including Satkhira. In the cyclone history, the highest occurring month was found June (8 times) followed by October (7 times). On the other hand, sea level rise and salinity intrusion are already affecting the coastal communities in Bangladesh. It is projected that the possible SLR may severely affect the coast of the country. The research delves into all these aspect through the lens of livelihood options which needs to be ensured against the devastating climate hazards within the coastal zones of Bangladesh.

This research shows the results of the district Satkhira but the most marginalized and poor populations situated in disaster prone areas of Bangladesh are often victims of extreme climatic conditions. **Table 2** shows impacts on the life and livelihoods is also jeopardized through these uncertain series of hazards and vulnerabilities, environmental migration from riverbank erosion, inundation, sea level rise etc. along with the impact on agriculture, a number of other sectors are impacted which includes water resources, forestry, food security, human health, infrastructure, settlements including displacement of inhabitants and loss of livelihood, coastal management and even sustainable disaster response and recovery plans. Therefore, resulting adaptation need of the vulnerable communities remain wide-ranging. Needs and demands of the vulnerable poor range from financial, technological needs to capacity building, administration, research and development, health, infrastructure etc. There is a dire requirement for serious intervention in the areas of food security, water security and related aspects to build a resilient community of poor and marginalized communities. Additionally, effort is also needed in comprehensive disaster management, flood control, enhanced rural and urban resilience of vulnerable groups, migration and other critical issues. Innovative, well-communicable, transparent and climate-smart solutions are needed to combat these challenges for the poor.

4. Conclusions

The currently existing challenges of climate change has been unanimously recognized by the international community, as a result of which Bangladesh, small country in terms of territory, ranks high in the list of most vulnerable countries on earth- as stated. The geographical characteristics of Bangladesh are heavily dependent on local and regional hydrological characteristics, which rely on climatic processes incusing dimensions of seasonality. Despite being a nano-emitter of the harmful greenhouse gases that contribute to increase the global average temperature, Bangladesh is the hardest hit as a result of climate change impacts. Larger and more integrated and innovative solutions are required to truly tackle the problem. Being one of the most climate vulnerable countries in the world, Bangladesh is at risk of natural disasters such as riverine and flash floods, tropical cyclones, storm surges, droughts, salinity intrusions, sea level rise and riverbank and coastal erosions. The trend analysis in this study shows that the changing patterns of temperature and rainfall of the country has a number of implications on agriculture and food security, water sector, health sector, livestock and fisheries and overall livelihood of the communities. Both the rural and the urban sectors are affected due to the unpredictable implications within these sectors. The effects of climate change are ubiquitous and all encompassing, causing the magnitude, frequency and duration of such natural disasters to increase, thus making marginalized communities extremely vulnerable. While the country has made a number of significant progress in reducing death burden from natural disasters, the remaining challenge is to protect the livelihoods which push people below the poverty line, force them to migrate to the urban areas in search of work and also excludes them from a rightful participation in shaping the local level decision-making and service provisioning in favor of them. To that, this study concludes that further data science applications to socio-environmental studies like these should be in continuation for deeper assessment of vulnerability assessment to take right national and global policy uptakes.

Author details

Md. Golam Rabbani[1*], Md. Nasir Uddin[2] and Sirazoom Munira[3]

1 Climate Bridge Fund (CBF), Dhaka, Bangladesh

2 Bangladesh Centre for Advanced Studies, Dhaka, Bangladesh

3 North South University, Dhaka, Bangladesh

*Address all correspondence to: golam.rabbani72@gmail.com

IntechOpen

References

[1] General Economic Division. Bangladesh Delta Plan 2100, Baseline Studies on Disaster and Environmental Management, *volume 2,* Bangladesh Planning Commission; 2018.

[2] General Economic Division. 8th Five Year Plan, July 2020-June 2025. Bangladesh Planning Commission; 2020.

[3] IPCC. Summary for Policymakers. In: J. J. McCarthy, O. F. Canziani, N. A. Leary, D. J. Dokken, & Kasey S. White (eds.). *Climate Change 2001: Impacts, adaptation and vulnerability. Contribution of working group II to the third assessment report of the intergovernmental panel on Climate change.* Cambridge, UK: Cambridge University Press, 2001. pp. 1-18.

[4] Wong P.P, Losada J.P, Gattuso J. Hinkel A, Khattabi K.L, McInnes Y, Saito and Sallenger A. Coastal systems and low-lying areas. In: *Climate Change 2014: Impacts, Adaptation, and Vulnerability. Part A: Global and Sectoral Aspects. Contribution of Working Group II to the Fifth Assessment Report of the Intergovernmental Panel on Climate Change* [Field C.B, V.R. Barros D.J, Dokken K.J, Mach M.D, Mastrandrea T.E, Bilir M, Chatterjee K.L, Ebi Y.O, Estrada R.C, Genova B. Girma, E.S. Kissel, A.N. Levy S. MacCracken P.R, Mastrandrea, and L.L. White (eds.)]. Cambridge University Press, Cambridge, United Kingdom and New York, NY, USA, 2014. pp. 361-409.

[5] IPCC. Summary for Policymakers. In: Global warming of 1.5°C. An IPCC Special Report on the impacts of global warming of 1.5°C above pre-industrial levels and related global greenhouse gas emission pathways, in the context of strengthening the global response to the threat of climate change, sustainable development, and efforts to eradicate poverty [V. Masson-Delmotte, P. Zhai H. O, Pörtner D. Roberts, J. Skea, P. R. Shukla, A. Pirani, W. Moufouma-Okia, C. Péan, R. Pidcock, S. Connors, J. B. R. 2018.

[6] CCC. Assessment on sea level rise on Bangladesh coast through trend analysis. Climate Change Cell, Department of Environment, Ministry of Environment, Forests and Climate Change, Dhaka, Bangladesh. 2016.

[7] McGranahan G. D. Balk, and B. Anderson. The rising tide: Assessing the risks of climate change and human settlements in low elevation coastal zones. Environment and Urbanization, 2007; **19**, 17-37.

[8] Rabbani M.G, RAHMAN S.H, MUNIRA S. Prospects of pond ecosystems as resource base towards community based adaptation (CBA) to climate change in coastal region of Bangladesh. Published by Journal of Water and Climate Change March 2018, *9 (1) 223-238.*

[9] Ali A. Climate change impacts and adaptation assessment in Bangladesh. Climate Research. Clim Res. Vol. 12: 109-116, 1999.

[10] Islam T. and Peterson R. E. Climatology of land falling tropical cyclones in Bangladesh 1877-2003. Nat. Hazards 48, 2008;115-135.

[11] Karim MF, Mimura N. Impacts of climate change and sea-level rise on cyclonic storm surge floods in Bangladesh. Global Environmental Change. 2008;**18**(3):490-500.

[12] IWM, 2005. Impact assessment of climate change on the coastal zone of Bangladesh. Final report, Institute of Water Modeling, Dhaka, Bangladesh, 37p. In Karim M. F. & Mimura N. 2008. Impacts of climate change and sea level

rise on cyclonic storm surge floods in Bangladesh. Glob. Environ. Change 18, 490-500.

[13] Lal M. Tropical cyclones in a warmer world. Curr. Sci. India, 80, 1103- 1104. 2001.

[14] Cruz R.V, H. Harasawa M. Lal S. Wu Y. Anokhin B. Punsalmaa Y. Honda M. Jafari C. Li and N. Huu Ninh. Asia Climate Change 2007: Impacts, Adaptation and Vulnerability. Contribution of Working Group II to the Fourth Assessment Report of the Intergovernmental Panel on Climate Change, M.L. Parry, O.F. Canziani, J.P. Palutikof, P.J. van der Linden and C.E. Hanson, Eds., Cambridge University Press, Cambridge, UK, 469-506. 2007.

[15] SMRC. The Vulnerability Assessment of the SAARC Coastal Region due to Sea Level Rise: Bangladesh Case. SAARC Meteorological Research Center, Dhaka SMRC-No. 3; 2003.

[16] Wilkinson C. Ed. Status of Coral Reefs of the World: 2000. Australian Institute of Marine Science, Townsville; 2000.

[17] Arceo H.O., M.C. Quibilan, P.M.Alino, G. Lim and W.Y. Licuanan. Coral bleaching in Philippine reefs: coincident evidences with Mesoscale thermal anomalies. B. Mar. Sci., 2001; 69, 579- 593.

[18] Wilkinson C., Ed. Status of Coral Reefs of the World: 2002.Australian Institute of Marine Science, Townsville; 2002.

[19] World Bank. The Impact of Sea Level Rise on Developing Countries: A Comparative Analysis, World Bank Policy Research Working Paper 4136, Washington, DC, 2007: World Bank.

[20] Hare, W. Assessment of knowledge on impacts of Climate change – Contribution to the specification of article 2 of the UNFCCC: Impacts on ecosystems, food production, water and socio-economic Systems.2003; Berlin, Germany.

[21] CEGIS-IWM-BUET. Assessment of Sea Level Rise and vulnerability of the coastal zone of Bangladesh using trend analysis; Report prepared for the Department of Environment under the Ministry of Environment, Forests and Climate Change, Government of Bangladesh; Centre for Environment and Geographic Information Services-Institute of Water Modeling-Bangladesh University of Engineering and Technology, 2015; Dhaka, Bangladesh.

[22] Rahman A, Alam M, Alam S, Uzzaman M.R, Rashid M, Rabbani M.G. risks, vulnerability and adaptation in Bangladesh. A background paper prepared for UNDP human development report 2007. Dhaka, Bangladesh.

[23] Ericson J.P, C.J. Vorosmarty S.L, Dingman L.G.Ward and M.Meybeck. Effective sea-level rise and deltas: causes of change and human dimension implications. Global Planet. Change, 2005; 50, 63-82.

[24] Rabbani M.G, Rahman A.A, Bhadra S. Climate Change Impacts on Water Resources and Connected Ecosystem in South Asia, with special reference to Bangladesh. In Gunawardena, E.R.N., Gopal, B., Kotagama, H (Eds). Ecosystems and Integrated Water Resources Management in South Asia. Published by Routledge in 2012, New Delhi, India and London, United Kingdom.

[25] Bronmark C, Hansson, L.A. Environmental issues in lakes and ponds: current state and perspectives. Environmental Conservation 29 (3): 290-306. Foundation for Environmental Conservation, Lund, Sweden. 2002.

[26] Karafistan A. & Arik-Colakoglu F. Physical, chemical and microbiological

water quality of the Manyas Lake, Turkey. Mitig. Adapt. Strat. 2005; Glob. Change 10 (1), 127-143.

[27] Erwin, K.L. Wetlands and global climate change: The role of wetlands restoration in a changing world. WCM, 2009; 17, 71-84.

[28] Husman, A.M. de Roda., Schets, F.M. Climate change and recreational water related infectious diseases. National Institute for Public Health and the Environment. 2010; Netherlands.

[29] Lal, M.climate change: India's monsoon and its variability. Journal of Environmental Studies and Policy, 2003; 6, 1-34.

[30] Rabbani, G., Huq, S. & Rahman, S. H. 2013a. Impacts of climate change on water resources and human health: Empirical evidences from a coastal district (Satkhira) in Bangladesh. In: Impact of Climate change on water and health (V. I. Grover, ed.). CRC press, Taylor & Francis Group, Boca Raton, FL,2013; pp. 272-285.

Chapter 7

Climate Change Risks in Horticultural Value Chains: A Case Study from Zimbabwe

Nqobizitha Dube

Abstract

Increasing frequency and severity of droughts and floods, shift in onset and cessation of the rainfall and increasing intensity of mid-season dry spells in the last 50 years have been identified in Zimbabwe. This paper presents an assessment of risks from climate change to the horticulture sector of Zimbabwe with the aim to provide mitigatory actions that could alleviate climate change risks in the horticultural sector of Zimbabwe. Specifically the chapter seeks to outline the climate change risks facing the horticulture sector in Zimbabwe, propose actions to reduce risks and assess financing and policy options for climate change adaptation in Zimbabwe. The study followed the approach taken by the International Fund for Agricultural Development (IFAD) which analyses climate risks at each stage of the horticulture value chain. The stages used by Vermeulen are input supplies (seeds, fertilisers, pest management, etc.,); agricultural production (water use, soil management, skill base, etc.,) and postproduction processes (storage, processing, transport, retail, etc.,). Data was collected from multiple stakeholders in areas with notable horticultural production across Zimbabwe using semi-structured interview guides. The study population composed of horticulture farmers, produce processing firms, value chain support organisations and government arms related to horticulture.

Keywords: Climate change risk, Horticulture, value chains, Zimbabwe

1. Introduction

Zimbabwe has a sub-tropical climate with four seasons: cool dry season from mid-May to August; hot dry season from September to mid- November; main rainy season from mid-November to mid-March; and the post rainy season from mid-March to mid-May [1]. The mean monthly temperature varies from 15°C in July to 24°C in November whereas the mean annual temperature varies from 18°C in the Highveld to 23°C in the Lowveld [2]. The lowest minimum temperatures (on average 7°C) are recorded in June or July and the highest maximum temperatures (on average 29°C) are recorded in October [2]. The climate is moderated by altitude with the Eastern Highlands enjoying cooler temperature compared to the low-lying areas of the Lowveld. In their research on agroecological conditions of Zimbabwe, Vincent and Thomas [1] argued that Zimbabwe was generally a semi-arid country with low annual rainfall reliability. The average annual rainfall is 650 mm but geographically it ranges from around 350 to 450 mm per year in the Southern Lowveld to above 1,000 mm per year in the Eastern Highlands. The rainfall pattern of Zimbabwe is variable with years below and above normal rainfall [1, 3]. **Figure 1** divides the country into five agro-ecological regions on the basis of soil type, rainfall, temperature and other climatic factors. The darker colours are indicative of higher rainfall, better soils and other positive climatic indicators. These regions also represent the agricultural potential for the production of crops and livestock [1, 2, 4]. Region 1 has the highest rainfall, followed by region 2a whose rainfall amounts average the upper limits —1000 mm—of region 2 while those of region 2b average the lower limits —750 mm—of region 2. Region 5 is the most arid agro-ecological region of Zimbabwe and is the second largest agro-ecological region after region 4 (see **Figure 1**). From a climate hazard perspective, the country experiences some relatively frequent drought years which are more frequent in region 4 and 5 (see **Figure 2**).

According to Zimbabwe's Third National Communication on climate change [5], climate change in the country is characterised by high temperature and rainfall variability and extremes. The increasing frequency and severity of droughts and floods, shift in onset and cessation of the rainfall and increasing intensity of mid-season dry spells in the last 50 years have been identified in Zimbabwe's Third National Communication [5] as a major consequence of climate change. The next sections consider the temperature and rainfall changes that have occurred over the years in Zimbabwe. The annual-mean temperature in Zimbabwe has increased

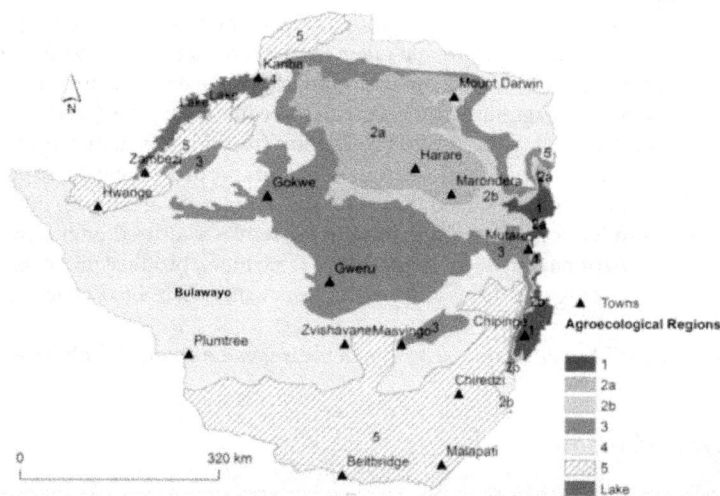

Agro-ecological region	Area covered as a percentage of total national are	Description
1	7000km/ 2%	<1000m rainfall, tea, coffee, plantation farming, macadamia, fruits, intensive livestock production
2a and 2b	58 600km2/ 15%	750-1000mm rainfall, Intensive crop and livestock production
3	72 900km² / 19%	650-800mm of rainfall. Severe mid-summer droughts but maize, tobacco, cotton and other cash crops grown
4	147 800km² / 38%	650- 800mm of rainfall .Livestock and drought resistant crop production
5	104 400km² / 27%	<450mm rainfall supports extensive cattle or game protection

Figure 1.
Agro-ecological regions of Zimbabwe. Source: Chikozi et al. [3].

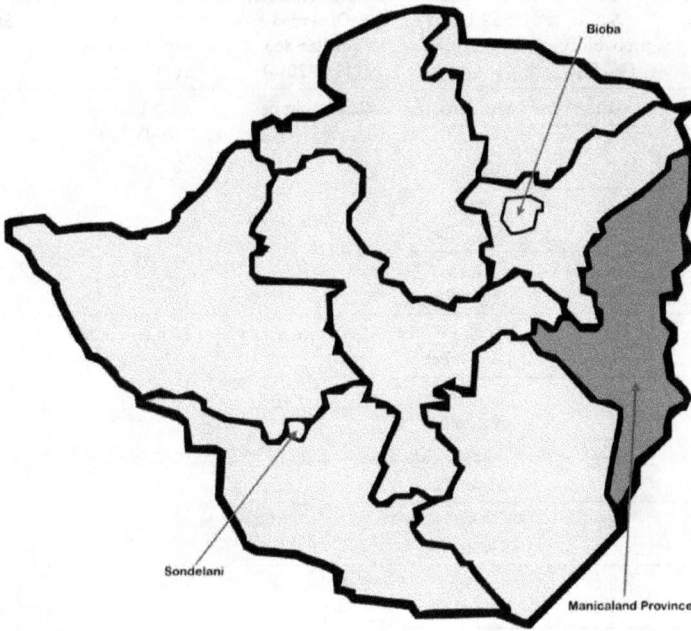

Figure 2.
Study areas.

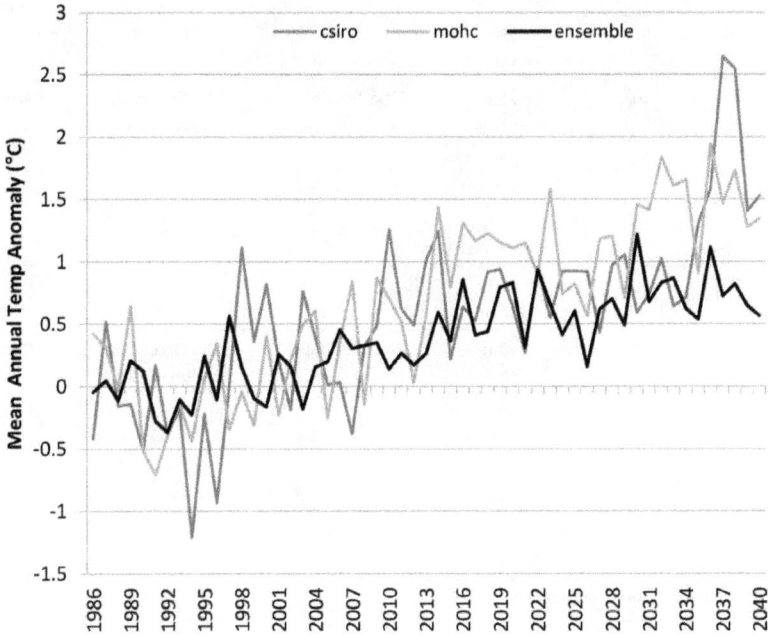

Figure 3.
Predicted mean annual temperature anomalies for Zimbabwe using various forecasting models. Source: Unganai [6].

by about 0.4°C since 1900 [2]. The 1990s decade was the warmest experienced in Zimbabwe during the last century. Between 2000 and 2020 temperatures have also followed and upward trend with the highest average annual temperatures recorded

Region	SoS probability (%) [1]	SoS date [1]	Observed median SoS (1997–2014)	SoS extremes and year observed (1997–2014)	Station used
V	20	31st October	21st November	6th January 2005–2006	Buffalo Range
	40	10th November			
	60	20th November			
	80	25th November			
IV	20	5th November	23rd November	28th December 2012–2013	Plumtree
	40	20th November			
	60	25th November			
	80	5th December			
III	20	25th October	19th November	13th December 2012–2013	Gweru
	40	5th November			
	60	10th November			
	80	25th November			
II	20	25th October	22nd November	19th December 2003–2004	Darwendale
	40	10th November			
	60	10th November			
	80	25th November			
I	20	25th October	22nd November	30th December 2006–2007	Chisengu
	40	5th November			
	60	15 November			
	80	25th November			

Source: Third national communication [5].

Table 1.
Changes in start of growing season.

between 2004 and 2005 (see Figures A2.3, A2.4 and A2.5 in Annex 1). There has been an overall rainfall decline of nearly 5 percent across Zimbabwe during the 20th century with the early 1990s witnessing probably the driest period in the past century [2]. Model experiments suggest that annual rainfall will continue to decrease across Zimbabwe in the future [2] (also see **Figure 3**). Despite the expected rainfall reductions, there have also been substantial periods —the 1920s, 1950s, 1970s— that have been much wetter than average [2].

According to the Third National Communication [5], the Start of Season (SoS) dates for the five representative meteorological stations in each of the agro-ecological zones showed a delayed onset of the rainy season. **Table 1** shows the SoS for representative stations in each agro-ecological zone historically by Vincent and Thomas [1] together with the observed dates for contemporary Zimbabwe. According to **Table 1**, in agro-ecological region V, the rainy season in 1960 was expected to start at the end of October but had shifted to the latter parts of November between 1997 and 2014. The same applies to agro-ecological region 1 where the start of the 1960 rainy season was in late October compared to late November in 2014.

This paper presents an assessment of risks from climate change to the horticulture sector of Zimbabwe with the aim to provide mitigatory actions that could alleviate climate change risks in the horticultural sector of Zimbabwe. Specifically the paper seeks to outline the climate change risks facing the horticulture sector in Zimbabwe, propose actions to reduce risks and assess financing and policy options for climate change adaptation in Zimbabwe.

2. Methodology

The study followed the approach taken by the International Fund for Agricultural Development (IFAD) [7] which analyses climate risks at each stage of the horticulture value chain. The stages used by Vermeulen [7] are input supplies (seeds, fertilisers, pest management, etc.,); agricultural production (water use, soil management, skill base, etc.,) and postproduction processes (storage, processing, transport, retail, etc.,). The methodology included a review of relevant

Large scale horticulture farmers	Small scale farmers	Agro-processors	Support organisations
Makoni farms (flower producers) (1)	Tomato out growers (5)	Schweppes (tomato) (1)	ZimTrade (1)
Chomodzi (sugar snap peas) (1)		Bioba (the African food hunter) (organic indigenous foods) (1)	IFAD (3)
		Nhimbe fresh foods (blueberry, bananas and sugarsnap peas) (1)	HIVOS (1)
		Sodelani (tomatoes) (1)	Palladium (1)
			HDC (1)
			Department of climate change (1)
			DR&SS department (1)
			Technoserve (1)
Total number of respondents: (2)	**Total number of respondents: (5)**	**Total number of respondents: (4)**	**Total number of respondents: (10)**

Table 2.
Representatives of firms and organisations interviewed for the study.

peer reviewed, technical and 'grey' literature on environmental and climate challenges in Zimbabwean agriculture. Examples of such documents include, *inter-alia* the Zimbabwe climate policy (ZCP), national climate change response strategy (NCRS), climate-change national communication documents and pre-liminary information on the National adaptation plan (NAP). Data was collected from multiple stakeholders in areas with notable horticultural production across Zimbabwe using semi-structured interview guides. The study population composed of horticulture farmers, produce processing firms, value chain support organisa-tions and government arms related to horticulture. **Table 2** summarises the selected study respondents —total 21 respondents— in accordance with the discussed study categories. All respondents save for Sondelani and Bioba have operations in Manicaland province (see **Figure 1**).

3. Findings: Zimbabwe climate change risk outline

Climate predictions in Zimbabwe related to temperature and precipitation have to be done in light of the possible representative concentration pathways (RCPs)[1]. The climate change information fact sheet on Zimbabwe of (2015) predicts that in 2030 the mean annual temperatures in the country will increase by 0.46°C, 1.04°C, and 1.83°C for the 10th, 50th, and 90th percentiles for the RCP4.5. Similarly, the 10th, 50th, and 90th percentiles for the RCP8.5 will witness increases of 0.62°C, 1.25°C, and 1.83°C. In 2050 the climate change information fact sheet for Zimbabwe predicts that the mean annual temperature in the country will increase by 0.95°C, 1.68°C, and 2.66°C for the 10th, 50th, and 90th percentiles for the RCP4.5. Similarly, the 10th, 50th, and 90th percentiles for the RCP8.5 will witness increases of 1.43°C, 2.17°C, and 3.13°C. The greatest increases are projected for the months June through to September. Unganai [6] also predicts an overall increase in annual mean temperature anomalies —tem-perature changes out of tune with the 1986–2005 average—using three different models shown in **Figure 4**.

Unganai [6] goes further in **Figures 3** and **5** showing the likely temperature changes in the country in accordance with the major river catchments. According to **Figures 3** and **5**, at RCP4.5 most of the country —save for the Eastern highlands— will see an increase in temperature anomalies of above 1.41°C while at RCP 8.5 the general rise in temperatures will be more pronounced in the northern than southern part of the country.

Regarding rainfall, the climate change information fact sheet on Zimbabwe of (2015) predicted changes in the scale of the rainfall probability distribution, indicating that extremes on both sides (floods and droughts) may become more frequent in the future. Furthermore, the climate change information fact sheet on Zimbabwe of (2015) projected mid-century decline of groundwater recharge, decrease in soil moisture and annual runoff. Still on future rainfall patterns and

---

[1] An RCP is a greenhouse gas (GHG) concentration (not emissions) trajectory adopted by the IPCC for its fifth Assessment Report (AR5) in 2014 [8]. Four pathways that describe different climate futures, all of which are considered possible depending on the volume of GHGs emitted in the years to come have been selected for climate modelling and research. The four RCPs are RCP2.6, RCP4.5, RCP6, and RCP8.5 and all are consistent with a wide range of possible changes in future GHG emissions. RCP 2.6 assumes that global annual GHG emissions —measured in CO-equivalents— peak between 2010 and 2020, with emissions declining substantially thereafter. Emissions in RCP 4.5 peak around 2040, then decline. In RCP 6, emis-sions peak around 2080, then decline. In RCP 8.5, emissions continue to rise throughout the 21st century.

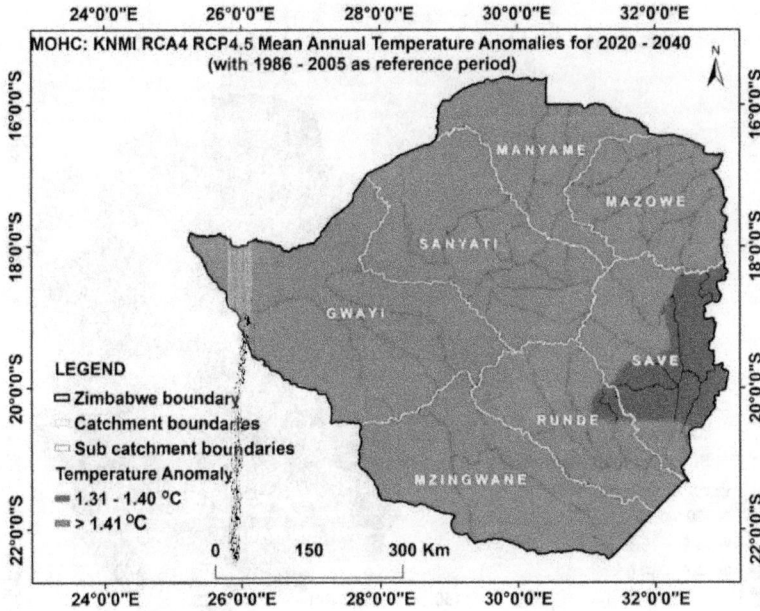

Figure 4.
Predicted mean annual temperature anomalies for Zimbabwe using various forecasting models. Source: Unganai [6].

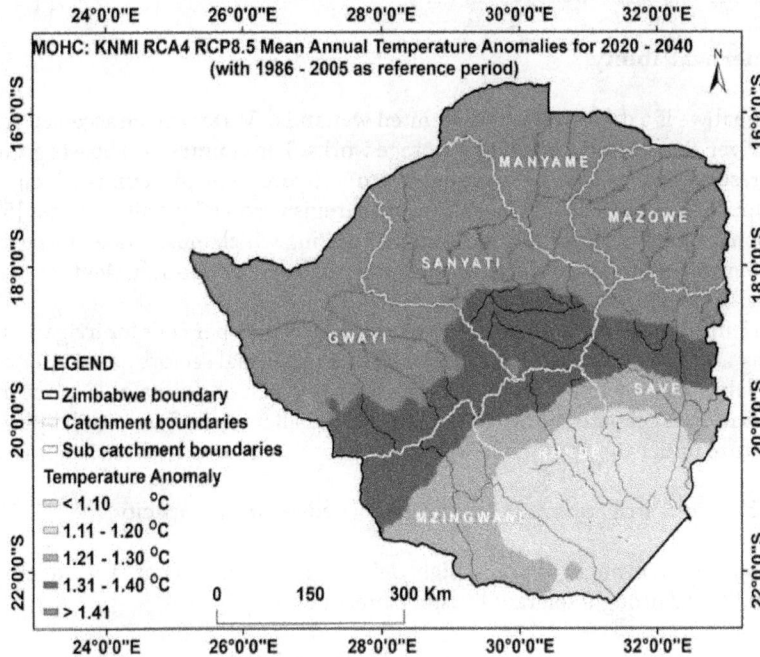

Figure 5.
Temperature predictions for Zimbabwe at RCP 4.5. Source: Unganai [6].

using 1985–2005 as a benchmark, Unganai [6] predicted rainfall reductions in most of the country under the RCP 4.5 (see **Figure 6**) and rainfall increases in most of Zimbabwe under the RCP 8.5 (see **Figure 7**).

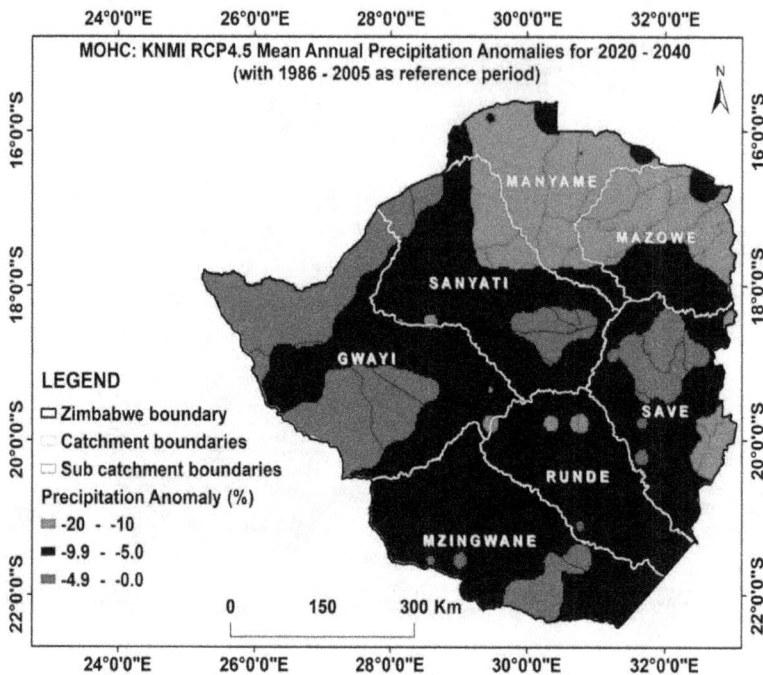

Figure 6.
Precipitation predictions for Zimbabwe under RCP 4.5. Source: Unganai [6].

3.1 Water availability

Zimbabwe is a dry country with limited wetlands[2]. Water can be accessed from direct river abstractions or through storage works. The country also boasts groundwater reserves which have been grouped into 10 hydro-geological units which yield approximately 1.8×10^6 megalitres from registered and monitored uses [5]. Additional water can be obtained through recycling which mostly takes place in urban centres where, potentially, wastewater can be treated to sufficient standards for discharge into public river systems.

Agriculture uses most of Zimbabwe's water that is 81 per cent for irrigation, fish farming and livestock watering. The urban and industrial sectors uses 15 per cent of available water, while mining accounts for 2 per cent of the water [9]. According to the National Climate Policy (NCP) [10], key challenges in water availability for agriculture under the changing climatic scenarios are rooted in three major issues

- Absence of irrigation systems and associated technical capacity,

- Absence of, drought tolerant, high yield, high nutrient, water efficient crops and heat / drought tolerant livestock breeds and

- Failure to manage episodic floods and excess rainfall.

The three major issues point to the need for effective harvesting and management of water resources in order to sustain agriculture in the presence of variable

---

[2] Zimbabwe's long-term average annual surface run-off is estimated to be 23.7×10 megalitres. The distribution of average runoff varies from 21 mm per year in the Gwayi catchment to 126 millimetres per year in the Mazowe and parts of the Save catchment [5].

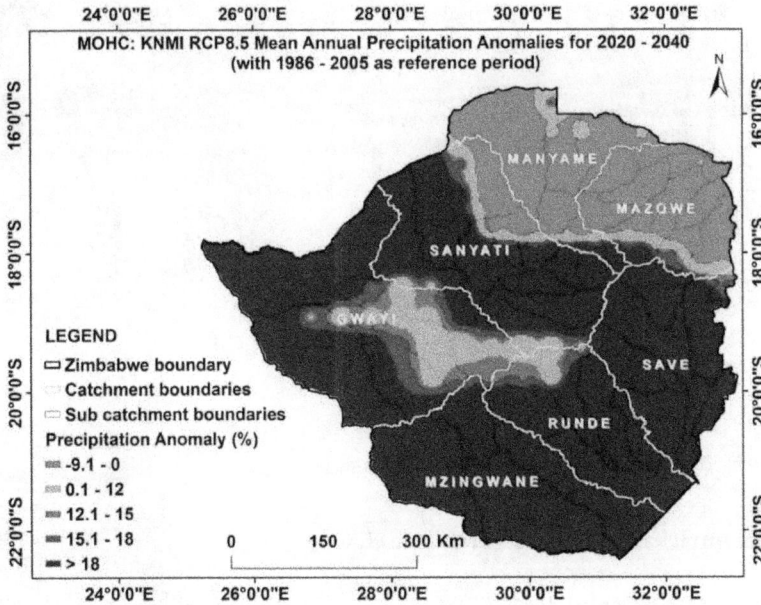

Figure 7.
Precipitation predictions for Zimbabwe under RCP 8.5. Source: Unganai [6].

rainfall and extreme temperatures. Nonetheless, Zimbabwe has the most dams in the Southern African Development Community (SADC) region after South Africa. The country has almost 40 medium-to-large dams and lakes including Lake Kariba as well as about 10,200 small dams. In 2001, about 152,000 hectares of land were under formal irrigation with a total of a further 600,000 hectares of land nationwide that can be made available for irrigation development [2].

3.2 Sectoral and development impacts

Zimbabwe is particularly vulnerable to climate change due to its heavy dependence on rainfed agriculture and climate sensitive resources such as hydro-electric power, wildlife tourism and other ecosystem goods and services. Zimbabwe has an agriculture-based economy with the sector contributing about 15 per cent each year to the GDP [11]. As a result of this linkage, climate is a major driving factor for most of Zimbabwe's socio-economic activities such that Zimbabwe's Gross Domestic Product (GDP) is tightly linked to rainfall patterns [9]. Robertson [12] further illustrated this point showing that in the years when Zimbabwe experienced droughts economic growth levels also declined (see **Figure 8**[3]).

Livelihoods of the poor, particularly women[4] who are highly dependent on climate sensitive sectors like agriculture, are likely to be impacted by climate change in various ways. Climate change impacts are also expected to disproportionately affect the young, elderly, sick, and otherwise marginalised populations who may not have the necessary livelihood capital assets —natural, financial, physical, human and social—to allow for adaptation or recovery when climate disasters strike.

---

[3] All years with red columns were drought years while those with orange columns were years of the fast track land reform programme. Blue years are considered normal years.

[4] Rural women had gender related duties that saw them increase their level of effort due to the negative effects of climate such as drought and extreme temperatures [13, 14].

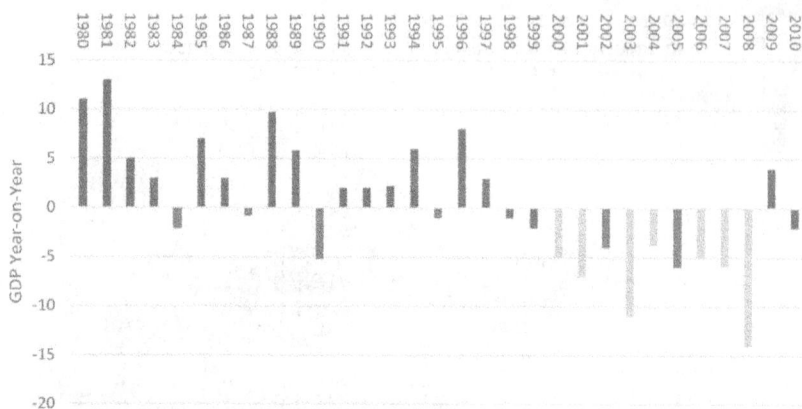

Figure 8.
Relationship between GDP and drought in Zimbabwe. Source: [12].

3.3 Climate risks and horticulture value chains

Zimbabwe's agricultural sector is divided into four major sub-sectors namely; large scale commercial farms, small scale commercial farms, communal and resettlement areas. The agrarian structure has changed with the recent land reform in Zimbabwe with 99 per cent of the farmers now being smallholder farmers (SHF). Of these 81 per cent are communal farmers, 18.7 per cent resettled farmers and 0.1 per cent large scale farmers [11].

Zimbabwe has a diverse horticultural subsector, producing vegetables —for export and domestic sales—, fruits —for export and domestic markets— and flowers —primarily for export—. The major horticultural exports from Zimbabwe are destined mostly to European and other African markets. At its peak in the late 1990s, horticulture was the second largest agricultural foreign exchange earner after tobacco, recording export figures in 1999 of up to US$144 million. Its trade balance however declined significantly over the years, influenced by the dollarization and the previously mentioned land holding changes. This negative trend reversed recently with exports recording significant growth in 2018 —more than $112 million against $50.9 million in 2017— [15]. The sector is also a significant earner of foreign currency thereby improving the country's terms-of-trade in addition to numerous downstream benefits in the packaging, processing, input suppliers and transport industries. According to the ITC [16], there is potential for the horticulture sector in Zimbabwe to contribute significantly to growth due to the following important factors:

a. Proven experience of commercial and SHFs in producing internationally competitive crops,

b. Short gestation period of some of its commodities,

c. Direct social impact on poor farming communities and

d. Creation of decent jobs for both men and women across the value chain.

The UKTP [15] together with Shone [17] noted a number of constraints inhibiting the growth of the sector that may be summarised as follows:

- *A lack of export diversification*: Zimbabwe is currently exporting on a few products compared to the pre-Land Reform Programme of 2000. It is mainly limited to fresh exports whereas in the past, Zimbabwe was exporting processed fruit (paste, ketchup, juices) and vegetables (fresh cut and dried). This is mainly due to the lack of expertise and infrastructure as some of the exporting companies closed shop.

- *Low production volumes:* since the Land reform of 2000, horticulture production for export has significantly gone down as systems that supported production —out-grower schemes, irrigation schemes, greenhouses and pack houses and cold chain transport— have been run down.

- *Lack of supply chain integration and linkages with UK/EU firms:* The supply chain is fragmented and the sector is disconnected. Its apex body, the Horticulture Promotion Council (HPC) which was affiliated to the Commercial Farmers Union (CFU) is now defunct. At its peak, the HPC was responsible for facilitating viable export linkages for producers.

3.4 Major risks and possible mitigation

Regarding inputs in the horticulture sector, production works best under drip-irrigation hence drought tolerant seed varieties may not be necessarily applicable in this case. However, seed varieties that are resistant to frost and heatwaves are crucial given the expected cooler winters and higher temperatures in future. Crops may also be prone to pests which will likely increase in the incidence of longer hotter summers. In this regard, the use of tissue culture and the development of varieties that are resistant to temperature extremes is essential for mitigating seed related climate risk. Tissue culture will assist the expansion of horticulture production and build resistance to water shortages, specific pests and temperature extremes.

Soils will require nutrient augmentation and the use of chemical fertilisers can have adverse impacts on agro-biodiversity and can result in eutrophication through erosion in the case of flash floods related to climate change. Use of fertilisers is linked to increased GHG emissions particularly in the case of methane and nitric-oxide. As such, in is critical to improve availability of fertilisers / build capacity for composting and to promote crop rotation to increase soil nutrients. Crops are also susceptible to attacks by pests and viruses. The increase in extreme temperatures can increase the prevalence of certain pests and viruses or reduce that of insects that attack pests. Thus, it is essential to explore organic / agrobiodiversity solutions for specific pests in order to maintain organic production. It is also useful to identify ideal crops for intercropping and integrated pest management.

Regarding information services in Zimbabwe, extension services may be considered as limited in resources. Nonetheless, there is scope for electronic messages on weather trends. Messaging is however limited by weak communication infrastructure and low forecasting capacity. As such, it is useful to strengthen early warning systems —including Geoinformation Science (GIS) and Earth Observation —on cropping season quality, rangelands conditions, droughts, floods, disease/pest outbreaks and wildlife movement; Strengthen capacity to generate new forms of empirical knowledge, technologies and agricultural support services that meet emerging development challenges arising from increased climate change and variability and Strengthen the capacity of farmers, extension agencies, and private agro-service providers to take advantage of current and emerging indigenous and scientific knowledge on stress tolerant crop types and varieties, including landraces that are adaptable to arising climatic scenarios.

In the Zimbabwean horticulture sector, SHFs are often constrained by the lack of adequate technology for production, harvesting, handling and storage. In some cases hostile temperature fluctuations hinder production or storage processes as better technology may be required to deliver the same quality output in the presence of *inter-alia* heatwaves, frost, hailstorms, droughts and floods. The absence of adequate technology for all SHF may open up opportunities in the public and private sector for sharing agricultural equipment using equipment pooling mechanisms that may see the emergence of pool tractors, trucks, ploughs, etc. This is already happening using the VAYA agricultural sharing tools platform. Such a service could be extended by the public sector. It is also necessary to develop appropriate storage which can handle high temperatures and maintain humidity will build resilience in the horticulture sector.

From and agricultural production perspective, the increased temperatures are likely to increase evapotranspiration and potentially salinity as more groundwater is extracted for irrigation purposes. Also, there is potential increased erosion due to long dry spells. Also, the erratic supplies of water rooted in the recurrent droughts have been the major negative effect of climate change. The commencement of the rainy season is often delayed and there is an overbearing need for water harvesting and irrigation if at all the value chain is to be viable. Drip irrigation is used in most of commercial horticulture production as it is more efficient and ideal. In the absence of water reservoirs, horticulture production would be seriously compromised. Given the above, it would be essential to identify latest technology for minimum tillage production models in order to conserve the soil. Also, the promotion of the regeneration of native species in and around growing areas could be encouraged together with water harvesting.

Horticulture tends to be dominated by monoculture. Monoculture leaves crops more vulnerable to pests and diseases as well as speeds up soil degradation. Monoculture also compromises biodiversity conservation efforts as it limits species diversity. In the presence of negative climate changes —e.g. temperatures that promote the breeding of particular pests— the overall ecosystem loses resilience and the risk of total loss of produce increases. Farmers can maximise productivity by selecting appropriate crops to be utilised for intercropping; this will not only provide shade for the soil but also maintain soil health and provide additional income.

From an energy dimension, it is important to note that Zimbabwe's electrical energy comes from hydro and thermal —coal fired power stations—generation mechanisms. The absence of regular dependable rainfall makes hydro-electricity generation a challenge. This in turn pushes the nation to depend on coal fired thermal power which increases GHG emissions. Furthermore, this negatively affects irrigation and other farm activities that require electrical energy given the energy deficit rooted in drought. Dwindling hydro-power potential and increasing emissions in coal fired power stations opens up new avenues in public and private investment in clean renewable energy. As such it is critical to promote and invest in the production of clean energy.

Harvest management often requires the use of complex energy consuming equipment. Previous sections in have already explained the climate risks that limit the access to hydro-power and ultimately energy. Furthermore, deliberations on access to finance also exposed challenges in accessing complex equipment. Furthermore, processing is undertaken by the large off-takers and climate related risk is entirely shouldered by them. Most of this risk is again related to energy availability and processing activities that also require energy. Water availability is also crucial in the processing phase. As such, the recurrent droughts again present risks to the effectiveness of the processing stage. This phase also generates a lot of waste that is often dumped into the natural environment creating negative externalities that are a great ecological cost to society.

Finally, this study exposed a number of climate risk factors to consider in the various horticulture value chain models. A significant number of the risk factors may be countered using the knowledge possessed by the farmers regarding climate change management. In the absence of such knowledge, farmers may fail to deal with the negative aspects of climate change resulting in higher climate risk.

3.5 Horticulture farming models in a changing climate

This section presents the models observed in various horticulture value chains in Zimbabwe with a focus on how the different models manage climate and environmental risk. The two prominent models were a model based on variations of contract farming and a lease-based model.

3.5.1 Contract farming based model

The first variation of the contract farming model was observed under Technoserve —an NGO operating within Zimbabwe—. The model is characterised by a lead firm —Linkflora/Lingfield Farm— working with 100 SHFs in 10 irrigation schemes in the Midlands region of Zimbabwe. This model has a field officer on each irrigation scheme who is responsible for production and the day-to-day running of the horticulture programme. The field officers from the Irrigation Schemes report to a manager who manages all the schemes. It is anticipated that after the programme the SHFs will pool funds and continue to hire the field officers for services which are critical in continued export production.

The second variation of the contract farming based model was observed at Sondelani ranching in the Matabeleland region. Sondelani Ranching is working with over 25,000 SHFs as a lead firm in a tomato contract farming scheme in Matobo district, Matabeleland south province. Sondelani provides inputs to farmers who pay back when they sell their produce to the lead firm. The SHFs are grouped into production capacity groups: the best being Platinum with 13,000 farmers followed by Gold with 13,000 farmers and then Silver with 1,000 farmers. The SHFs produce and feed into the Sondelani tomato processing plant that processes 150 tonnes of tomatoes into puree per day (commissioned by the President Mnangangwa in 2018). Sondelani also works with 2 commercial farms who produce about 10% of their raw produce while 90% comes from SHFs. The tomato puree is sold locally and exported into the SADC region. Sondelani key informants explained that Farmers are trained through a subsidised training programme (training costs 450 and farmers are asked to pay ZW$100 for a 3-day training) run by Sondelani's foundations for farming division. Sondelani has put in place a social capital based safeguard system where members —mostly women— are organised into groups that observe production rules. Any offender is dealt with by the group and has to make amends. Failure to which the group has power to dismiss the member who does not follow the rules. If a farmer sells to other markets they have to make sure repay their loans of inputs to Sondelani on time and in full.

3.5.2 Lease-based model

The alternative to the contract farming based model observed in value chains across Zimbabwe was a lease-based model. Schweppes —a large beverages manufacture in Zimbabwe— is implementing such a model through leasing land —150 hectares for tomato production —from SHFs in Darwendale close to the capital Harare. Schweppes plants tomatoes on 50 hectares at a time and the other 50 hectares are used for the next crop —usually peas—in a crop rotation system.

The 50 hectares that Schweppes does not use each season may be used by SHFs to produce crops of their own choice so long as they are in line with crop rotation of the export crop being produced by Schweppes. In this model, SHFs agree to lease their land to Schweppes on a 5-year basis, receive payment for leasing the land and a share of profit from production of their land. Schweppes has full control of the land use and production while SHFs are offered first refusal of offer to provide labour to Schweppes. Over the 5-year lease period, SHFs benefit from knowledge transfer, management partnerships, learning farming as a business and technical expertise. Other benefits include SHFs receiving constant income generated from working for Schweppes, land leasing, share profit and infrastructure development.

3.6 Development impacts of the models

Research on similar models presented by Scoones [18] at the British-Africa investment summit of 2020 compared three broad types of commercial agricultural investments —estates and plantations, medium-scale commercial farms and out grower schemes— in Ghana, Kenya and Zambia focusing on outcomes for land, labour and livelihoods. The cases included investments with some UK-linked companies, including the Blue Skies company in Ghana, which packages and exports fruit produced by smallholder out growers and an out grower scheme in Zambia, operated by Illovo, now largely owned by British Foods, whereby SHFs' land is incorporated into an estate, and they are paid revenues for the use of land. Scoones [18] concluded that the 'terms of incorporation' into business arrangements really mattered. Too often estates/plantations operated as 'enclaves' separated from the local community, possibly providing employment opportunities, but frequently with poor conditions. Those investments that had substantial linkage effects included those with smallholder-led out grower arrangements, where leverage over terms was effective. Meanwhile, consolidated medium scale farms potentially had positive spill over effects into neighbouring communities through labour, technology and skill sharing linkages. Scoones [18] further argues that private sector investment that has the most impact is usually small, often informal, and deeply linked into local economies. Clusters are usually spontaneous, not planned as part of grand corridor or investment hub schemes.

In relation to this study, perspectives raised by Scoones [18] generally favoured the variations of the contract farming-based models from a development perspective. Scoones' [18] arguments were centred on the yield and livelihood transformation potential of model in question. In addition to that, this study considers the pros and cons of each model —using information gathered from data collected—with respect to climate and environmental risk.

3.7 Environmental impacts of different models

In the lease-based model, the mitigation of climate and environmental risks related to production —e.g. composting, organic production, pest control, etc.,— is highly linked to market requirements and costs of implementing the initiatives. Thus, if the risk mitigation actions are beneficial for the natural environment but not required by the market, they less likely to be implemented. Regarding access to critical climate and environmental risk information, in the agricultural manager and lease-based model, the manager/lead firm are the source and distributers of information sourced at a lower transaction cost due to the centralised organisational structure. This allows the lease-based models to better access climate risk related information services. The same argument applies to climate and environmental risk financing within the agricultural manager and lease-based models. In the lease-based model, climate and environmental risk financing is not much of a challenge as the protagonist —e.g. Schweppes—is well placed to access required financing to deal with climate and environmental risks. The agricultural manager-based model is also similar though could change in a negative direction after the contract between the off-taker and manager lapses given that SHF would have to source their own financing. Regarding climate and environmental risks related to energy, the lease-based model allows for the easy use of alternative —to hydro-electric power—localised electricity generation processes such as solar mini-grids which are most effective and efficient in such centralised systems including the agricultural manager based model.

The contract farming model has better opportunities for mitigating climate and environmental risks related to production. For instance, the use of organic fertilisers may be easier mainstreamed together with minimum tillage principles because of the extensive training— silver, gold and platinum— that the various categories of farmers go through[5]. Also, in the contract farming model, social capital levels are high and the social rules and norms are not always driven by profit thus, may see farmers practicing what was taught even if it took a little more of their time and effort. In the contract farming-based model, farmer disaggregation may increase the transaction cost of accessing reliable and useful information thereby making such a model not as effective in dealing with information related climate and environmental risk. Nonetheless, the central role played by the lead firm —e.g. Sondelani ranching— allows for centralised information sourcing though dissemination of the information may be less efficient —due to the disaggregated nature of individual contract farmers— in comparison to the compact lease model. In the contract farming model, the access to financing may not always be extended to the disaggregated smallholder farmers as the risk of loan defaults changes with each SHF in question. Also, In the contract farming model, the decentralised nature of SHFs exacerbates the energy problem and makes a centralised solar PV system difficult to set up. This model therefore leaves the challenging issues related to energy in the domicile of the SHF who in most cases does not have adequate resources to make the investment required to avoid climate risks related to energy.

In all the models the large off-takers have the ability to develop suitable packaging for increasingly high temperatures, humidity and precipitation while maintaining food safety standards. The quality of the packaging however is likely to be a function of market preferences rather than environmental considerations given that the primary objective of the off-taker is like to be profit maximisation.

4. Climate policy and financing in Zimbabwe

There are a number of national policies in Zimbabwe that focus on climate and the broader environment. Zimbabwe has a National Climate Response Strategy (NCRS) supporting the National Climate Policy while consultations on the development of National Adaptation Plan (NAP) are on-going. The country has also submitted its Nationally Determined Contributions (NDCs) to the UNFCCC.

The National Climate Change Response Strategy provides a framework for a comprehensive and strategic approach on aspects of adaptation, mitigation, technology, financing, public education and awareness. It will help to inform Government on how to strengthen the climate and disaster risk management policies.

---

[5] Such training may not be as intensive in the lease based model given that the large firm already brings in the technical expertise.

The Zimbabwe National Climate Policy explains that Zimbabwe seeks to create a pathway towards a climate resilient and low carbon development economy in which "the people have enough adaptive capacity and continue to develop in harmony with the environment". To achieve this, the Climate Policy is supported by the National Climate Change Response Strategy, National Adaptation Plan, the Low Carbon Development Strategy, National Environmental Policy and Strategic Document as well as other policies aimed at achieving sustainable development. The actions envisioned in the Policy will "safeguard the Zimbabwean natural environment, sustain society, and support the economy for the years ahead". "Adequate financing, cross sectoral coordination, climate change science, research and systematic observations will form the backbone of actions towards a climate resilient Zimbabwe" (NCP, 2016:2).

It is the vision in the National climate policy to "climate-proof" all the socio-economic development sectors of Zimbabwe in order to reduce Zimbabwe's vulnerability to climate and climate related disasters, while at the same time developing along a low carbon pathway (NCP, 2016:2). Zimbabwe aims to reduce per capita emissions by 33% from "business-as-usual" baselines by 2030 (NCP, 2016:4). This ambition is based on the availability of financial resources and technology transfer from bilateral and multilateral funding mechanisms in addition to domestic financing.

4.1 Costs of climate change financing

The Development Bank of Southern Africa estimate the costs of climate change adaptation to be US$9.9 billion. The costs to agriculture are almost US$2.4billion [2]. Evidently, the climate risk mitigation and adaptation options discussed in the previous section require financial and technical resources to come to fruition. The potential sources of finance within and outside Zimbabwe are discussed below.

The Zimbabwean agro-food sector has undergone major changes ever since the fast-track land reform programme (FTLRP). These changes influenced new models of production, marketing and financing [19]. Access to external financial capital for the majority of participants remains limited. If available it is costly and inequitably distributed thus, severely limiting the productivity and competitiveness of the majority rural SHFs [20–22].

The changing agricultural environment in Zimbabwe has come with a drought of the typical financial products and services for agricultural production that were previously designed for large scale commercial farmers [23]. According to Sachikonye [24], financial intermediaries lack depth and understanding of the rural SHFs who in most cases have non-liquid assets that are not recognised by conventional financial markets. As such, conventional thinking is that the agricultural sector (particularly the SHF) is too costly and risky for lending [24–26].

Biyam [23] notes that commercial banks have traditionally shied away from the rural smallholder agricultural sector because of uncontrollable and systemic risks, higher costs and fear of the unknown. The cost of directly lending to rural smallholder farmers in hard-to-reach rural areas with less-educated and low-income populations has become prohibitive to most formal financial institutions [25, 27]. Microfinance institutions that loan cash to these low-income households do so at a high cost, with short-term loan products that are generally not able to address the full range of agricultural needs of the rural SHFs [25].

4.1.1 Financing for SHF and related stakeholders outside Zimbabwe

The most important multilateral sources of climate financing at the international level are the World Bank's carbon funds, the Global Environment Facility

(GEF)—recently supported the development of the Zimbabwe NAP—, the African Development Bank (AfDB), African Sustainable Forestry Fund, the United Nations Framework Convention on Climate Change (UNFCCC's) Adaptation Fund, the Green Climate Fund (GCF)—which is a financial mechanism of the UNFCCC—and the Kyoto Protocol's Clean Development Mechanism.

Pswarayi-Jabson [28] noted that there is limited funding that private entrepreneurs and CSOs can access in the global climate management mostly because most of these organisations tend to compete rather than work together to access such financial resources. Local private organisations and CSOs in Zimbabwe are often prevented from accessing climate funds directly due to the large size of available grants, donor partner preference and the absence of an enhanced direct access mechanism [28]. The above is mostly because Zimbabwe currently does not have institutions accredited as a National Implementing Entity, so as to access direct to the GCF. However, efforts are being made to have the Environmental Management Agency (EMA) and the Infrastructure Development Bank of Zimbabwe (IDBZ) accredited with the GCF as implementing entities. Such accreditation would deal with access to finance challenges such as failure to meet global fiduciary standards; misalignment of financing application with other institutional arrangements and requirements of climate funding such as the GCF, GEF and IFC/World Bank and lack of expertise in packaging of green projects, with particular focus on mainstreaming of climate issues. Nonetheless, the NCRS [2] notes that Zimbabwe has received modest funding from, for example, multilateral organisations (such as UNDP, UNICEF, UNEP; GEF; GEF Small Grants Programme and FAO); international organisations (such as the Global Water Partnership); regional organisations (e.g. COMESA); private organisations (such as the Evangelischer Entwicklungsdienst); and research funding organisations (e.g. IDRC and DFID). However, most of the climate finance has come from the Government [2].

Author details

Nqobizitha Dube[1,2]

1 Ali-Douglas Development Consultancy, Bulawayo, Zimbabwe

2 Institute of Development Studies, National University of Science and Technology, Bulawayo, Zimbabwe

*Address all correspondence to: nqobizithad@gmail.com and nqobizitha.dube@nust.ac.zw

IntechOpen

References

[1] Vincent, V. and Thomas. R. G., 1960. *Agro-ecological survey*. Government of Southern Rhodesia, Salisbury.

[2] National Climate Response strategy (NCRS)., 2014. Government of Zimbabwe, Causeway, Harare.

[3] Chikodzi, D., Zinhiva, H., Simba, F.M. and Murwendo, T., 2013. Reclassification of agroecological zones in Zimbabwe–the rationale, methods and expected benefits: The case of Masvingo Province. Journal of Sustainable Development in Africa, *15*(1), pp.104-116.

[4] Mugandani, R., Wuta, M., Makarau, A., & Chipindu, B. (2012). Re-classification of agro-ecological regions of Zimbabwe in conformity with climate variability and change. African Crop Science Journal, *20*(2), 361-369.

[5] Third national communication (2016). Department of climate change, ministry of environment, water and climate, Harare.

[6] Unganai L. S., 2020. Downscale climate projection for Zimbabwe. 2040 estimates. Oxfam, Zimbabwe, In Press.

[7] Vermeulen, S., 2015. How to do Climate Change risk assessments in value chain projects. International Fund for Agricultural Development (IFAD). Environment and climate change. Adaptation for Smallholder Agriculture Programme (ASAP). IFAD.

[8] Intergovernmental Panel on Climate Change (IPCC)., 2014. Working Group AR5 Climate Change 2014: Impacts, Adaptation, and Vulnerability. Online: https://www.ipcc.ch/report/ar5/wg2/. Accessed 12/12/19.

[9] National Water Policy, 2013. Government of Zimbabwe. Causeway, Harare. Online: https://www.google. com/url?sa=t&rct=j&q=&esrc=s&sourc e=web&cd=3&cad=rja&uact=8&ved=2 ahUKEwjUh83LmaTmAhVDMewKHUI KBlAQFjACegQIBBAC&url=http%3A% 2F%2Fwww.conscientiabeam. com%2Fpdf-files%2Feco%2F74%2 FIJPPAR-2015-2(3)-60-72rev. pdf&usg=AOvVaw0ziu- Le4vlO5qj_1H7xVdK. Accessed 9/12/19.

[10] National Climate policy, 2017. Government of Zimbabwe, Causeway, Harare.

[11] Zimbabwe Agriculture Investment Plan (ZAIP)., 2012. A Comprehensive Framework for the Development of Zimbabwe's Agriculture Sector 2013-2017. Causeway, Harare.

[12] Robertson, J, 2011. Statistics on the Zimbabwean economy. Www.Slidenet/ Sokwanele/ Zimbabwe-economy-october-2009, accessed 28 December 2009.

[13] Dube, T., Intauno, S., Moyo, P. and Phiri, K., 2017. The gender-differentiated impacts of climate change on rural livelihoods labour requirements in southern Zimbabwe. Journal of Human Ecology, *58*(1-2), pp.48-56.

[14] Muchacha, M. and Mushunje, M., 2019. The gender dynamics of climate change on rural women's agro-based livelihoods and food security in rural Zimbabwe: Implications for green social work. Critical and Radical Social Work, *7*(1), pp.59-72.

[15] UKTP, Draft inception report., 2019. International Trade Centre (ITC), Geneva.

[16] International Trade Centre (ITC)., 2019. Environmental Mainstreaming Strategy. UK Trade Partnerships Programme. ITC, Geneva.

[17] Shonhe, T., 2019. The agrarian question in contemporary Zimbabwe. Africanus, *49*(1), pp.1-24.

[18] Scoones I., 2020. UK-Africa trade and investment: is it good for development? Online: http://ianscoones.net/blog/. Accessed 29/01/20.

[19] Moyo, S., 2011. Land concentration and accumulation after redistributive reform in post-settler Zimbabwe. Review of African Political Economy 38 (128): 257-276.

[20] Miller, C. and Jones, L., 2010. Agricultural Value Chain Finance Tools and Lessons. Food and Agriculture Organization of the United Nations. Practical Action Publishing, Dunsmore, Rugby.

[21] Moyo, M., 2014. *Effectiveness of a Contract Farming Arrangement: A Case Study of Tobacco Farms in Mazowe District, Zimbabwe*. Unpublished MPhil. Thesis, University of Stellenbosch.

[22] Scoones, I., Mavedzenge, B. and Murimbarimba, F., 2016. Sugar, people and politics in Zimbabwe's Lowveld, Journal of Southern African Studies: 1-18.

[23] Biyam, S. T., 2013. *An Overview of Farming Financial Services- Sharing Decades of Farming Finance Experience and Ideas on Farming Finance Looking into the Future*. Agribusiness Development Service Provision Business Linkage, Harare.

[24] Sachikonye, L., 2016. Old wine in new bottles? Revisiting contract farming after agrarian reform in Zimbabwe. Review of African Political Economy, 43 (1): 86-98.

[25] Malaba, S.M.T., 2014. *State of funding for agriculture and the whole value chain as well as prospects of the fourth coming agriculture summer season*. 71st commercial farmers union of Zimbabwe, Harare.

[26] Parirenyatwa, K., and Mago S., 2014. Evolution and development of contract farming. Mediterranean Journal of Social Sciences 5 (20): 237-244.

[27] Masiyandima, N., Chigumira, G. and Bara, A., 2011. *Sustainable Financing Options for Agriculture in Zimbabwe.* Zimbabwe economic policy analysis and research unit (ZEPARU), working paper series, ZWPS 02/10, Harare.

[28] Pswarayi-Jabson, G., 2018. Opportunities for Accessing International Climate Adaptation Finance for Civil Society Organizations in Zimbabwe. Africa centre for climate change knowledge foundation. Harare.

Section 3

Adaptation to Climate Change

Chapter 8

The Impacts of Climate Change in Lwengo, Uganda

Shyamli Singh and Ovamani Olive Kagweza

Abstract

Climate Change has become a threat worldwide. Vulnerable communities are at foremost risk of repercussions of climate change. The present study aimed at highlighting a case study of climate change impacts on Lwengo District of Uganda. Out of the total geographical area of the district, 85% hectares are under cultivation and most of its population depends majorly on the rain- fed agriculture sector to meet the food requirement and as a major income source. With the changing climatic conditions, agriculture is the major sector which is being impacted. The region has experienced disasters from some time, usually the second seasons rains used to result in such disasters but since 2016 both seasons have occurred disasters, which majorly include hailstorm, strong wind, long dry spells, pests and diseases. The situation became more severe due to shortage of availability of skilled human resources, quality equipment for disaster management, limited financial resources and weak institutional capacity, which resulted in increasing vulnerability of small farm holders. Some of the adaptation strategies are being taken up by the government but there is a need to understand prospects of decision-making that are site specific and more sustainable for smallholder communities. Climatic changes possess many obstacles to farming communities which require sustainable adaptation to enhance the adaptive capacities of the communities through continued production systems, which are more resilient to the vagaries of weather. Farmers are practising such options which are location specific, governed by policy framework and dependent on dynamism of farmers. This study investigated how these drivers influence farmers' decision- making in relation to climate change adaptations.

Keywords: climate change, agriculture, disaster, vulnerability, adaptation

1. Introduction

The Fifth Assessment Report of the Intergovernmental Panel on Climate Change (IPCC 2014) anticipated quick changes in climate, even if greenhouse gas emissions are condensed [1]. The impact of future climate change can be reduced by mitigation but cannot be stopped altogether. Around 80 percent of the Ugandan population is dependent on agriculture. Uganda's rain fed agriculture is crucial to the masses for consumption and income generation [2]. It thus becomes paramount to scrutinize the change in climate, in terms of overall temperature and precipitation levels, or in terms of variations in seasonality of rainfall as it directly alters the state of revenue and services provided by the ecosystem. Of late, substantial amounts of variations and extreme events are being witnessed. The detrimental effects of anthropological and environmental impacts can surely be fended off by anticipating and getting accustomed to them. The most vulnerable however, are the marginalized smallholder farmers in Sub- Saharan Africa as their inadequate capacity to adapt keeps them at the receiving end of facing setbacks to farming livelihoods [3]. Hence, it becomes essential to categorize and understand possible adaptation to cope with the impacts of it.

Climate change has a global impact which is evident in the contemporary adversarial changes in climate and can be seen in some of its manifestations in the form of prolonged drought spells, temperature variations [4, 5]. Countless instances of climate change impacting biomes, livelihoods, and human development were recorded and recognized by several studies [6–8]. A risk in the form of biomass loss and runoff is posed due to prolonged drought spells and floods [8–10] that consequently affects the agricultural sector. Moreover, it has been established that Global warming has direct consequences on Global Food security and Agricultural production. [11–14]

As the population of Lwengo, Uganda is dependent on agriculture which is a Climate Sensitive sector. Any change in the Climatic conditions would definitely reflect on the agriculture and would effect the food security of the masses hence it is very important for the population of Lwengo to see the manifestations of Climate Change and to ponder upon the various dimensions of the same.

This in turn affects people's basic livelihood enterprises amounting to great uncertainty [15] especially in places like Sub- Saharan Africa that is predominantly dependent on rain-fed agriculture [16]. Agriculture sector is the backbone of Uganda's economy as the agriculture sector supports over 70% of the people, so the variations in temperature and Climate Change is a real cause of concern for them [17]. Surrounded by a small scale and mixed crop system, the agriculture sector contributes 70% to the country's GDP - wherein 75% of employment is provided to the national workforce by the ago -enterprise niche [18]. The agriculture is however suffering in the hands of hydrometeorological disasters contributing to hunger and death of the livestock. In recent times, new challenges in the form of precipitation have been noticed which makes it urgent to make preparations for necessary and possible adaptation practices in order to avoid economic shocks that could possibly pour out from the agricultural sector [19, 20].

2. Impact of climate change on Uganda

A large population depends on agriculture for their food security and livelihood and eventually becomes susceptible to climate change related consequences [21]. It is likely that Climate change will have an adverse effect on African countries owing to the fact that it is dependent on rain-fed agriculture and there is a shortage of skilled human resources in the domain of disaster management besides limited financial resources and weak institutional capacity [22]. Other than this, poor condition of the soil in Sub-Saharan Africa, along with poor production techniques and lack of appropriate policies with respect to use of inputs (fertilizer) and access to credit only exacerbates the situation [23].

To top it all, Africa's warmer climate increases the chances of pandemic recurrence (e.g., HIV/AIDS), crop and animal pests and other such diseases [24]. The negligent governance of the alarming issue in these countries has widened the income disparity making more and more people economically weak [22].

In the eastern region, the agriculture production supply has started falling lower than the demand to a great extent, worsening the already delicate food security. This has also resulted in increasing vulnerability and rural poverty, further amplifying the impact of droughts that appear to have taken a grave turn in the recent years. [25].

3. General climatic characteristics of the study site: Lwengo district, Uganda

Map of Uganda showing Lwengo

A discussion on hot and cold seasons (**Figure 1**) of Lwengo would not be meaningful as the variation in the temperature of the study area is too little. The type of climate found in Lwengo is moderately hot, humid and misty. The temperature alters over the year ranging from 58°F to 81°F and is seldom below 55°F or above 87°F. A steady fall of water from the sky in significant quantities happens during 31 days in and around April 17 with an average total accumulation of 5.6 inches (**Figure 2**). The lowest rain falls around the month of June 27 accumulating an average total of 1.0 inches [26].

3.1 Key geographical characteristics

Located in the central region of Uganda, its coordinates as 00-24S, 31 25E. With an average altitude of 1150 m above sea level, it is spread over 1,024.3 sq. kms where

Temperature

The temperature in Lwengo varies so little throughout the year that it is not entirely meaningful to discuss hot and cold seasons.

Average High and Low Temperature

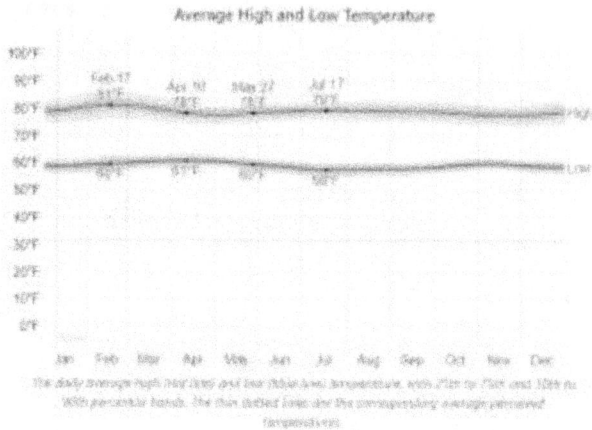

Figure 1.
Average high and low temperature, Lwengo. Source: https://weatherspark.com/y/96871/Average-Weather-in-Lwengo-Uganda-Year round as accessed on 13 January, 2020.

Average Monthly Rainfall

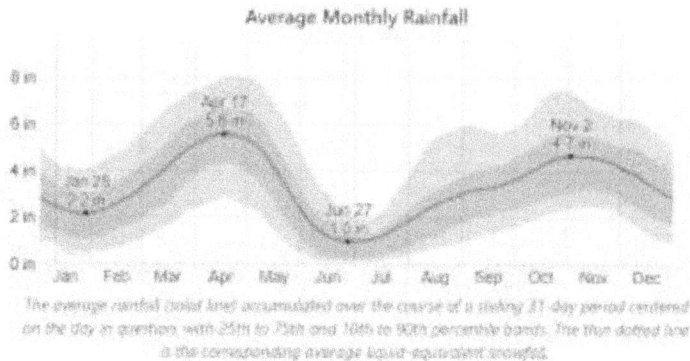

Figure 2.
Average monthly rain. Source: https://weatherspark.com/y/96871/Average-Weather-in-Lwengo-Uganda-Year round as accessed on 13 January, 2020.

the land occupies area of 1,013.46 sq. km. The population density is 269 sq. kilometers per person.

The district adjoins Sembabule District to the North, Bukomansimbi District to the North -East, Masaka District to the East, Rakai District to the South, and Lyantonde District to the West. (Lwengo district, 2020). Its district headquarters is situated in Nyenje zone Church ward Lwengo Town council. The nearest largest city is Masaka, which is 45 kilometers away from the Lwengo.

Lwengo became a separate district on 1st July 2010, prior to which it was a part of the Masaka district. According to the National Population and Housing Census released in 2014, Lwengo has a population of approximately 275,450 which comprises of 53% males and 47% females. The district has total three constituencies, four town councils *viz.* Kyazanga, Kinoni, Lwengo and Katovu, six sub counties namely, Kkingo, Lwengo, Kisekka, Ndagwe, Kyazanga and Malongo, one town board, forty-two parishes and four hundred and fifty-four villages.

In the sub counties of Kkingo, Kisekka and Lwengo, the water flows through rivers and streams of swamps, predominantly in rainy season. On the other hand, the sub counties of Kyazanga, Malongoand, Ndagweare face dry spells as they are situated in a dry corridor. Other than this, no permanent lakes exist in the region.

The total area of Lwengo District is about 1,023.7 sq. kilometers, most part of which are dotted with bear hills. The landscape and topography are by and large rolling and falling with valley bottom swamps including streams flowing to swamps. The texture of the soil is different at different places and is largely productive. It ranges from red literate to sandy loam and loam.

3.2 Climate: rainfall, temperature, humidity and winds

Being adjusted by help and proximity to Lake Victoria, the climate of the district is tropical in nature. The region witnesses long droughts between May and August, and January to March making the precipitation design bimodal. The two periods of downpour happen in the long stretches of March and April, and September to December. Chiefly, the region lies in dry-cattle passage with low dampness levels and wrecking twists consequently delayed times of dry spell.

3.3 Vegetation

85% hectares are under cultivation out of the absolute geological zone of the District (1,023.7 sq. km.) The area gazetted to the forest estate is about 21 hectares (Lwengo Local Forest Reserve) comprising about 0.021% of the absolute land territory of the region.

Summary of administrative units in Lwengo District

	Name of Sub-County / Town Council	Parish Name	Number of Villages
Bukoto	Kkingo sub county	Kagganda	10
		Kasaana	11
		Kiteredde	10
		Kisansala	9
		NkoniSsenya	9
Bukoto	Kisekka sub county	Busubi	8
		Kikenene	10
		Kinoni	9
		Kankamba	7
		Ngereko	10
		Kiwangala	9
		Nakateete	7
		Nakalembe	10
Bukoto	Lwengo sub county	Kito	18
		Kyawagonya	17
		Mbirizi	3
		Musubiro	11
		Nkunyu	13
		Nakyenyi	16
		Kalisizo	12
		Lwengo	03

	Name of Sub-County / Town Council	Parish Name	Number of Villages
Bukoto	Ndagwe sub county	Makondo	14
		Nanywa	20
		Ndagwe	26
		Mpumude	18
Bukoto	Malongosubcounty	Kalagala	18
		Katovu	20
		Kigeye	16
		Malongo	15
Bukoto	Kyazanga sub county	Kakooma	18
		Bijaaba	14
		Lyakibirizi	16
		Katuulo	14
Bukoto	Lwengo Town council	Kabalungi ward	3
		Church ward	3
		Central ward	2
		Lwengo ward	4
		Mulyazaahoward	3
Bukoto	Kyazanga Town council	Kitooro ward	4
		Lwentale ward	4
		Central ward	2
		Nakateete ward	3

4. Detailed demographic profile

The demographic of Lwengo District stands at 275,450 people, with an annual population growth rate of 3.1 percent, as per the 2014 National Population and Housing Census results. Out of the total population, 48% are men, while 52% are women. The number of inhabitants is 61,923 and 54 percent of its population is over 16 years of age. The density of population was then 192 people per square kilometers, 96% of the populace lived in rural areas, 5 people were the average fami ly size, 22% of the households were headed by women while 51.1% of the population was below the age of 18 years. The allocation of the population across sub-counties and town councils with their respective households and demographics over 16 years is shown in the following table (**Figure 3**).

Demographic characteristics of Lwengo District

Sub-county	Male	Female	Total	Households	Above 16 years
Kisekka	23,467	25,718	49,185	11,710	26560
Kkingo	16,674	17,673	34,347	8,061	18547
Kyazanga	16,953	17,794	34,747	7,269	18763
Kyazanga Town Council	7,366	8,165	15,531	3,745	8387
Lwengo	24,092	25,840	49,932	11,094	26963
Lwengo Town Council	7,259	8,268	15,527	3,561	8385
Malongo	18,030	19,875	37,905	8,151	20469
Ndagwe	18,356	19,920	38,276	8,332	20669
District	132,197	143,253	275,450	61,923	148,743

Source: UPHC 2014.

Figure 3.
Population by parliamentary constituency, LWENGO district. Source: https://www.ubos.org/wp content/
uploads/publications/2014CensusProfiles/LWENGO.pdf as accessed on 13 January, 2020.

5. Natural resources endowment

5.1 Wetlands

Wetlands are a vital part of theecosystem and one of the most important natural resources that contribute to the district's environmental health and socio-economic stability. They retain a tremendous amount of fresh water and provide buffering capacity against pollution, flooding and siltation. In order to assess their area of coverage, a final inventory and demarcation of wetlands in the Lwengo District becomes essential. In Kkingo and Kyazanga sub-County, there are Kyojja and Kiyanja Wetlands, both of which can be created for various economic benefits, such as through craftmaking and bird watching activities.

5.2 Social economic infrastructure

The district of Lwengo is blessed with a healthy atmosphere as well as an industrious population. The District's key economic profile includes agriculture and husbandry, fishing and trade, and pit sawing. Life standards indicators; employment patterns; patterns of human settlement; productive capital and district economic activities are a part of social economic infrastructure of the place.

Other than this, a lot of economic activities take place as well. Agriculture, livestock keeping, and trade are the major one s, as mentioned, with agriculture being the most significant element in terms of the District's revenue contribution. Employment income amounts to the District's second highest portion of revenue, followed by trade, property income tax, and the cottage industry eventually contributes the least.

6. LWENGO district and disaster

For some time, Lwengo district has suffered disasters, normally the second seasons of rain used to cause such disasters, but disasters have occurred since 2016. In Lwengo, hailstorms, heavy winds, long dry spells, pests and diseases are the main types of disasters (**Figures 4–7**).

Figure 4.
After a short heavy down pour these are patches of snow that were left behind.

Figure 5.
Banana plants destroyed by strong wind.

On 9 January 2020, a powerful hailstorm hit several villages in Lwengo sub county inflicting the impact on several villages that further affected numerous crops, including coffee, mangoes, passion fruits, banana, maize and cassava. Leaves of innumerable plants werebroken, stems uprooted further mangoes, bananas, cherries and other fruits in a detrimental state (**Figure 8**).

7. Conclusion

There is a missing link for a socio-economic assessment of what defines the capacity and initiative for adaptation among smallholder farming communities to

Figure 6.
Banana plants destroyed by heavy wind.

Figure 7.
A heap of snow around a banana plant after a heavy down pour.

adopt climate change adaptation practices. Since, farming communities are very sensitive to climate change in the Ugandan region, it was noticed over the past three decades, that the temperature rises and low rainfall have been a fundamental issue for farmers trying to boost seasonal farm yields.

Other barriers in the form of knowledge shortages, lack of finance for implementing improved technologies, the absence of irrigation and the short supply of labour are problematic for climate change management and adaptation in the area. Owing to this, Smallholder farmers can come up with adaptation strategies such as soil co nservation, mixed cropping, change of planting dates, tree planting/

Figure 8.
Maize plants destroyed by strong wind.

agroforestry, and furrow irrigation. This sustainable adaptation would help the ongoing on -farm practices that farmers modify on a seasonal basis to cope with climate change with the goal of improving yield and land productivity. Such activities are location-specific, regulated by policy structures, have a temporal aspect, and are based on farmers' dynamism. On one hand, the use of adaptation practices and behavior, which are imposed without recognizing the historical environmental characteristics of a region, is often presented as re-active to the immediate response to climate change events. For example, sometimes urgent steps taken by the authorities and others in cases of droughts and f loods to reinstate an 'equilibrium' within that environment. Higher-level organizations are also required to prepare for adaptation in an anticipatory manner by formulating legislation, proposing adaptation projects, such as Climate Smart Agriculture, and most recently, through the National Action Plans for Adaptation (NAPAs). The NAPAs have for the most part been developed by'countries by giving limited attention to factors that drive farmers' decision-making around implementation [27]. As a result, smallholder farmers are therefore practicing different adaptation strategies on their farms, including sub -optimal ones, with the option of adaptation measures that are known to be influenced by different factors including, sex, age, farming experience, family size (household members), membership (Group affiliation of members), Shock floods, Land size, Farm inputs, Landscape position, Level of Education, Crop yield, and Farm income [28–31]. This study explores how these drivers affect the decision-making of farmers in relation to adaptation to climate change. Understanding the relative value of these parameters would allow farmers to use viable adaptation practices easily and to address famine and major constraints on crop production.

Author details

Shyamli Singh[1*] and Ovamani Olive Kagweza[2]

1 Indian Institute of Public Administration, New Delhi, India

2 Entomology Officer, Lwengo District Government, Uganda

*Address all correspondence to: shyamli.env@gmail.com

IntechOpen

References

[1] Change IC. Mitigation of climate change. Contribution of Working Group III to the Fifth Assessment Report of the Intergovernmental Panel on Climate Change. 2014 Apr;1454.

[2] Asiimwe JB, Mpuga P. Implications of rainfall shocks for household income and co nsumption in Uganda.

[3] Lemma WA. Analysis of smallholder farmers' perceptions of climate change and adaptation strategies to climate change: the case of Western Amhara Region, Ethiopia. Ethiopia Doctoral Thesis University of South Africa. 2016 May.

[4] Easterling DR, Meehl GA, Parmesan C, Changnon SA, Karl TR, Mearns LO. Climateextremes: observations, modeling, and impacts. science. 2000 Sep 22;289(5487):2068–74.

[5] Howden SM, Soussana JF, Tubiello FN, Chhetri N, Dunlop M, Meinke H. Adapting agriculture to climate change. Proceedings of the national academy of sciences. 2007 Dec 11;104(50):19691–6.

[6] Ainsworth EA, Ort DR. How do we improve crop production in a warming world?. Plant physiology. 2010 Oct 1;154 (2):526–30.

[7] Rosegrant MW, Cline SA. Global food security: challenges and policies. Science. 2003 Dec 12;302(5652):1917–9.

[8] Parry ML, Rosenzweig C, Iglesias A, Livermore M, Fischer G. Effects of climate change on globalfood production under SRE S emissions and socio-economicscenarios. Global environmental change. 2004 Apr 1;14 (1):53–67.

[9] Schmidhuber J, Tubiello FN. Global food security under climate change. Proceedings of the National Academy of Sciences. 2 007 Dec 11;104(50):19703–8.

[10] Majaliwa JG, Omondi P, Komutunga E, Aribo L, Isubikalu P, Tenywa MM, Massa-Makuma H. Regional climate model performance and prediction of seasonal rainfall and surface temperature of Uganda. African Crop Science Journal. 2012;20:213–25.

[11] STATISTICS UB. Uganda Bureau of Statistics 2014 statistical abstract.

[12] Hardee K, Mutunga C. Strengthening the link between climate change adaptation and national development plans: lessons fr om the case of population in National Adaptation Programmes of Action (NAPAs). Mitigation and Adaptation Strategies for Global Change. 2010 Feb;15(2):113–26.

[13] Zizinga A. Viability Analysis of Climate Change Adaptation and Coping Practices for Agriculture Productivity in Rwenzori Region, Kasese District.(2013) Makerere University: Kampala, Uganda, 2013.

[14] Kraybill D, Kidoido M. Analysis of relative profitability of key Ugandan agricultural enterprises by agricultural produc tion zone. Uganda Strategy Support Program (USSP) Brief No. 7.

[15] Almeida JP, Montúfar R, Anthelme F. Patterns and origin of intraspecific functional variability in a tropical alpine species along an altitudinal gradient. Plant Ecology & Diversity. 2013 Dec 1;6(3–4):423–33.

[16] Challinor AJ, Ewert F, Arnold S, Simelton E, Fraser E. Crops and climate change: progress, trends, and challenges in simulating impacts and informing adaptation. Journal of experimental botany. 2009 Jul 1;60(10):2775–89.

[17] Nhemachena C, Hassan R. Micro-level analysis of farmers adaption to climate change in Southern Africa. Intl Food Policy Res Inst; 2007.

[18] Cooper PJ, Cappiello S, Vermeulen SJ, Campbell BM, Zougmoré RB, Kinyangi J. Large-scale implementation of adaptation and mitigation actions in agriculture.

[19] Deressa TT, Hassan RM, Ringler C, Alemu T, Yesuf M. Determinants of farmers' choice of adaptation methods to climate change in the Nile Basin of Ethiopia. Global environmental change. 2009 May 1;19(2):248–55.

[20] Ruijs A, de Bel M, Kononen M, Linderhof V, Polman N. Adapting to climate variability: learning from past experience and the role of institutions. World Bank; 2011 Aug.

[21] McCarthy JJ, Canziani OF, Leary NA, Dokken DJ, White KS, editors. Climate change 2001: impacts, adaptation, and vulnerability: contribution of Working Group II to the third assessment report of the Intergovernmental Panelon Climate Chan ge. Cambridge University Press; 2001 Jul 2.

[22] Bagamba F, Bashaasha B, Claessens I, Antle J. Assessing climate change impacts and adaptation strategies for smallholder agricultural systems in Uganda. African Crop Science Journal. 2012;20:303–16.

[23] Scholes RJ, Biggs RA. Ecosystem services in Southern Africa a regional assessment. CSIR; 2004.

[24] McCarthy JJ, Canziani OF, Leary NA, Dokken DJ, White KS, editors. Climate change 2001: impacts, adaptation, and vulnerability: contribution of Working Group II to the third assessment report of the Intergovernmental Panel on Climate Change. Cambridge University Press; 2001 Jul 2.

[25] Funk C, Dettinger MD, Michaelsen JC, Verdin JP, Brown ME, Barlow M, Hoell A. Warming of the Indian Ocean threatens eastern and southern African food security but could be mitigated by agricultural development. Proceedings of the national academy of sciences. 2008 Aug 12;105(32):11081–6.

[26] Weatherspark.com. 2021. Average Weather in Lwengo, Uganda, Year Round - Weather Spark. https://weathe rspark.com/y/96871/Average-Weather-in-Lwengo-Uganda-Year-Round (Accessed 13 January 2020).

[27] McCarthy JJ, Canziani OF, Leary NA, Dokken DJ, White KS, editors. Climate change 2001: impacts, adaptation, and vulnerability: contribution of Working Group II to the third assessment report of the Intergovernmental Panel on Climate Change. Cambridge University Press; 2001 Jul 2.

[28] Hassan RM, Nhemachena C. Determinants of African farmers' strategies for adapting to climate change: Multinomial choice analysis. African Journal of Agricultural and Resource Economics. 2008;2(311-2016-5521):83–104.

[29] Average Weather in Lwengo, Uganda, Year Round - Weather Spark. Weatherspark.com. 2021. [cited 13 January 2020]. Available from : https://weatherspark.com/y/96871/Average-Weather-in-Lwengo-Uganda-Year-Round

[30] Lwengo district. (2020). [13th January, 2020] Retrieved from https://www.lwengo.go.ug/lg/overview

[31] Panda A, Sharma U, Ninan KN, Patt A. Adaptive capacity contributing to improved agricultural productivity at the household level: Empirical findings highlighting the importance of crop insurance. Global Environmental Change. 2013 Aug 1;23(4):782–90.

Chapter 9

Opportunities and Challenges of Mitigation and Adaptation of Climate Change in Indonesia

Gun Mardiatmoko

Abstract

The impacts of climate change are changes in rainfall patterns, sea level rise and extreme weather or extreme meteorological events. This impact will further provide dangers that threaten the sustainability of human life. The main causes of climate change are deforestation and forest degradation and the growth rate of industry and transportation modes that are not environmentally friendly. Therefore, Indonesia is participating in the Paris Agreement and implementing the Reducing Emissions from Deforestation and Forest Degradation program, role of conservation, sustainable management of forest and enhancement of forest carbon stocks in developing countries (REDD+). In an effort to increase the prosperity of the State, many forests have been transferred to other uses such as the development of oil palm plantations, agricultural land and urban expansion etc. In fact, many agricultural lands have changed their function into settlements. If this happens, the forest area will continue to decrease again because after the agricultural land has turned into residential land, the forest land is converted again for agricultural expansion, this happens continuously. When viewed from the CO_2 flux, there will also be changes in the basic CO_2 flux from forest land, plantation land, agriculture and urban areas. The problem of deforestation and forest degradation is inseparable from the large number of forest conversion functions into oil palm plantations, expansion of agricultural areas and other uses such as urban development and infrastructure. Opportunities for climate change mitigation and adaptation include the implementation of the REDD+ program, financing of climate change mitigation and availability of climate information. The challenges faced include the lack of synergy in the policy framework and implementation of climate change control, recognition of indigenous peoples' rights and uncertainty in the implementation of the REDD+ program.

Keywords: flux CO_2, REDD+, climate change, mitigation, adaptation, oil palm plantations

1. Introduction

This paper discusses the implications of climate change in Indonesia and discusses the challenges to and opportunities for climate change mitigation and adaptation within Indonesia.

It is widely known that one of the reasons for climate change is global warming which is marked by an increase in air temperature. Climate change is associated with increased atmospheric temperature caused by the "Green House Effect" which occurs due to the increase in green house gases (GHG) in the atmosphere. Carbon dioxide (CO_2) is one of the gases that causes global warming. According to the IPCC [1], the average temperature of the earth's surface over the past century has increased by 1.3^0 F. The presence of CO2 is related to the condition of forests in an area. The trees that make up forests of various types and growth rates, known as forest structure and composition, have a role in storing CO_2. Forests are dominated by vegetation that has chlorophyll which functions in the photosynthesis process by requiring light energy, water and CO_2 to form carbohydrates. Thus the forest will absorb carbon from the air and accumulate in the plant body in the form of stems, branches, twigs, leaves, flowers, fruit and roots and soil. In general, this process is known as Carbon Sequestration [2, 3]. Thus the forest can function as a carbon sink. Therefore, well-well-maintained forests can increase carbon sequestration or reduce the amount of carbon in the atmosphere. In addition, by expanding the forest area, of course, its ability to absorb carbon will be higher. The development of various ecosystems over millions of years has resulted in certain patterns of carbon flow in global ecosystems. However, human (anthropogenic) activities in the use of fossil fuels, conversion of forest land and others have resulted in changes in the exchange of carbon in the atmosphere, land and marine ecosystems. As a result of these activities, there was an increase in the concentration of CO_2 into the atmosphere by 28% from the CO_2 concentration that occurred more than 150 years ago.

Indonesia's swamplands, which are around 33 million ha, 20.6 million ha of which are peatlands. Most of the peatlands are spread across three major islands, namely Sumatra (35%), Kalimantan (32%), Papua (30%), Sulawesi (3%), and the rest (3%) is spread over a narrow area [4]. The role of peatland is important because it has a high carbon sequestration and is a natural resource that has a hydrorological function. The existing peatlands need to be protected from fire. Because if a fire occurs in the peat forest, it will cause large CO_2 emissions and the resulting smoke will disrupt airlines and cause shortness of breath, etc. Peatlands play a major role in the development of agriculture, oil palm plantations or industrial plantations. For this reason, peatlands are managed with the principle of sustainable peatland management so that they can minimize environmental damage. Apart from peatlands, there are also mangrove forests that are found on the coast of the Indonesian archipelago which have a high carbon content known as blue carbon.

Peatlands planted with oil palm and acacia function as a carbon sequester through the photosynthetic process and carbon is stored as plant biomass. The carbon tethering process through the photosynthesis process is able to offset the loss of carbon stocks in the soil which are oxidized to CO2 gas emissions. However, if the expansion of oil palm plantations is excessive to the point where many natural forests are converted, it will have a negative impact on the biodiversity of the peatlands. The existing mangrove forests have also suffered a lot of damage because the area is used for the construction of ponds, excessive mangrove wood extraction and the large number of mangrove forests that have turned into settlements in coastal areas. The area of mangrove forests in Indonesia reaches 3.49 million ha but 52% or 1.82 million ha is in a damaged condition [5].

Carbon emissions from forest land including peat and mangrove forests generally fluctuate depending on many factors including climate, soil and hydrology. Environmental factors that greatly influence the amount of carbon emissions, especially from peatlands, are temperature, soil moisture and electrical conductivity (EC) [6]. These three factors fluctuate greatly from day to day depending on climatic and hydrological factors, resulting in high fluctuations in carbon emissions [7, 8]. High carbon content in natural and plantation forests is usually found in forests where the potential for wood or wood volume unit (m3/ha) is also very high. Therefore, if an area is only used for seasonal plant development, of course the carbon content is low. The lowest carbon content is when forest land has been converted into urban areas with the development of housing, markets, offices, development of road networks and infrastructure. Even with the construction of urban areas with various tall buildings, it has triggered the emergence of a heat island. One of the safety valves so that forest areas can maintain carbon content is the application of the agroforestry system. This system is a cultivation in an area with a mixture of perennials and seasonal plants.

In an effort to increase the prosperity of a country, a lot of forest is transferred to other uses such as the development of oil palm plantations, agricultural land, livestock grazing and urban expansion etc. In fact, many agricultural lands have changed their function into settlements. If this happens, the forest area will continue to decrease again because after the agricultural land has turned into residential land, the forest land is converted again for agricultural expansion, this happens continuously. In other words, deforestation and forest degradation have triggered climate change.

If viewed from the CO_2 flux, there will also be changes in the basic CO2 flux from forest land, plantation land, agriculture and urban areas. It is certain and inevitable that the forest area will decrease and be used for non-forestry development. One of the reasons is the increase in population which is difficult to control every year. Thus, changing a forest area to non-forest will have an impact on the lack of carbon sequestration as shown in **Figure 1**.

The conversion of forest land to non-forest land actually occurs as a result of economic motivation. For example, more forest land will be converted into oil palm plantations if the results of oil palm management turn out to be more profitable from an economic perspective. Therefore, forest management must endeavor to be able to generate more tangible benefits from non-forest uses.

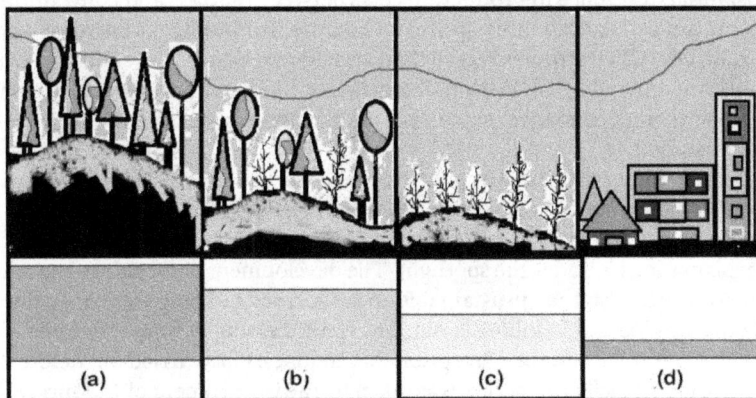

Figure 1.
The lower carbon sequestration of forest to non-forest areas. (a) Forests: very high carbon sequestration, (b) Agroforestry: high carbon sequestration, (c) Agricultural crops: low carbon sequestration, (d) Cities with infrastructure: very low carbon sequestration.

2. Expansion of oil palm plantations and the issue of deforestation

Indonesian oil palm plantations have grown rapidly in large parts of Indonesia. Sumatra and Kalimantan are two large islands which are the main centers of oil palm plantations in Indonesia. About 90% of oil palm plantations in Indonesia are located on these two oil palm islands, and the two islands produce 95% of Indonesia's crude palm oil (CPO) production. In the period 1990–2015, there was a revolution in the exploitation of oil palm plantations in Indonesia, which was marked by the rapid growth and development of smallholder plantations, namely 24% per year during 1990–2015. During this period, the forest land changed into oil palm plantations. This marks the end of the logging era and the drastic reduction of the plywood industry. The Ministry of Forestry has revoked many HPH licenses and an increasing number of plywood industries have closed due to a shortage of log raw materials. So in addition to the development of oil palm plantations, it is also planting industrial tree plantations which encourage the construction of pulp and paper mills. The area of Indonesian oil palm plantations in 2015 reached 11.3 million ha [9]. In 2017 it has reached 16 million ha. The largest proportion of oil palm plantations is smallholder plantations 53%, large private plantations 42%, and state plantations 5%. The rapid development of the palm oil industry has attracted the attention of the world community, particularly the world's major vegetable oil producers. In 2019 the area of oil palm plantations has reached 14.6 million ha [10]. Indonesia has become the world's largest palm oil producing country since 2006. Indonesia managed to surpass Malaysia in 2016 where Indonesia's CPO production share has reached 53.4% of the world's total CPO. Meanwhile, Malaysia only has a share of 32%. Likewise in the global vegetable oil market, palm oil has also managed to outperform soybean oil since 2004. In 2004, total CPO production reached 33.6 million tons, while soybean oil was 32.4 million tons. In 2016, the share of world CPO production reached 40% of the world's main vegetable products, while soybean oil had a 33.18% share [11].

Indonesia with its enormous reserves of oil palm plantations needs to ensure that these resources contribute to its national energy plan. Therefore the central government has compiled a Biodiesel Mandate which is among the most ambitious in the world. By 2016, liquid fuels must contain at least 20 percent of biofuels (and by 2025, 30 percent). A subsidy program has also been established to account for the substantial difference in production costs between biofuels and conventional diesel. One can feel considerable optimism because this funding is based on taxes on Crude Palm Oil (CPO) exports rather than on national budget expenditures which are negotiated annually [12]. With the mandate of biodiesel, in an effort to achieve national energy independence, expansion of oil palm plantations is something that cannot be avoided.

The rapidly increasing share of palm oil in the world vegetable oil market has influenced the dynamics of competition between vegetable oils and has even led to a negative / black campaign against palm oil. In addition, the sustainability aspect of oil palm plantations is under the spotlight. The development of oil palm plantations in Indonesia is perceived as unsustainable and is accused of being the main cause of deforestation and loss of wildlife habitat. The rapid clearing of forest land into oil palm plantations has led to the perception that Indonesia has carried out deforestation on a large scale. Actually this action was taken by the Government of Indonesia in carrying out national development in order to improve the welfare of its people. So there are stages for a country to deforest for the welfare of its people. When viewed from the development history of a number of major countries in the world, both the United States and Europe have deforested their countries. Therefore, it is unfair if the issue of deforestation is used to suppress the growth of Indonesian oil palm plantations.

So far there have been many accusations stating that 67% of oil palm plantations are obtained from forest conversion [13]. Gunarso et al. [14] tried to examine
the truth of forest conversion in Indonesia for oil palm plantations. This is done
by using data from disturbed and undisturbed forest land cover classes according
to the carbon stock sequence published by the Forestry Planning Agency in 2011.
Carbon stock of natural/production forests, either undisturbed forest or disturbed
forest, contains carbon stocks higher than carbon. Oil palm plantation stock. Thus,
if there is conversion of production forest to oil palm plantations, there will be
a decrease in land carbon stock or deforestation. Meanwhile, timber plantation,
agricultural land (mixed tree crops, dry cultivation land) and shrubs/ abandoned
land (schrub) contain lower carbon stocks than oil palm plantations. Thus, the conversion of scrub agricultural land/abandoned land, including industrial plantation
forest land, into oil palm plantations is categorized as an increase in land carbon
stock or reforestation. This study turns out to provide conclusions that are different
from the allegations by Koh and Wilcove [13]. The Indonesian oil palm plantations
planted until 2010, namely 8.1 million ha, turned out to be 5.5 million ha of which
came from the conversion of agricultural land and abandoned land (reforestation).
While the rest, namely 2.5 million ha, comes from conversion of production forests
(deforestation). Because the area of deforestation for oil palm plantations is much
less than the area of reforestation for oil palm plantations, in net terms, the expansion of Indonesian oil palm plantations to reach 10.4 million ha in 2013 is a form of
reforestation and not deforestation. This means that the expansion of Indonesian
oil palm plantations to 10.4 million ha in 2013 on a net basis is to increase land
carbon stock or reforestation [15]. However, the conversion of forest land which
was converted into oil palm plantations in 2019 has reached 14.6 million ha, so that
deforestation cannot be avoided. This is what causes a huge source of CO_2 emissions
that actually triggers climate change.

3. The opportunities for mitigating and adapting to climate change in Indonesia

3.1 Mitigation and adaptation opportunities through the REDD+ Program

The remaining forest area in Indonesia in 2019 is 94.1 million ha or 50.1% of the
total land area [16]. These forests play an important role in climate change mitigation and adaptation, so various strategies are needed and identification of opportunities to strengthen the results for both. a logical step. Therefore, the existence
of the REDD+ Program will be very useful to support various steps that will help
reduce the vulnerability of forest communities to the impacts of climate change.
Reducing Emissions from Deforestation and Forest Degaradation (REDD+) is an
effort to reduce emissions from deforestation and forest degradation, the role of
conservation, sustainable forest management and increasing forest carbon stocks
using a national approach and sub-national implementation. In its implementation,
mitigation-adaptation synergy is needed which aims to find ways to take advantage
of the synergy between REDD+ and climate change adaptation. Thus there is cer-
tainty that REDD+ will have impacts that go beyond mitigation and are sustainable
in a climate that changes over time [17].

3.2 Climate change financing opportunities

Indonesia still dominantly uses fossil energy sources that are not environmentally friendly and contribute to the increase in GHG which has been scientifically

proven to change climate patterns with the emergence of global warming. Climate change will affect the duration of the dry and rainy seasons. This will certainly affect the yields in the agricultural-plantation sector and also the results of fishing in the sea. Therefore, people whose income depends on these two livelihoods will definitely be affected directly. To overcome this, it is necessary to implement climate change mitigation and adaptation programs. Here there are funding opportunities to carry out climate change mitigation and adaptation sourced from (1) public funds through the State Budget (APBN), (2) funds from abroad in the form of grants or loans (3) Funds from the private sector through Corporate Social Responsibility (CSR) and Green Bond [18].

3.3 Opportunities for providing climate information

The selection of the types of adaptation that can be carried out in various regions is basically a follow-up to the National Action Plan - Climate Change Adaptation (RAN-API). Understanding the impacts of climate change varies depending on location or region. In this condition, an assessment of the impacts and vulnerability of climate change specific to the economic sector in a location or region is required as a first step in selecting climate change adaptation options. Furthermore, an evaluation of adaptation options is carried out considering that the implementation of climate change adaptation requires additional costs [19]. One of the important elements needed in conducting a climate change impact and vulnerability assessment is climate information. This climate information plays a vital role in identifying the impact of global climate change on climate conditions in a region. The trend of climatic elements such as rainfall and air temperature observations is the earliest stage to see the effects of climate change in an area. Climate information is needed to (1) undergo impact models, for example: crop simulation models to assess the impact of climate variability in a region on the agricultural sector, (2) validate climate model outputs for projecting future climate conditions, compiling climate change scenarios. The uncertainty of future climate change is often approached by using more than one climate model or emission scenario. To understand the capabilities of climate models, validation of climate model outputs for the current period (control) is carried out using observational climate information. Compiling climate change scenarios also requires observational climate information, for example by changing (adjusting) observational climate information with differences between future climate projections and control periods [20].

4. The challenges to mitigating and adapting to climate change in Indonesia

4.1 A policy framework and implementation of climate change control have not yet gone hand in hand

Indonesia is the fourth largest country with GHG emissions in the world but does not make climate change a national priority agenda. At the international level, Indonesia has ratified the PARIS Agreement and has committed to reduce GHG emissions without conditions by 29% under a business as usual scenario in 2030 and up to 41% with international assistance. The government has established a policy framework such as RAN GRK SINCE 2011 and RAN API in 2014. These policies must be broken down to sub-national levels in the form of RAD GRK and RAD API. However, in practice, the policy framework and implementation often do not go hand in hand because local governments do not fully implement the policies set by the central government [21]. Addressing this challenge requires a strong synergy between the central government and local governments.

4.2 The rights of indigenous peoples to the REDD+ program

For indigenous community activists, fighting for community rights to support the implementation of REDD+ is very important. This is because the role of indigenous peoples is very real in protecting the forest and its environment. Those with local wisdom have the knowledge to protect and protect their territory with customary laws, customary institutions and tenure systems that are different from the Western system. In general, they apply communal ownership and do not understand property rights [22]. Tenure issues cannot be eliminated in forestry management in Indonesia. This is due to overlapping control of forest areas because there are claims of state blasphemy over customary forests which are controlled by customary law communities. State forest claims provide room for the State's unilateral control over the forest through the various companies it owns or granting permits on it with the authority of the regional government. This has resulted in legislation and policies that are not clearly formulated, uncoordinated granting of permits and denial of recognition of indigenous peoples and other local forest users [23]. Indigenous peoples have a special role in REDD+, especially from the policy context, namely their participatory role. They have long lived in the forest and are able to care for and protect the forest for their survival from generation to generation. In addition, their cultural and spiritual relationship with the land and forest where they live is very deep [24]. Actually, the existence of this tenurial conflict has been eliminated somewhat by the implementation of the Social Forestry Program. In general, indigenous peoples have been given access to be able to carry out activities and manage in State forests. Tenure conflicts do not only occur on land already owned by companies that have forest concession permits but also in forest areas that have implemented the REDD+ program. So this is a challenge that must be resolved in the future.

4.3 Uncertainty in the implementation of the REDD+ program

Before REDD+ is fully implemented, a Demonstration Activity (DA) is carried out in the early stages. The implementation of DA is based on international guidelines from COP's decision in the form of International Guidance for DA. The aim is to find out progress, evaluate the implementation of activities and lessons learned related to DA REDD+. In the implementation of DA REDD+, various activities carried out refer to the methodology issued by the IPCC but the mechanisms mostly follow the schemes issued by the Voluntary Standard such as VCS, CCBS and Plan Vivo. The implementation of REDD+ provides benefits and provides opportunities because it is in accordance with the principles of forest sustainability and provides benefits to the community and biodiversity preservation. The current conditions for DA REDD+ are various, many lessons learned have ended and are also results-based with varying progress which still needs further guidance [25]. A crucial implementation stage is the implementation of the Measuring, Reporting and Verifying (MRV) System. Developing country governments at the COP 16 meeting in Cancun 2010 were encouraged to carry out various mitigation activities, including: reducing emissions from deforestation and forest degradation, conserving forest carbon stocks, sustainable forest management, and increasing carbon stocks (FCCC/CP/2010/7/Add.1/C/Par. 70). In connection with these activities, a suitable and transparent measurement and reporting system needs to be established (FCCC/CP/2010/7/Add.1/C/Par.71). Specifically for activities funded by international or domestic sources, verification must be carried out based on the conventions/ guidelines that will be developed (FCCC/CP/2010/7/Add.1/C/Par.71; FCCC/ CP/2010/7/Add 1/B/Par. 61 and 62). During its development, the Monitoring, Reporting and Verification system was changed to Measurement, Reporting and Verification (MRV) at the Subsidiary Body for Scientific and Technological Advice (SBSTA) 36 in Bonn, 2012. MRV system is the basic and main requirement of implementing the REDD+ program using the principles incentives that are assessed based on performance or pay for performance [26].

MRV activities include measuring and reporting the effectiveness of GHG reduction or absorption quantitatively using methods and procedures that are reliable, transparent and accountable. MRV is part of a monitoring system where measurement methods and results are conveyed using standard and consistent scientific principles. These activities will serve as the basis for payment for the performance of reducing emissions. Each MRV activity must be in line with the reporting principles of the IPCC (Intergovernmental Panel on Climate Change), which must be transparent, accurate, consistent, complete, comparable and have minimal uncertainty. The MRV system implementer is an independent body but still coordinates with the REDD+ Agency as a governing council. The UN-REDD Program has recommended a set of key considerations for the development of a national MRV system. As a system, MRV can be applied to several scales, namely national, subnational (province, district) and projects. The MRV system can also be reported to certain agencies and verified or validated by certain agencies or associations related to carbon. The use of MRV at the local and national levels is highly recommended. At the international level, reporting to the UNFCCC is a must or a requirement. Because the MRV system reporting must be based on scientific principles, this is a challenge for scientists and foresters in implementing the MRV system [27]. This is very important because the MRV principle is applied to collect data on each type of forest, forest cover and the amount of carbon content contained therein. Forest conditions in Indonesia are very diverse and categorized as mega-biodiversity. Of course, there will be many difficulties in implementing MRV. The challenge that is often faced is the calculation of the biomass present in each forest type. Ideally, biomass calculations are carried out by developing an allometric equation for each tree species which is very expensive. If this is done per tree type in each forest type, it certainly requires a large biomass measurement fund. The REDD+ program is known for leakage, additionality and uncertainity. In REDD+ activities, forest land which is designated as the location for REDD+ implementation according to the stipulated time period must be able to prevent leakage from occurring [28]. Here it is necessary to take intensive care for the location of the implementation of REDDD + so that there is no leakage originating from the work area and the surrounding area. Thus, year after year additionality must be guaranteed. Given the prevalence of forest conversion to non-forest, unresolved tenurial conflicts and illegal logging, etc., it will definitely be difficult to avoid uncertainty.

5. Conclusion

Global warming has caused climate change around the world. The impact of climate change is very large which affects the joints of life from an economic, ecological, and social perspective. The main cause is deforestation and forest degradation which releases CO_2 emissions into the atmosphere. Thus, if deforestation and forest degradation cannot be controlled, the earth's temperature will get warmer. The warming of the earth's temperature is also triggered by the use of fossil energy which is not environmentally friendly. Nowadays there is awareness from each country to start replacing fossil energy with biofuels that are more environmentally friendly. Indonesia has planned the production of biofuels to be independent of national energy that is environmentally friendly. One of them is by converting forest land for expansion of oil palm plantations and of course it will cause deforestation. So on the one hand developing environmentally friendly energy but on the other hand, sacrificing the area of the forest so that it becomes a contributor to CO_2 emissions that trigger climate change. Therefore, it requires a strong determination from the Government to be able to find the best way that can benefit both of them in controlling climate change. In every program that is executed, there are always opportunities and challenges that must be faced. One of them is the implementation of climate change mitigation and adaptation programs, such as opportunities for implementing the REDD+ program, financing climate change management, and the availability of climate information. There are also challenges faced, such as the lack of synergy in the policy framework and implementation of climate change control, recognition of indigenous peoples' rights, and uncertainty in the implementation of the REDD+ program.

Author details

Gun Mardiatmoko
Forestry Department, Faculty of Agriculture, Pattimura University, Ambon, Indonesia

*Address all correspondence to: g.mardiatmoko@faperta.unpatti.ac.id

IntechOpen

References

[1] IPCC. 2007. Climate Change 2007: The Physical Science Basis. Contribution of Working Group I to the Fourth Assessment Report of the Intergovernmental Panel on Climate Change [Solomon, S., D. Qin, M. Manning (eds)]. *http://ipcc- wg1.ucar. edu/wg1/wg1-report.html.* (accessed 17 December 2013)

[2] EPA. 2008. Carbon Sequestration in Agriculture and Forestry. *www.epa.gov/ sequestration/index.html* (accessed 19 February 2017)

[3] Jana, B.P., S. Biswas, M. Majumder, P. K. Roy and A. Mazumdar, 2009. Carbon sequestration rate and aboveground biomass carbon potential of four young species. *Journal of Ecology and Natural Environment* Vol. 1(2), pp.015-024.

[4] Wahyunto S, Heryanto B. 2005. *Sebaran Gambut di Papua.* Bogor (ID): W etland International Indonesia Programme.

[5] KKP. 2020. Hari mangrove sedunia, KKP targetkan rehabilitasi 200 ha lahan mangrove di 2020. Kementrian Kelautan dan Perikanan. https://kkp.go.id/djprl/ artikel/21996-hari-mangrove-sedunia-kkp-targetkan-rehabilitasi-200-ha-lahan-mangrove-di-2020 (accessed 7 August 2020)

[6] Setia R, Marschner P, Baldock J, Chittleborough D, Verma V. 2011. Relationships between carbondioxide emission and soil properties in saltaffected landscapes. Soil Biology and Biochemistry. 43(3): 667-674.

[7] Hirano T, Kusin K, Limin S, Osaki M. 2014. Carbon dioxide emissions through oxidative peat decomposition on burnt tropical peatland. Global Change Biology. 20(2): 555-565.

[8] Marwanto S, Agus F. 2014. Is CO flux from oil plam plantations on peatland controlled by soil moisture and/or soil and air temperatures?.*Mitigation Adaption Strategy Global Change.* 19(6): 809-819.

[9] Kementerian Pertanian Republik Indonesia. (2015). *Statistik perkebunan kelapa sawit Indonesia 2013-2015.* Jakarta: Kementerian Pertanian.

[10] BPS. 2019. Indonesian Oil Palm Statistics 2019. BPS-Statistics Indonesia. https://www.bps.go.id/publication/ 2019/11/22/f9ad9da6bac600960802c85f/ direktori-perusahaan-perkebunan-kelapa-sawit-indonesia-2018.html (accessed 27 August 2020)

[11] United States Department of Agriculture (USDA). (2016). *Indeks mundi, agricultural statistic.* Washington D.C.: USDA.

[12] Pirard, R. 2016. Indonesia: Biofuels from palm oil and power from tree plantations?. http://blog.cifor. org/43268/indonesia-biofuels-from-palm-oil-and-power-from-tree-plantations?fnl=en, (accessed 31 August 2017)

[13] Koh, L.P., Wilcove, D. S. 2008. Is oil palm agriculture really destroying tropical biodiversity? Conservation Letters 1(2):60 - 64.

[14] Gunarso, P, Hartoyo, M. E., Nugroho, Y., Ristiana, N.I., & Maharani, R. S. (2013). Analisis penutupan lahan dan perubahannya menjadi kebun kelapa sawit di Indonesia (Studi Kasus di 5 Pulau Besar di Indonesia periode 1990-2010). Jurnal Green Growth dan Manajemen Lingkungan, *1*(2), 10-19.

[15] Purba, J.H.V., Sipayung, T. 2017. Kelapa sawit Indonesia dalam perspektif pembangunan berkelanjutan. *Academic Forum on Sustainability I.* Masyarakat Indonesia, Vol. 43 No.1, Juni 2017.

[16] PTKL-KLHK. 2020. Hutan dan deforestasi Indonesia tahun 2019. http://www.ppid.menlhk.go.id (accessed 21 November, 2020)

[17] CIFOR-CGIAR. 2014. Sinergi mitigasi dan adaptasi. https://www.cifor.org/publications/pdf_files/factsheet/4515-factsheet.pdf (accessed 17 October 2018)

[18] ICCTF. 2019. Peluang pembiayaan perubahan iklim di Indonesia. https://www.icctf.or.id/2019/07/12/peluang-pembiayaan-perubahan-iklim-di-indonesia/ (accessed 10 August 2020)

[19] Tamirisa, N. 2008. Climate Change and the Economy. Finance & Development. March 2008.

[20] Perdinan. 2014. Perubahan Iklim dan Demokrasi:Ketersediaan dan Akses Informasi Iklim,Peranan Pemerintah,dan Partisipasi Masyarakat dalam Mendukung Implementasi Adaptasi Perubahan Iklim di Indonesia. JURNAL HUKUM LINGKUNGAN 1(1): 109-132.

[21] ICCTF. 2020. Yang terabaikan dalam perubahan iklim. https://www.icctf.or.id/2020/01/21/yang-terabaikan-dalam-perubahan-iklim/ (accessed 16 May 2020)

[22] Satriastanti, F. E. 2016. Menjawab tantangan REDD+. https://forestsnews.cifor.org/40246/menjawab-tantangan-upaya-nyata-redd-di-indonesia?fnl= (accessed 19 July 2019)

[23] HuMa. 2014. Konflik Kehutanan di Indonesia: apakah REDD peluang atau ancaman. https://web.huma.or.id/kehutanan-dan-perubahan-iklim/konflik-kehutanan-di-indonesia-apakah-redd-peluang-atau-ancaman.html (accessed 9 May 2018)

[24] Cusworth, C.C. 2017. Ini alasan hak adat menjadi penting untuk REDD+. https://forestsnews.cifor.org/53264/ini-alasan-hak-adat-menjadi-penting-untuk-redd?fnl= (accessed 22 May 2019)

[25] Wibowo, A. 2016. Implementasi Demonstration Activity (DA) REDD+ untuk mendukung kebijakan pengendalian perubahan iklim. P3SEKPI-Research Data. http://puspijak.org/Myfront/indexUsulan

[26] REDD+ Task Force MRV Working Group. 2012. Strategy and Implementation Plan for REDD+ Measurement, Monitoring, Reporting, and Verification (MRV) in Indonesia. https://static1.squarespace.com/static/566f0f00d8af100a22fb3bfd/t/58540e10e4fcb56428732a47/1481903646370/Indonesia+REDD%2B+Task+Force.+National+Strategy.pdf (accessed 21 March 2019)

[27] Jaya, I.N.S., Saleh, M.B. 2013. Peta Jalan (*Road Map*)MRV Kehutanan. Direktorat Jenderal Planologi Kehutanan, Kementrian Kehutanan. Jakarta.

[28] Mardiatmoko, G. 2018. From Forest Biomass to Carbon Trading. https://www.intechopen.com/books/renewable-resources-and-biorefineries/from-forest-biomass-to-carbon-trading (accessed 7 December 2017)

Chapter 10

Integrating Local Farmers Knowledge Systems in Rainfall Prediction and Available Weather Forecasts to Mitigate Climate Variability: Perspectives from Western Kenya

Daniel Kipkosgei Murgor

Abstract

This chapter examines relevant studies and examples on integrating farmer's traditional knowledge systems in rainfall prediction with available weather forecasts to mitigate impact of changing climate among rainfall dependent farmers in Western Kenya. The chapter combines the results of a study conducted in Western Kenya among maize and wheat growing farmers in Uasin Gishu County and perspectives from other related studies within the Eastern and Southern part of Africa. The chapter details how farmers have navigated the impact of changing climate on the farming enterprise that is largely dependent on rainfall. The findings reveal that farmers in western Kenya have experienced crop losses during planting and harvesting seasons due to prevailing variations in weather patterns. This is corroborated by over 340 (87.8%) of farmers in Uasin Gishu county of Kenya who agreed so and further stated that they had experienced changes in rainfall patterns and even the timing for maize and wheat growing had become uncertain and contrary to what they have known over time in the recent years. Similarly, like other findings in the reviewed studies in this chapter, the Kenyan farmers (84.9%) agreed strongly that they applied their local indigenous knowledge and experience gained over time to predict rainfall onset and cessation dates thus making key farming decisions. Relying heavily on traditional weather forecasting by farmers is catastrophic now due to changes on the environment associated to environmental degradation; ecosystem disturbance and changing climate which have seen important traditional predictor indicators disappear or lost completely from the environment. Although over 90% of the Kenyan farmers in average belief in use of weather forecast information, integration of this information is not effective because of its adaptability, format and timing challenges. The same is true for farmers in some countries within the region. Importantly, provision of context-specific and downscaled weather forecast information to support farmer's resilience is crucial. Most studies and programmes reviewed in this chapter agree that there is synergy in integrating local knowledge systems and available weather forecast information for better weather prediction. It is critical that policymakers, practitioners or key stakeholders and forecasters (both from the meteorological services and indigenous groups) converge and agree on weather prediction if they are to support farmers in managing climate risk or uncertainties.

Keywords: indigenous knowledge systems, farmers, changing climate, weather forecasts, climate information, rainfall prediction, Western Kenya, Eastern and Southern Africa

1. Introduction

Due to the prevailing climate variability brought about by a changing climate global phenomena, the devastating impact on various sectors have been felt at regional and local levels more so, the developing countries. The changing climate is having a growing impact on the African continent, hitting the most vulnerable hardest, and contributing to food insecurity, population displacement and stress on water resources. Further, the latest decadal predictions, covering the five-year period from 2020 to 2024, shows continued warming and decreasing rainfall especially over North and Southern Africa, and increased rainfall over the Sahel [1]. Extensive areas of Africa will exceed 2 °C of warming above pre-industrial levels by the last two decades of this century under medium scenarios as reported in the Intergovernmental Panel on Climate Change Fifth Assessment Report [2]. Some countries in the sub Saharan Africa have recently witnessed increased flood disasters, invasion of desert locusts that endanger food security and livelihoods and now face the looming danger of drought due to the likelihood of La Niña event [3].

The Horn of Africa region experienced a combination of very dry conditions, floods and landslides associated with heavy rainfall between 2018 and 2019 period while the Southern Africa region faced drought phenomena. A report by WMO explains that extensive flooding occurred over large parts of Africa in 2020 as rainfall was mainly above average in most of the Greater Horn of Africa region during the March–May season. This followed a similarly wet season in October– December 2019. Sudan and Kenya thus were affected more with 285 deaths reported in Kenya and 155 deaths and over 800,000 people affected in Sudan in addition to disease impacts associated with flooding. Heavy rains in the Arabian Peninsula and East Africa resulted in the largest desert locust outbreak in 25 years across the Horn of Africa [4, 5]. In Ethiopia alone, 200000 hectares of cropland were damaged and over 356000 tons of cereals were lost, leaving almost one million people food insecure [6]. In Somalia, floods were associated with the displacement of over one million people in 2020, mostly inside the country, while drought-related impacts induced a further 80,000 displacements [7].

Based on Remote Sensing data and other observations, Kenya's Lakes that include Naivasha, Elementaita, Nakuru, Bogoria, Baringo, Turkana, Logipi have been rising since 2018. Most of the lakes have over flown displacing thousands of people from their homes, leaving behind submerged schools, health facilities and social amenities and rendering some of the public infrastructure like roads impassable [8]. The agricultural sector in Eastern and some Southern Africa region remain dominantly a rainfall dependent system and majority of farmers work on a small-scale or subsistence level and have few financial resources, limited access to infrastructure, and disparate need for access to agronomic data and information [9]. For the farmers, any abnormal variation in rainfall onset and cessation dates in addition to extreme temperatures result in serious crop loss or damage. With the prevailing changes in weather patterns due to changing climate, African traditional knowledge systems in weather prediction for agriculture otherwise reliable for centuries could or have been rendered ineffective to some extent. Agriculture, being the backbone of

Africa's economy and accounting for the majority of livelihoods across the continent is both an exposure and vulnerability "hot spot" for climate variability and change impacts [10]. IPCC projections have suggested that warming scenarios risk having devastating effects on crop production and food security. Farmers therefore have suffered and will continue to suffer the most in their farming enterprise due to crop losses related to heat stress and pest damage, crop diseases and food system destabilization due to floods brought about by variations in weather and climatic conditions [10]. To mitigate this phenomena, most farmers have continued to rely on their indigenous knowledge systems in weather prediction in addition to integrating available weather forecasts to create a synergy to enable them navigate such uncertainties related to climate variability. This has created a great need for climate and weather information to be delivered to those farmers engaged in farming activities at the farm level in rainfall dependent farming systems. The information thus needs to be context specific, timely and delivery through the most reliable and accessible modes.

The objective in this chapter is to determine how farmer's indigenous knowledge systems in rainfall prediction influence farm level planning and decisions in Kenya and other countries. Perspectives from some Eastern and Southern Africa region on utilization of indigenous knowledge system in rainfall prediction together with available weather forecasts to mitigate climate variability in the agricultural enterprise is captured for evidence based practice.

2. Indigenous knowledge systems and meteorological forecasts use among farmers dependent on rainfall in western Kenya and some countries in Eastern and Southern Africa – some examples

Indigenous knowledge (IK) is generally defined as "knowledge of a people of a particular area based on their interactions and experiences within that area, their traditions, and their incorporation of knowledge emanating from elsewhere into their production and economic systems" [11]. The terms indigenous, traditional and/or local knowledge as commonly referred to is knowledge and know-how that is accumulated over generations and guides human societies in their innumerable interactions with their surrounding environment [12]. Indigenous Knowledge is still important among local communities in many parts of Africa and the global scientific community do acknowledge its value [12]; however, most of this knowledge except a few captured by researchers has not been well documented hence the risk of losing some of this tacit knowledge when holders of such knowledge are incapacitated in any way. Over the years, communities have developed their own systems for monitoring climate conditions, but this information may not be adequate to inform adaptation if the changing climate continues in unprecedented way. An example from Burkina Faso illustrates that farmers who traditionally relied on observation of environmental indicators to predict climate patterns have now lost much confidence in their ability to predict rainfall accurately given increased changing climate thus increasingly seek to incorporate weather forecast information [13]. Socio–cultural changes also account for the shift away from traditional practices such as the use of bio–indicators for agricultural production, even when such practices continue to provide useful information [13, 14]. It is thus important to reflect on the traditional knowledge systems of communities as this provides an important entry point to scientist, researchers and key stakeholders in water, climate change, agriculture, food security and livelihood sectors into understanding how a new type of information about the climate/weather might be accepted and used by the local people to counter prevailing adverse weather conditions [15].

Climate information is a valuable resource for confronting and living with an increasingly uncertain future. Availability of climate information or weather forecasts provides a basis on which people whose livelihoods are affected by climate can make forward looking and flexible plans that are adapted to a range of climate possibilities. Consequently, climate information allows us to move from strategies which react to conditions as or after they occur, to those which seek to build resilience under all possible conditions and ultimately, to proactive strategies informed by forecasts and forecast probabilities [16]. In the agricultural enterprise, crop growth, or crop yield, requires appropriate amounts of moisture, light, and temperature at its correct time hence timing of farmer activities is critical at the local farm level. Detailed and accurate historical, real-time and forecast weather information can help farmers better understand and track the growth status of crop hence being able to make informed decisions. Having access to this critical agronomic data and information can guide farmers in making significant and potentially costly decisions, such as when to start tilling the land and subsequent planting period of crops as this is directly related to rainfall onset and cessation dates in rainfall dependent agriculture [16]. The most useful weather forecast information that can assist farmers in making decisions on agricultural management is the early indication of the characteristics of the rainy season. It should include: onset date of the main rains; quality of the rainy season (rainfall amount); cessation date of the main rains; temporal and spatial distribution of the main rains; timing and frequency of active and dry periods (wet and dry spells) and probability of extreme weather events. These decisions relate to the choice of crops to be made, cultivar (early or late flowering), mix of crops fertilizer use, pest and disease control and also timing of the harvest period [17].

There is evidence that substantial gain to sustainable food security and national development adaptation strategies can be achieved in Africa through provision and integration of improved climate information and prediction products into decision-making systems [18]. Accurate monitoring, prediction and early warning of seasonal rainfall performance can be used to improve planning and management of various rainfall dependent socio-economic activities like agriculture and the same can be used to enhance the livelihoods of the communities and services and support their resilience to adverse weather conditions. According to [13], access to climate information and technologies for adaptation is essential to enable actors to anticipate long–term risks and make the appropriate adjustments to increase their resilience. However, despite significant scientific gains in predicting the climate, often there is a lack of climate information available at the local level due to uncertainty in climate projections and seasonal forecasts, or due to lack of information on particular climate indicators, such as rainfall variability. Even when climate information is available, incorporation of scientific climate information into local decision making may not often occur because of the way such information is communicated and disseminated [19]. Several studies have shown that there is a need to make climate information more accurate, accessible, and useful for rural communities [20].

Adapting to climate change requires improved understanding of the linkages between climatic conditions and the outcomes of climate sensitive processes or activities; agricultural production for example in a certain region could be influenced by the availability of water resources and their management ways. Information from literature according to [21] explains that adequate use of climate and weather information conditions by farmers' results in at least 30% increase in crop yields. The utilization of this information reduces farmers' vulnerability to weather related risks, ensures that informed decisions are made on time, and reduces the risk of agricultural losses as well as indicating to farmers the most marketable crop in respective times. The analysis in the study on economic value of climate forecasts for livestock production in the Northwest Province of South Africa demonstrated that, for the commercial farmers, long term average annual income could potentially be increased through using ENSO predictions [22].

During a World Bank funded workshop in Dar-es-Salaam in 1999 on users responding to seasonal climate forecasts in southern Africa and the lessons learned then, it became apparent that there were communication barriers between the generators of the information and the users of the same information thus there was a need to develop appropriate information channels to relay such information. The second was that there were bottlenecks in the effective use of seasonal climate forecasts by farmers [23–25]. Users of seasonal climate forecasts have not been able to decode the information disseminated and therefore, users could not make use of the information provided if they did not understand the information provided in the first place [25]. Field studies conducted in the southern part of Africa reveal the existence of a considerable gap between information needed by farmers and that provided by meteorological services. There existed a communication barrier as the two parties have been interacting for a long time but to some extent, they have not been able to communicate effectively. The farmers know what they want and the meteorological services know what they need to give to the farmers, but there is no "shared meaning" [23, 24]. Without a shared meaning in communication, the value attached to particular information availed to the user (farmer) is diminished and may not serve the intended purpose.

In the light of such challenges, adopting an integrated approach where weather forecasts are combined with indigenous knowledge systems locally have shown to be effective in mitigating variability in the farming enterprise. According to [26–28], the use of scientific weather forecasts and indigenous climate forecast information for farm level decision making has been reported in Kenya and Mozambique. Previous studies in East Africa indicate that both IK and scientific weather forecasts are used for making crop and livestock production decisions, conserve the environment, and deal with other natural disasters. In Malawi and Zimbabwe, communities have combined scientific and indigenous climate forecast information for farm-level decisions hence being able to cope with prevailing drought [29]. Therefore indigenous knowledge system in weather forecasting is crucial in complementing available weather forecast information for improved decision making by farmers.

The Climate Change Adaptation in Africa (CCAA) program, funded by the International Development Research Centre (Canada) and the Department for International Development (United Kingdom) have supported projects that inves- tigated how seasonal climate forecasts might be better integrated into agricultural and pastoral decision-making to strengthen livelihoods and food security [30]. Through these projects, it is apparent that indigenous knowledge forecasts, which have been used by communities for decades, provide information that is comple- mentary to the meteorological forecasts. Indigenous seasonal forecasting and weather forecast information from the meteorological services help assist develop- ment challenges related to climate variability brought about by changing climate. Many farmers already use indigenous forecasts in their farm-level decisions and may only need certain information, such as total rainfall expected in the season, to complement what they already have. It is, therefore, important that policymakers, practitioners, and forecasters (both from meteorological services and indigenous groups) target existing gaps and take advantage of opportunities if they are to sup- port farmers and pastoralists in managing climate risk in Africa [30].

Many societies and communities have their own ways of interpreting climate and weather patterns developed over years of experience. Traditional rainfall forecasts/predictions differ across communities, cultural background, and environ- ment around the farm. According to [31], in South-Western Free State and Kwa- Zulu Natal of South Africa, as well as Western Kenya, inhabitants use birds, toads, and white ants to predict the dry season and onset of rains as well as temperatures. In Tanzania, they look at the behavioral patterns of birds and mammals. In a study on climate forecasting among the Basotho in Lesotho, they were asked if there were any ways to predict the coming weather and climate from what they know traditionally and a lot of answers were given that touched on weather conditions (hours and days) rather than climate conditions (weeks to months). The indicators were both environmental and cultural beliefs. According to [15], birds and insects were the most common environmental indicators. People mentioned the 'squawk or the *Makara*' bird as being indicative of rain in the coming days. Winds that blow from a certain direction were thought to bring rain. Plants flowering at certain time, the amount and color of clouds gathering, rising groundwater and frisky animals were all mentioned as indicators of imminent rain.

Among the Nganyi community in Western Kenya, the traditional weathermen observe the flora and fauna in the Nganyi forest shrine to predict weather conditions. The forest, which lies on just one acre of land, has pristine biodiversity that has helped the local Bunyore community predict weather conditions for generations. According to [32], researchers with the Climate Change Adaption in Africa (CCAA) program, a collaboration between international organizations and Kenyan scientists recorded data from a meteorological weather station near Bunyore in western Kenya for two seasons. They then compared its results with predictions made by indigenous forecasters who use the forest shrine as their main tool. The two findings were similar in all aspects. Based on this outcome, the researchers recommended the combination of both meteorological data and indigenous knowledge to facilitate accurate predictions that are acceptable scientifically and by the local community [32]. The Nganyi shares the consensus forecast for the coming season with the local community, in local languages, through radio, in churches and other community gathering points. The Nganyi community weather predictions are based on close observation of natural phenomena, like the budding or flowering of specific plant species and the behavior of local insects, birds and animals, associated with seasonal changes. A colony of bees migrating from downstream to the upper land clearly means that long rains are approaching and the vice versa symbolizes dry season [32].

Among farmers in the Kalenjin community in the North Rift region of Kenya, there are indicators or rainfall predictors that have been observed over the years and have been perceived as very important. The indicators can be classified as those related to the plant species, meteorological, animal and universe indicators. According to [33], some of the meteorological indicators include wind direction blowing eastwards signifying rainfall near onset, clouds thickening at the horizon and wind veering or breaking towards the east and cloud darkening in color; this signify rainfall onset and also cloud movement from eastern to western side of their farms all indicating rainfall onset. High sunshine intensity during the day and warm nights or high temperatures at night and low temperature in the evening signify onset of rains. Lightning strikes in near vertical position in three specific locations indicate near onset of rainfall.

3. Methodology and data

3.1 Study site and study design

The study was carried out in Uasin Gishu County; one of the 37 counties in Kenya. The county is made up of six sub counties namely:-Turbo, Soy, Moiben, Anapkoi, Kesses and Kapseret. According to [34], the total area of the County is 3327.8 Km2 with arable land covering 2603.2 Km2 and non-arable land covering

682.6 Km2. The County extends between longitude 34° 50′ and 35 ° 37′ east and 0° 03′ and 0° 55′ north. The headquarters of Uasin Gishu County is Eldoret town located on the main highway serving Kenya, Uganda and other countries in the great lakes region areas that include Uganda, Tanzania, Rwanda, Burundi, Democratic Republic of Congo, Zambia, and stretching all the way to Cape Town in the Republic of South Africa. The town is located at an altitude of 2085 m above sea level with a relatively cool climate experiencing daily mean maximum temperatures of 23.7 ° C and a mean minimum of 9.5 ° C. Eldoret town is traversed approximately latitude 00° 30' North and Longitude 35° 15′ East of the Equator [35]. Uasin Gishu County is basically an agricultural district producing more than a third of the total wheat production in Kenya. Similarly, maize ranks second both as food and cash crop. A report compiled by [36] shows that the 2009 long rain maize production season was about 1.84 million Metric Tonnes, which was 28 percent below normal. There was a growing apprehension that the production could further be revised downwards due to insufficient and erratic rains in some parts of the main maize producing areas in North Rift including Uasin Gishu County due to the changing

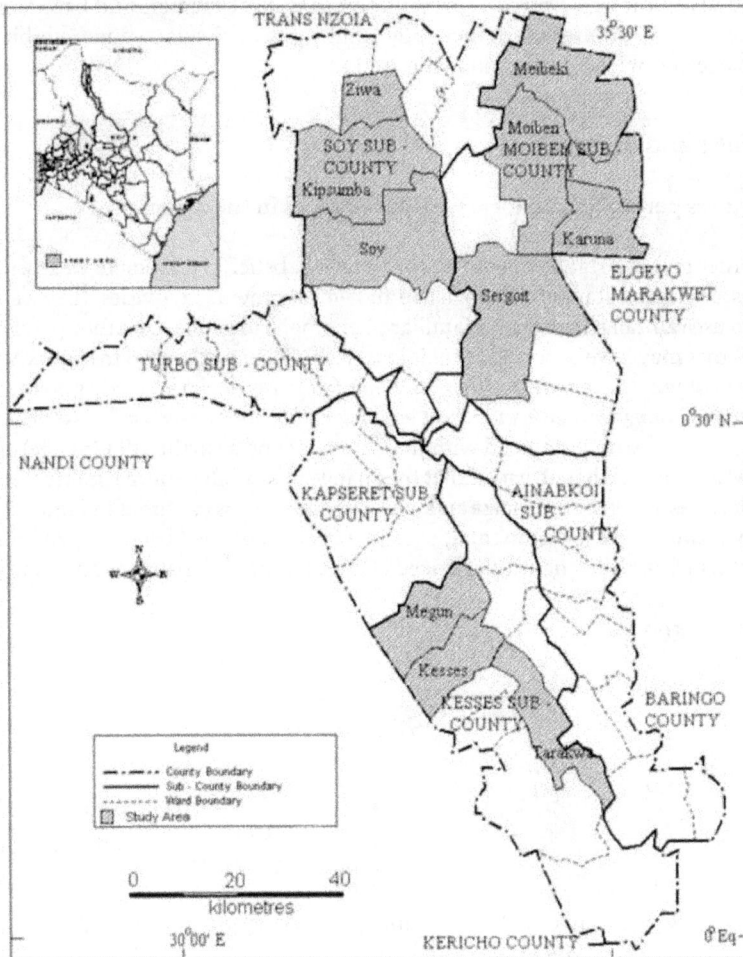

Figure 1.
Map of Uasin Gishu County showing the study area.

climate. The crop production has never been steady with each year having different challenges related to rainfall variations hence impacting positively or negatively to overall maize and wheat production.

The study utilized both qualitative and quantitative techniques. The use of both quantitative and qualitative research techniques are known to complement each other especially where exploration of indigenous knowledge among farmers is important. The qualitative aspect helped consolidate the themes emerging from the interview or survey. The target population was all farmers engaged in maize and wheat production resident in Uasin Gishu County at the time of study. Because of its nature of utilizing both qualitative and quantitative techniques, a representative sample was picked considering the levels of stratifications. The sampling frame for the study was 129, 384 farmers distributed as follows: - Soy Sub-County = 61, 138, Moiben Sub-County = 38, 950 and Kesses Sub-County = 29, 296. A minimum of 399 farmers were included in the study. In addition 12 key informants were interviewed; one from each ward totalling to 9 and also 2 from Directorate of Agriculture and Directorate of Meteorology in Uasin Gishu County respectively. One other key informant from the Kenya Meteorological Services in Nairobi was interviewed as well. Purposive, stratified and random sampling procedure was adopted to be able to capture a representative sample of farmers and based on this criterion, 3 sub-counties of Moiben, Soy and Kesses were selected for exhibiting both maize and wheat production (**Figure 1**).

4. Results and discussion

4.1 Farmers perception on local weather changes in the recent years

In order to understand what farmers perceive or belief in relation to weather changes and its resultant effects to maize and wheat growing activities, they were asked to use rainfall parameter; a familiar phenomena to gauge what they percieved as changes that may have occurred at their local level. They were asked to state whether *"in the recent years there were any changes in rainfall patterns experienced by them and that even the timing for maize and wheat growing had become uncertain".* A total of 142 (36.7%) farmers strongly agreed with the statement and a further 198 (51.2%) agree-ing as well. They further affirmed that the change in rainfall pattern experienced had led to declines and losses in maize and wheat production as captured in **Figure 2**.

The findings reveal that in total, over 340 (87.8%) of the farmers agreed that in the recent years they had experienced changes in rainfall patterns and even the

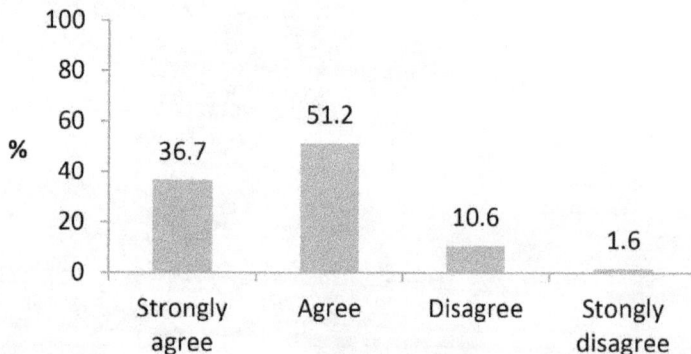

Figure 2.
Farmers perception on prevailing local weather changes affecting their crop production in Uasin Gishu County.

timing for maize and wheat growing had become uncertain and contrary to what they have known over time. This clearly show that local farmers can identify with the fact that the changing climate is areal phenomena and has been experienced locally at their farm level. The poor maize crop shown in **Figure 3** as captured during the field study help to illustrate what farmers are reporting in relation to losses they have incurred during their farming activities.

Figure 3.
Poor maize crop due to erratic rains in Kesses Sub-County of Uasin Gishu County, Kenya.

The finding in this study are similar to finding from [37] where farmers indicated that climate has been changing in the previous five years, in that rains start earlier, rains end latter and the maximum number of dry spells has increased. A comparison of the meteorological records with farmers' assessment of climate change showed a large disparity, with few of the stated changes being evident in the long term record. It is apparent that changes occurring in local and global weather patterns as a result of the changing climate will challenge to a great extend the indigenous knowledge systems in weather prediction otherwise reliable for centuries. A great need for climate and weather information to be delivered to those engaged in farming activities to supplement indigenous knowledge systems is critical in mitigating this phenomena.

4.2 Factors influencing a farmer's decision at the farm level in maize and wheat production in Uasin Gishu, county, Kenya

In an effort to understand what influences a farmers decision to commence an activity in the maize and wheat growing calendar (when to start land preparation, planting, type of crop to grow, weeding, top dressing, spraying of weeds, fungus or bacterial infection on crops and even harvesting), the following responses were gathered from the farmers:-Those who commence their activities just because those are the dates known to them through their experience over time in maize and wheat growing were 333 (84.9%). Another group of farmers 141 (36%) rely on looking around for *"signs that rains are about to fall (wind direction from east to western side, cloud movement from eastern to western side of their farms and high sunshine intensity during the day and warm nights)"*. Advice from the Ministry of Agriculture Livestock and Fisheries officials was mentioned by 29 (7.4%) of the farmers as having assisted them make decisions on the start of farming activities in their respective farms as shown in **Figure 4**.

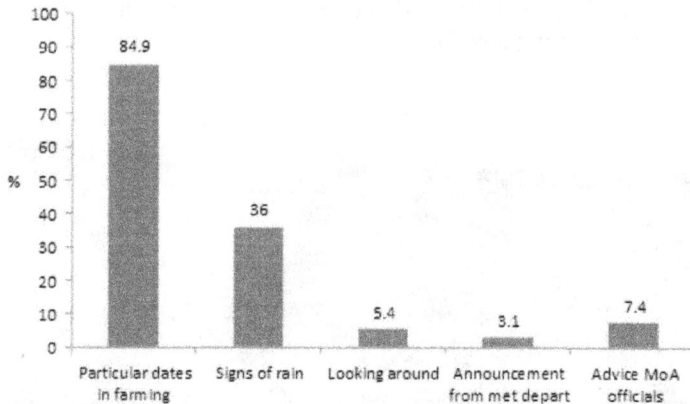

Figure 4.
What factors influence a farmer's decision at the farm level in maize and wheat production in Uasin Gishu, county, Kenya.

The results reveal a farming population solely dependent on their own indigenous knowledge systems and experience gained over time in maize and wheat growing to make certain farming decisions. The fact that 333 (84.9%) farmers commence their activities just because those are the dates known to them through their experience over time and another 141 (36%) looking around at some indigenous knowledge system indicators that rains are about to fall is a clear indication that framers do base their farming decisions on their own indigenous knowledge systems and experience gained over time. However, depending solely on indigenous knowledge indicators to predict onset dates of the rains and experience gained over time practicing maize and wheat creates a big challenge to the farmer especially with the prevailing changing climate that keeps on distorting seasons known to farmers. This makes farmers vulnerable and may incur much losses related to unpredictable weather patterns brought about by climatic changes. The findings in this study correspond with findings in [24] which asserts that in different parts of the world, farmers depending on rain-fed cultivation have developed complex cultural models of weather and may be able to cite local predictors of seasonal climate. Similarly, [13] argue that even when climate information is available, incorporation of scientific climate information into local decision making may not often occur because of the way such information is communicated and disseminated. There is need to identify clear channels of information delivery and also downscale the information to make sense for the farmer for effective uptake of such information.

4.3 Indigenous knowledge system indicators in rainfall prediction for farmers in Uasin Gishu County, Kenya

In an effort to understand what farmers use as their local indigenous knowledge indicators for weather prediction, farmers were asked to state what they thought were signs that rains were about to fall in their respective areas based on their experiences gained over time as maize and wheat farmers. The farmers who affirmed and believed that the real sign for rain commencement in their farms is when there are consistent lightning flashes around the Lake Region or Tindiret area in Nandi South as affirmed by 245 (62.7%). Those farmers who belief that heavy cloud cover and intense sunshine during the day was a sign that rains were about to commence were 178 (45.5%). The farmers relying on wind direction (eastwards and sometimes westwards) to alert them that the rains were about to commence were 159 (40.7%). Farmers who belief that very warm nights was a sign that rains were about to fall in their area were 195 (49.9%). Those who affirmed that in the recent years, it has been quite difficult to predict when the rains are about to fall were few farmers 42 (10.7%). The responses are as shown in **Figure 5**.

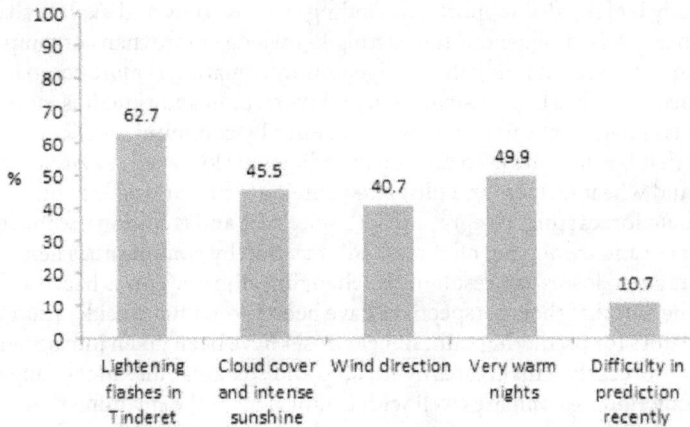

Figure 5.
Indigenous knowledge system indicators for predicting rainfall in Uasin Gishu County, Kenya. (*multiple responses - farmers were allowed to note as many reasons as were applicable).

The findings in this study reflect a farming community that has developed its own indicators in the traditional knowledge system on rainfall prediction. The set of indicators of rainfall used by farmers include consistent lightning flashes around the Lake Region or *Tindiret* area in *Nandi* South, heavy cloud cover and intense sunshine during the day, blowing of wind towards eastern side and sometimes westwards and very warm nights. During the key informant interview to supplement and corroborate some of the response, the following were identified as other key indicators used by farmers especially among the *Nandi* community and include examining the behavior of certain plants or trees to determine rainfall near onset. Among the plant species known include the tree *Erithrina Abyssinica (Kakarwet)*. The tree starts flowering red with full leaves regained, *Vernonia Auriculifera (Tepengwet)* tree starts flowering, *Flacourtia indica (Tungururwet)* start budding is an indicator for rainfall near onset and farmers prepare to commence dry planting. The Fig tree (*Ficus sycomorus*) known locally by farmers as (*Mogoiywet*) start shading leaves is an indicator for rainfall near onset. One other small plant (herb) growing in thickets or bushes *Scadoxus multiflorus* of the *Amaryllidaceae* family (*Ngotiotet*) starts flowering red around March period and found in thickets or bushes near river banks is a real indicator for rainfall near onset.

Other indicators include migratory birds among them the White Stock (*Kaptalaminik*) moving or flying towards the north side signifying near rainfall onset and when flight changes towards the south, this signifies rainfall cessation. In addition to the indicators, the community had a unique way or prayer asking their God for rainfall or water for their domestic use during drought manifestation. According to the key informants interviewed, the traditional prayer song (*ingoo*) was sang by mature women and those at child bearing age at night and in several groups (3 to 4 groups of women) from various locations. They could join in the prayer song at night converging at one place in the open near a watering point away from homesteads as participants sang without cloths on them. They carried with them cooking items that include cooking sticks (*mukanget ak Kurbet*) and some finger millet flour (*Beek ab Kipsongik*). This form of prayers according to key informants and other farmers participating in this study all agreed that the women prayers at night were answered almost instantly there and then because before mid-night during the singing, heavy rains would fall on the singing women as they retreat back to their homesteads happy.

The study by [18] also supports the findings in this study and explain that majority of farmers prefer indigenous forecasting knowledge more than contemporary forecasting. The reasons being that indigenous information is more compatible with local culture and it has been tested, tried and trusted. In addition, it is more specific and is in a language that can be understood better by communities. There is a clear view here that it is important to document indigenous knowledge system used by maize and wheat farmers and ultimately integrate the knowledge with science based climate forecasting. Integrating both scientific and traditional knowledge will enable uptake and ownership of climate information by communities hence helping farmers avoid losses as a result of the changing climate. This is because farmers consider the fact that their perspectives have been taken into consideration when their indicators for predicting rainfall near onset have been taken into account by the weather forecasts. This creates ownership and sustainability in the long run.

The study finding compares well with findings by [38] explaining that in Uganda, indigenous knowledge forecasters associate the onset of rainfall with the appearance of clouds. The appearance of nimbostratus and cumulonimbus clouds indicates a high probability of rainfall. Biological indicators focus on the behavior and activities of domestic and wild animals, insects and different species of plants for weather forecasting. For instance, in Uganda, the *Mvule* tree indicates onset of the rainy season. In Kenya like Ethiopia, the intestines of sheep and goats are used to forecast about the magnitude, severity, and duration of drought, drought-affected places, disease outbreak, the prospect of peace, and/or conflict. In Tanzania, the occurrence of large flocks of swallows and swans, roaming from the South to the North during the months of September to November, is an indication of onset of short rains [38]. From the afore going, to effectively mainstream access to climate and weather information in key sectors, it is important to understand the local perspectives in relation to the changing climate and coping strategies in place. Ignoring this fact might hinder uptake of weather forecasts by the farmers hence leaving them completely vulnerable to the changing climate as they would rely only on their indigenous knowledge systems to mitigate prevailing weather conditions.

4.4 Do famers belief in the use of climate information/weather forecasts or their indigenous knowledge system indicators in rainfall prediction

Farmers were further tested on whether they relied on the use of climate and weather information or their indigenous knowledge/practice in their farming decisions and the responses were varied as outlined in **Table 1**.

More than half of the farmers 219 (55.7%) sometimes belief in seasonal climate and weather information, 163 (43.7%) rely on experience/indigenous knowledge in wheat production most of the time and 148 (37.8%) rely on experience/indigenous knowledge in maize production most of the time as outlined in **Table 1**. The results show a group of farmers who tend to belief in the use of climate and weather information to conduct their business of maize and wheat crop growing however, they remain influenced by their traditional knowledge system which they have relied upon over the years. This arrangement becomes very challenging at present

Statement	All the time	Most of the time	Sometimes	Not at all
Often belief in seasonal climate and weather information	68 (17.3)	80 (20.4)	219 (55.7)	26 (6.6)
Often rely on experience/ indigenous knowledge in maize production	89 (22.7)	148 (37.8)	141 (36)	14 (3.6)
Often rely on experience/ indigenous knowledge in wheat production	85 (22.8)	163 (43.7)	111 (29.8)	14 (3.6)

Table 1.
Belief in use of climate and weather information or indigenous knowledge among farmers in Uasin Gishu County.

with the prevailing changing climate. There is need to balance access to the two sets of information by farmers as stated by farmers in this study. The findings in the study agree with those by [13] in a paper on community based adaptation in climate change which reveal that farmers in Burkina Faso traditionally rely on observation of environmental indicators to predict climate patterns, but they lost confidence in their ability to predict rainfall given the increased climate variability and increasingly seek to incorporate meteorological forecasts/climate information.

To understand further the usage of climate and weather information and the influence of farmer's indigenous knowledge systems in maize and wheat production over time, indicators used in rainfall prediction were examined together with usage of climate and weather information as shown in **Table 2**. The results portray existence of a significant relationship between indigenous knowledge system indicators for rainfall prediction and the use of climate and weather information in maize and wheat production activity as shown in **Table 2**.

A higher proportion of the farmers relying on their indigenous knowledge system indicators to predict rainfall hence making decision on farming activities do not use climate and weather information as indicated in **Table 2**. A greater negative influence is created on utilization of climate information by farmers if they are left alone to decide whether to embrace climate information or stick to their indigenous knowledge systems in their farming decisions. Farmer's indigenous knowledge indicators and experience gained over time practising maize and wheat growing influence a lot their ability to use climate information. There is need therefore for the County Directorates of Agriculture in Kenya to undertake greater sensitization

Indigenous Knowledge indicators	Climate & Weather Information Use		Chi	P
	Yes	No		
Lightning flashes around lake region/Tindiret area in Nandi	110 (44.7%)	136 (55.3%)	0.476	0.490
Cloud cover and intense sunshine	70 (39.1%)	109 (60.9%)	6.391	0.011
Wind direction	83 (52.2%)	76 (47.8%)	4.059	0.046
Very warm nights	70 (35.7%)	126 (64.3%)	16.832	<0.001
Difficulty in prediction recently	23 (53.5%)	20 (46.5%)	1.074	0.300

Table 2.
Relationship between climate information and indigenous knowledge indicators usage in rainfall prediction in Uasin Gishu County, Kenya.

meetings during farmers village meetings, agricultural shows and farmer's field day outlets to educate farmers on the need to consider using climate information in addition to their great local knowledge and experience learned over time in farming decisions as this will cushion them against extreme weather variations that may lead to crop losses hence impacting negatively on their livelihoods.

The traditional weather forecasters form part of the decision making process sometimes when farmers experience delayed onset of rainfall when they look at their farming calendar. Farmers listen to what known traditional weather forecasters can deduct from reading the arrangement of stars in the solar system and the "positioning of male and female star". During the key informant interview with the traditional elder shown in **Figure 6**, he predicted that *"there was going to be plenty of rain in December 2013 and that rains would continue to March 2014. He advised the farmers to grow short duration crops like beans that take 3 months to mature"*.

Figure 6.
Explanation of weather phenomena by a traditional weather forecaster in Moiben Sub County of Uasin Gishu, Kenya.

From the diverse responses by the farmers, it is evident that indigenous knowledge system plays a crucial role in farmer's decision making process; ultimately, it has a significant impact on their activities. With changing climate being real in the region, farmers relying on "known signs" or meteorological indicators derived in their traditional knowledge systems are at a greater risk of losing out on benefiting from prevailing positive conditions or avoiding bad weather leading to crop loss. Farmers and key stakeholders in the agricultural sector need to incorporate climate and weather information or weather forecasts in their decision making process. Indigenous knowledge weather prediction methodologies are now facing serious challenges related to environmental degradation and interference of the natural ecosystem balance by man. According to the Director of Meteorology in Uasin Gishu County and some traditional weather forecasters interviewed, most indigenous trees have disappeared completely and are being replaced by exotic trees which are alien to indigenous weather predictors. They have thin leaves thus their ability to sequestrate carbon in the atmosphere is reduced hence does not help much in arresting greenhouse gasses. Burning of farms has not only destroyed micronutrients and shrubs but also destroy insects and their migratory paths which traditional weather forecasters use to predict near onset of rains.

Some bird species have migrated elsewhere where they can still find a natural habitat hence rendering traditional weather forecasters using birds to predict onset of rains helpless. It is however important to integrate both scientific and traditional knowledge system as the scientific forecast diverges from traditional farmer's prediction in scale and to some extent on predictors [30]. In any case, some of the principles of the predictors like wind flow, temperature changes converge with the scientific forecast. Farmers have been using combination of various biological, meteorological and astronomical indicators to predict the rainfall. While the scientific forecast are developed using the predictors such as wind, sea surface temperatures and others which are primarily meteorological indicators.

4.5 Value attached to utilization of climate and weather information

If weather forecasts/weather and climate information were to be integrated with indigenous knowledge system for synergy prediction among farmers, it is critical to understanding how such farmers perceive as value attached to utilization of such climate and weather information in relation to their maize and wheat growing. Farmers were asked to rank the value they attached to climate and weather information use in their farming enterprise and their responses were captured either as very important, important, unsure, and not important. As indicated in **Figure** 7, more than half of the farmers view climate and weather information as very important in both maize and wheat growing 215 (54.8%) and 206 (54.4%) respectively. Similarly, a large number of farmers still considered the same information as important 143 (36.7%) for maize growing and 136 (36.1%) for wheat growing.

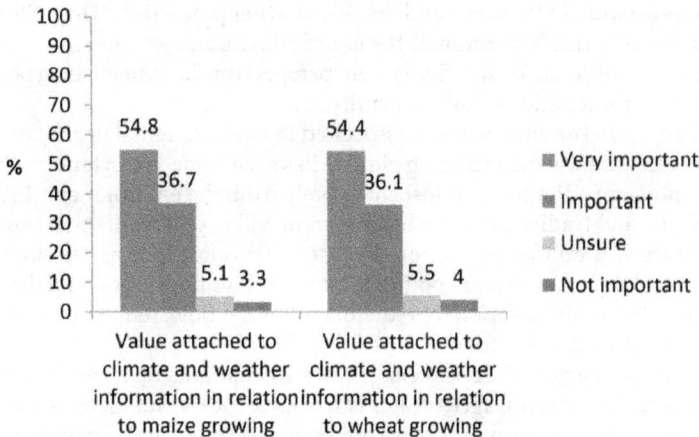

Figure 7.
Value attached to utilization of climate and weather information in Uasin Gishu County, Kenya.

As seen earlier farmers in this study have agreed that climate and weather information is important but whether this translates into action or use remains to be tested. A mere belief in some variable (climate and weather information) does not necessarily translate into adoption of the same especially where traditional knowledge systems influence decisions as seen elsewhere in the results of this study. This calls for greater efforts to educate farmers to be able to embrace weather forecasts as well in their farming decisions as mere belief in it would not alter losses they will incur especially with the prevailing climate variability which has distorted order of activities and seasons known to farmers in maize and wheat growing areas of Kenya. It is important to emphasize that climate information or weather forecasts is key to understanding climate as a major influence on lives, livelihoods, resources, ecosystems and development. According to [39], weather forecasts supports decision making on which option to invest in, when and how much to invest. Flexible and proactive planning enabled by climate information helps vulnerable communities especially the farming communities to cope with the risks inherent. It provides a way of analyzing the nature and scale of impacts due to past and current climate and the potential future impacts as the climate continues to vary and change. In the end, actors can then make informed and appropriate decisions and plans to deal with climate-related impacts through adaptation, risk reduction and development actions.

5. Conclusions

Farmers in Kenya and some countries in eastern and southern Africa who depend on rain-fed cultivation have developed complex cultural models of weather prediction and are able to cite local indicators of seasonal weather. From this study majority of farmers prefer indigenous forecasting knowledge because they have a belief that indigenous information is more practical as it has been tested, tried and trusted over time.

The dependency on traditional indicators alone without much uptake of available weather forecasts can and has predisposed farmers to crop losses as they try to navigate the risks of changing climate. Although farmers do trust climate and weather information, its uptake will require more sensitization and demonstration if farmers are to navigate risks associated to the changing climate in the farming enterprise.

There is need to integrate both traditional knowledge systems and weather forecast information for synergy and reliable weather prediction. This will act as a motivator to farmers to embrace the use of climate and weather information because their traditional knowledge system perspectives have been incorporated hence creating ownership and sustainability.

The changes on the environment associated to environmental degradation; ecosystem disturbance and changing climate has seen some important traditional predictor indicators disappear or lost completely from the environment. Integrating both scientific and traditional knowledge system will supplement the loss on traditional indicators of rainfall prediction that used to support farmer decisions hence mitigating against losses that are bound to occur. It is important to note that some of the principles of the traditional predictors like wind flow, temperature changes converge with the scientific forecasts.

A national policy guideline on a closer working relationship should be established between the farmers, agricultural extension officers and meteorological scientists including traditional weather predictors. A strong link, including feedback loops between scientists, advisory agents and farmers will help in communicating downscaled climate information and facilitating access by local farmer communities. Format, delivery mode and timing of the information is key. All key stakeholders in agriculture, livelihoods and climate change sectors during their participatory weather scenario planning action should be aware of farmers needs whether they are real or perceived because farmers do know what information they need at particular point in time at their local farm level.

5.1 Recommendation

This study suggests that there is need to integrate both traditional knowledge systems and weather forecasts at local farmer levels for synergy forecasting that will generate reliable climate information useful to farmers. Incorporating farmers perspectives by slightly modifying and using them to meet current needs and situations will help address the needs of farmers and may be a motivator in embracing the use of climate and weather information as their traditional knowledge system perspectives have been incorporated. Climate information services must be embedded in local, national and regional processes to enable scaled-up support for widespread adaptation activities.

Acknowledgements

Gratitude to University of Nairobi and IDRC Collaborative project on Innovative Application of ICTs in Addressing Water-related Impacts of Climate Change (ICTWCC) in The School of Computing and Informatics and the Department of Land Resource Management and Agricultural Technology (LARMAT) for providing the much needed research grants led by Prof. Timothy Waema and team members Dr. laban Mc'Opiyo and Anuradha Khodha. Prof. Grace Cheserek, Professor Gilbert Nduru and Dr. Victor Odenyo for their valuable consultation.

Conflict of interest

The Author declare that there is no conflict of interest to state. I certify that the submission is original work and is not under review at any other publication.

Thanks

Thank you Lord for granting me abundant life and good health in all times. Thanks to Professor Vincent Sudoi and University of Eldoret for the space and moral support. Kiprop Murgor, Chemutai Murgor and Brian Murgor for simply their loyalty.

Author details

Daniel Kipkosgei Murgor
University of Eldoret, Uasin Gishu County, Kenya

*Address all correspondence to: danmurgor2@gmail.com

IntechOpen

References

[1] WMO. State of the Climate in Africa [Internet]. 2019, 37p. Available at https://library.wmo.int/doc_num. php?explnum_id=10421 [Accessed 2021-01-20]

[2] Pachauri R K, Meyer L. A. editors. IPCC Climate Change 2014: Synthesis Report. Contributions of working groups I, II and III to the fifth Assessment Report of the Intergovernmental Panel on Climate Change, IPCC, Geneva, Switzerland [Internet] 2014. Available at http://www.ipcc.ch/site/assets/ uploads/2018/02/SYR_ARS_FINAL_ full.pdf [Accessed 2021-01-12]

[3] UNFCC. Climate Change Is an Increasing Threat to Africa [Internet]. 2020. Available at https://unfccc.int/ news/climate-change-is-an-increasing- threat-to-africa#:~:text=Rising%20 temperatures,than%20global%20 mean%20surface%20temperature [Accessed 2021-01-14]

[4] WMO. State of the Global Climate 2020-Provisional report [Internet]. 2020. Available at https://library.wmo. int/doc_num.php?explnum_id=10444 [Accessed 2021-01-14]

[5] WMO. State of the Global Climate 2020-Provisional report [Internet]. 2020. Available at https://library.wmo. int/doc_num.php?explnum_id=10444 [Accessed 2021-01-14]

[6] FAO Ethiopia. Desert Locusts drive one million to food insecurity [Internet]. 2020. Available at http:// www.fao.org/ethiopia/news/detail- events/en/c/1270924 [Accessed 2021-01-14]

[7] UNHCR. UNHCR Somalia Factsheet, 1-30 September [Internet]. 2020. Available at (https://data2.unhcr.org/en/ documents/details/83089) [Accessed 2021-01-10]

[8] Regional Centre for Mapping of Resources for Development (RCMRD). The Rising Water Level and Expansion of the Rift Valley Lakes from Space [Internet]. 2020. Available at https:// www.rcmrd.org/the-rising-water-level- and-expansion-of-the-rift-valley-lakes- from-space [Accessed 2021-01-10]

[9] Pereira L. Climate Change Impacts on Agriculture across Africa, USA; Oxford University: 2018, DOI: 10.1093/ acrefore/9780199389414.013.292

[10] UNFCC. Climate Change Is an Increasing Threat to Africa [Internet]. 2020. Available at https://unfccc.int/ news/climate-change-is-an-increasing- threat-to-africa#:~:text=Rising%20 temperatures,than%20global%20 mean%20surface%20temperature [Accessed 2021-01-14]

[11] Boef W, Amanor de Kojo, Wellard K, Bebbington A. Cultivating Knowledge; Genetic Diversity, farmer experimentation and crop research. London: Intermediate Technology Publications. [Internet]. 1993. Available at https://wordpress.clarku.edu/ abebbington/1993/book/cultivating- knowledge-genetic-diversity-farmer- experimentation-and-crop-research/ [Accessed 2018-01-30]

[12] Mafongoya P. L, Ajayi O C. editors, (2017). Indigenous Knowledge Systems and Climate Change Management in Africa. Wageningen, the Netherlands: CTA, 2017. 316pp [Internet]. Available at https://publications.cta.int/media/ publications/downloads/2009_PDF.pdf [Accessed 2020-12-15]

[13] Bryan E, Behrman J. Community– based adaptation to climate change: A Theoretical framework; overview of key issues and discussion of gender differentiated priorities and participation. Washington: International Food Policy Research Institute. CAPRI

Working Paper No. 109. 2013 [Internet] Available at http://dx.doi.org/10.2499/CAPRiWP109 [Accessed 2019-11-20]

[14] Gilles J. *[et al]*. Laggards or leaders: Conservers of traditional agricultural Knowledge in Bolivia. Rural Sociology: 78, Issue 1, pp. 51-74, 2013[Internet]. Available at http://onlinelibrary.wiley.com/doi/10.1111/ruso.12001/abstract [Accessed 2019 -09-20]

[15] Ziervogel G. Global science, local problems: Seasonal climate forecast use in Basotho village, Southern Africa. *Prepared for presentation at the open meeting of the Global Environmental Change Research Community,* Rio de Janeiro, October 6-8, [Internet], 2001. Available at http://cedac.ciesin.columbia.edu [Accessed 2012-01-12]

[16] Eilts M. The Role of Weather and Weather Forecasting in Agriculture [Internet]. 2018. Available at https://www.dtn.com/the-role-of-weather-and-weather-forecasting-in-agriculture/ [Accessed 2021-01-18]

[17] FAO. Handbook on climate information for farming communities: what farmers need and what is available FAO, ROME [Internet]. 2019. Available at Http://www.fao.org/3/ca4059en/ca4059en.pdf?Eloutlink=imf2fao [Accessed 2019-01-18]

[18] Kadi M, Njau L. N, Mwikya J, Kamga A. The State of Climate Information Services for Agriculture and Food Security in East African Countries. *CCAFS Working Paper No. 5.* Copenhagen, Denmark: CGIAR Research Program on Climate Change, Agriculture and Food Security (CCAFS) pp 1-49 [Internet]. 2011. Available at www.ccafs.cgiar.org [Accessed 2019-12-05]

[19] Vogel C, O'Brien K. Who can eat information? Examining the effectiveness of seasonal climate forecasts and regional climate–risk management strategies. Climate Research 33, *2006: p*111-122.

[20] Hansen J. W, Baethgen W, Osgood D, Ceccato P, Ngugi R K. Innovations in climate risk management: protecting and building rural livelihoods in a variable and changing climate. 2007; J. Semi-Arid Trop. Agric. Res. 4, 1-38. doi: 10.7916/D8ZW1S4C

[21] Anuforo A C. Nimet's agro-climate information services: A Vital tool for managing food crisis in Nigeria. A Paper presented by Dr. Anthony C. Anuforo; DG/CEO (NIMET) at the 2009 World Food Day Symposium: "*Achieving Food Security in Times of Crisis. Abuja, Nigeria.* 2009. Available at http://www.nignationalagricshow.biz/applications/Vital_Tool_For_Managing_Food_Crisis_in_Nigeria.pdf [Accessed 2012-11-05]

[22] Gunasekera D. Use of climate information for socio-economic benefits. *Abstract Presented at the World Climate Conference-3,* Geneva, *p*1-3. 2009 Available at http://www.wmo.int/wcc3/sessionsdb/documents/ [Accessed 2013-04-21]

[23] Walker S, Mukhala W J, Van Den Berg, Manley C. Assessment of Communication and use of climate outlooks and development of scenarios to promote food security in the Free State province of South Africa: Final Report submitted to the Drought Monitoring Centre (Harare, Zimbabwe) on the pilot project DMCH/WB/AAA 6.02/13/1. 2001. p. 4-6. Available at http://www.wmo.int/pages/prog/wcp/wcasp/clips/modules/documents/mod5_assessment_%20commun.pdf [Accessed 2014-01-06]

[24] Blench R. Seasonal climate forecasting: Who can use it and how should it be disseminated? World Bank funded workshop titled "*Users responses to seasonal climate forecasts in Southern Africa: what have we learned convened*

in Dar es Salaam, Tanzania. London: Overseas Development Institute, Natural Resource perspectives No. 47. 1999. Available at http://www.odi. org.uk/resources/download/2107.pdf [Accessed 2014-02-20]

[25] O'Brien K, Sygna L, Naess L, Kingamkono R, Hochobeb B. Is information enough? User responses to seasonal climate forecasts in southern Africa: Report to the World Bank, AFTE1- ENVGC. *Adaptation to climate change and variability in sub-Saharan Africa, Phase II, No. 3.* 2000. Oslo; CICERO Centre for International Climate and Environmental Research. Available at http://brage.bibsys.no/xmlui/ bitstream/handle/11250/191938/-1/ CICERO_Report_2000-03.pdf [Retrieved 2014-04-30]

[26] Mahoo H, Mbungu W, Yonah I, Recha J, Radeny M, Kimeli P, Kinyangi J. Integrating Indigenous Knowledge with Scientific Seasonal Forecasts for Climate Risk Management in Lushoto District in Tanzania. CCAFS Working Paper no. 103. CGIAR Research Program on Climate Change, Agriculture and Food Security (CCAFS) [Internet]. Available at www. ccafs.cgiar.org [Accessed 2021-01-19]

[27] Lucio F. Use of contemporary and indigenous forecast information for farm level decision making in Mozambique. Consultancy report. UNDP/UNSO p. 72. 1999.

[28] Ngugi R. Use of Indigenous and contemporary knowledge on climate and drought forecasting information in Mwingi district, Kenya. Consultancy report. UNDP/UNSO p. 28. 1999.

[29] Radeny M, Desalegn A, Mubiru D. et al. Indigenous knowledge for seasonal weather and climate forecasting across East Africa. Climatic Change 156, 509-526. 2019. Available at https:// doi.org/10.1007/s10584-019-02476-9 [Accessed 2021-01-10]

[30] Ziervogel G, Opere A. (editors). Integrating meteorological and indigenous knowledge-based seasonal climate forecasts in the agricultural sector: Climate Change Adaptation in Africa learning paper series Ottawa, Canada: International Development Research Centre; 2010, 24p. Available at https://www.idrc.ca/sites/default/files/ sp/Documents%20EN/CCAA-seasonal- forecasting.pdf [Accessed 2021-01-20]

[31] Zuma-Netshiukhwi G, Stigter K, Walker S. Use of Traditional Weather/ Climate Knowledge by Farmers in the South-Western Free State of South Africa: Agrometeorological Learning by Scientists. Atmosphere 2013, 4, 383-410. https://doi.org/10.3390/atmos4040383 [Accessed 2021-01-04]

[32] Esipisu I. Nganyi: The tiny forest in Kenya that predicts the weather [Internet] 2016. Available at https:// news.mongabay.com/2016/02/ nganyi-the-tiny-forest-in-kenya-that- can-predict-the-weather/ [Accessed 2021-01-20]

[33] Kipkorir E C, Mugalavai E M, Songok C K. Integrating Indigenous and Scientific Knowledge Systems on Seasonal Rainfall Characteristics Prediction and Utilization Kenya Science, Technology and Innovation Journal: ISSN 2079-5440, 2013. 35p. Available at https://www.researchgate. net/publication/236289850_ Integrating_Indigenous_and_Scientific_ Knowledge_Systems_on_Seasonal_ Rainfall_Characteristics_Prediction_ and_Utilization [Accessed 2012-12-16]

[34] Government of Kenya. Uasin Gishu County profile 2013. Ministry of Agriculture, (2013). Uasin Gishu County Directorate of Agriculture, Eldoret.

[35] Nyakaana J B. Kenya development center policy: the case of Eldoret. An assessment of its implementation and impact [Thesis], Amsterdam University, Department of Human Geography,

Faculty of Environmental Sciences, Amsterdam, 1996.

[36] Government of Kenya. Assessment of Long rains rapid food security in Uasin Gishu District 22nd – 25th July, 2008. Kenya Food Security Steering Group (KFSSG), Nairobi, Kenya. Available at http://www. kenyafoodsecurity.org/longrains08/ district_reports/uasin_gishu.pdf [Accessed 2010-10-30]

[37] Chagonda, Mugabe F T, Munodawafa A, Mubaya C P, Masere P, Murewi C. Engaging smallholder farmers with seasonal climate forecasts for sustainable crop production in semi-arid areas of Zimbabwe. African Journal of Agricultural Research. 2015; Vol. 10(7), 668-676. DOI: 10.5897/AJAR2014. 8509 [Accessed 2021-01-16]

[38] Desalegn, A., Maren, R. & Mungai, C. Indigenous knowledge in weather forecasting: Lessons to build climate resilience in East Africa [Internet] 2017. Available at https://ccafs.cgiar.org/ news/indigenous-knowledge-weather- forecasting-lessons-build-climate- resilience-east-africa-0#.W-qS7B9oTIU [Accessed 2021-01-15]

[39] Ambani M, Percy P, White E, Ward N. Facing uncertainty: the value of climate information for adaptation, risk reduction and resilience in Africa. [Internet] Care International, 2014. Available at https://careclimatechange. org/wp-content/uploads/2014/08/C_ Comms_Brief.pdf [Accessed 2021-01-27]

Chapter 11

Sustainable Carbon Management Practices (CMP) - A Way Forward in Reducing CO$_2$ Flux

Biswabara Sahu, Snigdha Chatterjee and Ruby Patel

Abstract

Asian agriculture sector contributes about 44% of greenhouse gas (GHG) emission. Predominantly paddy rice cultivation couples with indiscriminate use of agrochemicals, burning of fossil fuels in farm machinery majorly causes GHG emissions from farmlands in Asia. Presently, Asian soils have 25% cropland soil organic carbon (SOC) content but with moderately to highly vulnerability towards land degradation. To make up the soil carbon losses which has occurred due to continuous cultivation and tillage, it is recommended to adopt suitable carbon management practices to sequester carbon in soil through their physio-chemical protection. Conservation agriculture (CA), cover crop, crop diversification, integrated nutrient management (INM) and balanced fertilisation promotes better soil structure formation, stabilisation of aggregate associated carbon, microbial polymerisation of organic matter as well as a better root architecture. Carbon management practices not only improve soil fertility but also supports improved grain and straw yield. More the yield more biomass addition occurs to the soil. Soil carbon sequestration may not be the only panacea of climate change related issues, but is certainly a way forward to enriched soil fertility, improved agronomic production as well as adaptive- mitigation for offsetting anthropogenic GHG emission.

Keywords: GHG, Paddy rice, CA, INM, Microbial polymerisation, Adaptive mitigation

1. Introduction

The ever-increasing population growth of the world has resulted in putting more and more pressure on a piece of arable land demanding higher and higher production. The world statistics shows reduction of per capita arable land from 0.23 ha in 2000 to 0.19 ha in 2015. While the per capita arable land in North America is still 0.55 ha, the numbers for South Asia and East Asia- Pacific are 0.12 ha and 0.11 ha respectively (5–6 times lower than that of North America) [1]. The shrinkage of arable land compels the farmers to go for over dose of fertiliser application which is a main source of many kind of pollutions and emission. Food sector contributes to around quarters (26%) of the global greenhouse gas (GHG) emission out of which solely crop production practices cause **27% emission share of food sector.** The fields associated with food sector in Asian countries are also under threat as the current situation of vulnerability and their less reliance to changes are affecting their ecosystem function and services.

Sub-region	Food and fibre	Biodiversity	Land degradation
North Asia	Moderately resilient	Highly vulnerable	Moderately vulnerable
Central and West Asia	Highly vulnerable	Moderately vulnerable	Highly vulnerable
Tibetan Plateau	Moderately resilient	Highly vulnerable	Moderately vulnerable
East Asia	Highly vulnerable	Highly vulnerable	Highly vulnerable
South Asia	Highly vulnerable	Highly vulnerable	Highly vulnerable
South east Asia	Highly vulnerable	Highly vulnerable	Highly vulnerable

Table 1.
Sectoral vulnerability in food, land and biodegradation in sub-continents of Asia.

The food, biodiversity and land degradation condition in every sub-section of Asia are moderately to highly vulnerable and less resilient [2] which has a gradual effect on climate change. Before we get caught in catastrophic climate change impact, required management practices are to be adopted (**Table 1**).

2. Scene of Asian agriculture in GHG emission and carbon storage potential

According to IPCC (Intergovernmental Panel on Climate Change) 2014 [3] record, the scenario of GHG emission is very critical in Asia as Asian agriculture causes an average of 44% of global agricultural GHG emission (**Table 2**).

The agricultural GHG emission contributors such as enteric fermentation and paddy rice cultivation are the major source of methane emission whereas the major sources of nitrous oxide emission are application of manures and fertilisers. The worldwide contribution of paddy rice cultivation towards GHG emission (CH4) is 11%. For higher crop production farmers rely on synthetic fertiliser application which is a rapidly growing source of emission having the increase rate of around 37% since 2001 [5]. Along with that the use of large number of machineries are the source of CO_2 emission due to burning of fossil fuel. The imbalanced fertilisation is another reason for the release of soil carbon to the atmosphere (**Figure 1**).

To meet the daily food requirements, the agricultural stakeholders must make two kind of assessments in order to understand the impact of climate change on food and crop production i.e., mitigation and adaptation. Mitigation will reduce the emission of GHG from agricultural sources whereas adaptation will enable the agricultural sectors to perform well in the existing climate change situation through modified management and production systems. Both the approaches can be regulated through various policies e.g., ensuring the economic value of carbon and its sequestration will be an important development in the agriculture sector [7]. The adaptive-mitigation techniques to capture carbon in soil in organic form is a potential factor for controlling CO_2 emission as well as a factor for improving soil quality and health.

Carbon pool	Carbon changes		Rate of carbon increase in the atmosphere
	Fossil fuel use	Land Use	
750 Gt	+ 5.5 Gt yr.$^{-1}$	+1.6 Gt yr.$^{-1}$	+3.3 Gt yr.$^{-1}$

Table 2.
Carbon pool size and changes due to human activities [4].

PERCENTAGE CONTRIBUTION FROM AGRICULTURE
SECTORS IN ASIA.

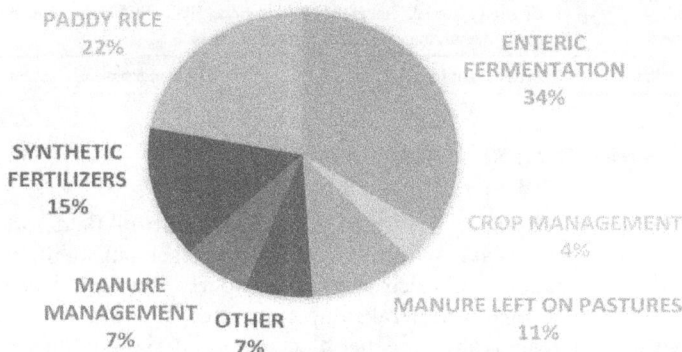

PADDY RICE
22%

ENTERIC
FERMENTATION
34%

SYNTHETIC
FERTILIZERS
15%

CROP MANAGEMENT
4%

MANURE
MANAGEMENT   OTHER
7%   7%

MANURE LEFT ON PASTURES
11%

Figure 1.
Contribution of various agricultural sectors towards GHG emission in Asia. See [6].

Carbon storage in terrestrial system is important as soil can hold three times more carbon than vegetations that they support. The Soil carbon pool which is the largest reactive carbon in terrestrial ecosystem [8], is estimated to be 2500 Pg (10^{15}) up to 1 mt depth, of which soil organic carbon is about 1500 Pg. This stock accounts for about 3.2 times the size of atmospheric carbon pool and 4 times that of biotic pool [6, 9]. Thus, capturing the carbon from agricultural lands in stable form can reduce CO_2 content of the atmosphere.

Again, the global distribution of carbon and its storage potential is highly influenced by climatic conditions such as temperature and precipitation [10]. The higher decomposition rate controlled by higher oxidation of organic matter result in lower Soil Organic Carbon (SOC) in the tropics as compared to higher SOC of cooler regions. Though all the parts of Asian croplands contain moderate amount of carbon, and all together they account for about 25% of global cropland carbon [11]. But the regions of South Asia with low level of SOC and with serious degradation problems are global highest in carbon storage per hectare basis (0.62–1.28 t C/ha/ yr) over 2.9 million km^2 of land which all together turns out to be 2.2 to 4.5 Pg C storage/yr. in South Asia [11]. Thus, the management practices which are proved to be potential drivers of SOC enrichment must be encouraged as mitigative measure in agricultural soils.

3. Understanding role of SOC and carbon management practices

Soil Organic carbon (SOC) is the controlling factor for soils physical, chemical, biological and ecological functionality and wellbeing. Not only soil's health and productive capacity but soil carbon can also mitigate hazardous climate change. Quality and quantity of SOC; its dynamics/turnover is the main governing factor of soils ecosystem functions. A huge loss (50 to 75% and with magnitude of loss of around 30 to 60 Mg C/ha) of antecedent soil C pool has occurred due to land conversion, cultivation and erosion associated with it in most agricultural ecosystems [12]. Generally, agricultural soils contain considerably less SOC than soils under natural vegetation, hence, these lands are deprived of C than their ecological potential.

Carbon management practices (CMP) aim to sequester i.e., to capture and secure storage of carbon that would otherwise be emitted to, or remain, in the atmosphere. In other words, CMP is enhancing and/or maintaining soil carbon

Attributes	Mechanism
Physical stability	Depth distribution, Aggregate stabilisation, Organic macromolecules.
Chemical recalcitrance	Charred materials, Interaction with cations, Hydrophobicity, Complexation with clay minerals, intermolecular interaction.
Biotic mechanisms	Recalcitrant fractions, Structural composition, Condensation reaction.

Table 3.
Mechanism of increasing MRT of SOC for its stabilisation.

not allowing it to escape out to the atmosphere. In agricultural fields, addition of biomass carbon and organic manure is a direct approach but stabilisation of the soil carbon is through its physico-chemical property. Physical mechanism includes formation of organo-mineral complexes, encapsulation in microaggregates within macroaggregates, deeper placement of carbon in the soil profile away from natural and anthropogenic perturbation zone [12]. At the same time, the producer must seek for those practices which will promote sequestration of SOC in croplands without compromising the provision of ecosystem services such as food, fodder, fibre or other agricultural products. Thus, it is very crucial to understand the mechanism of carbon stabilisation by improving the mean residence time (MRT) and by offsetting anthropogenic emissions [13] which is vary according to the climatic condition and soil properties and also on existing soil carbon content of the particular region. For example, the same management practice which are proved to increase SOC can result in high amount of loss and unintended consequences in those soils which are already saturated with organic carbon [14] (**Table 3**).

4. Carbon management practices (CMP)

A 4% increase in global agricultural soil carbon pool up to 1 m depth, 2–3 Gt C can be sequestered annually which would drawdown global anthropogenic GHG emission by 20–35% [15] but practicality has many constraints. For example, in countries with low (inherent) SOC like India, high rate of decomposition due to high temperature and the removal of crop residues does not allow this concept to work well [16]. Due to a greater surface area and charge density, organic matter can react with soil particles to form organo-mineral complexes. The mean residence time of carbon fractions are functions of their turnover rate which is dependent on the degree of protection within soil matrix [17]. Chemical protection involves formation of some recalcitrant compounds [18] like non- acid hydrolysable carbon fraction, aromatic compounds, double chained hydrocarbons and hydrophobic compounds which are not easily decomposed by microorganisms.

Change in soil carbon is a balance sheet of carbon input and output through mineralisation, loss, other emissions etc. [10]. So, the key for sequestering SOC is increasing carbon inputs and reducing carbon outputs. Cropping system biomass productivity has primary control over this carbon input through proper fertiliser, land, water management practices based on exiting soil and climatic condition. Integrated and balanced fertiliser application positively affect both above ground and below ground biomass and crop productivity. This adds more amount of organic matter to the soil directly in the form of straw returns, roots, exudates and organic manures directly. The organic carbon present in soil is very much prone to oxidation if neither biochemically protected (depends on its composition) not physically protected (in soil aggregates). So, researches focus on those practices which are helpful to protect pre-existing soil aggregates and/or to promote the genesis of new soil aggregates or to achieve both objectives of CMP.

Important carbon management practices are:

1. conservation agriculture (CA),

2. Cover crop

3. Crop rotation and diversification

4. Integrated and balanced nutrition (use of organic amendments viz. Crop residue, FYM, Compost, Biochar)

4.1 Conservation Agriculture (CA)

This is the technology of a set of management practices which aims at conserving the natural resources and biodiversity in the crop land and are characterised by the three principles e.g., i) No/minimum soil disturbance, ii) permanent organic cover or cover crops, and iii) crop diversification. Each principle individually and combinedly contribute towards carbon enrichment in soil. Build-up of carbon in soil can be successful through increased input, reduced decomposition and loss or both. Cultivation of previously uncultivated land can lead to 20%–40% loss in the native carbon in the initial years following initial cultivation [19]. Restoring that carbon in soil through addition and protection can be a potential carbon management practice. Every input like fertiliser, pesticide and irrigation has a carry a 'hidden carbon cost, thus optimising their quantity in a crop management practice should be estimated in the carbon balance sheet [20]. Historically, excessive cultivation operations like tillage can expose SOC for decomposition by microbes which further may cause many land degradation problems such as erosion and soil structural decline. Enhanced soil disturbance triggers carbon losses from soil system via increased decomposition and erosion of SOM. All these ultimately adds to the atmosphere as CO_2 fluxes or to the water resource [21]. Soil carbon levels of agricultural soils are lower than corresponding soils under natural vegetation or fallow that indicates the potential for soil carbon storage. In agricultural systems, soil carbon levels tend to be variable and dependent on management practices. Reducing soil disturbance can reduce rate of oxidation of organic matter and provide protection to the microbial habitat. Rate of decomposition can also be reduced by introducing slowly decomposing residues in the rotation. Intensifying crop rotation, legumes and green manure crops in crop cycle, elimination of fallow period, cover crop and residue mulch enhances soil carbon input in the form of both above ground and below ground biomass. The principles of conservation agriculture rotate around the concept of biomass addition and its protection through less soil disturbance. Soil C level and its composition under no-tillage and stubble retention (SOC = 2.5%) was more than the same soil under 3 pass tillage and stubble burning (SOC = 1.5%) after 19 years [4]. Reduced tillage increases the potential of soil c sequestration over conventional tillage practices as described in **Figure 2**. The concept of achieving steady state carbon status in cultivated soil through maximisation of organic input (residues, root biomass, organic amendments) is depicted in **Figure 3**. Conservation agriculture technology can be a potential method for conserving soil moisture, supplying plant nutrient and mitigate pathogen, peat and weed infestation there by cutting off fertiliser, pesticide requirement. Every input like fertiliser, pesticide and irrigation has a carry a 'hidden carbon cost, thus optimising their quantity in a crop management practice should be estimated in the carbon balance sheet [23].

A study conducted by [20] Sapkota et al. (2015) in the Indo-Gangetic region showed that conventional rice- wheat cropping system has 27% higher GHG emission (in terms of CO_2 equivalence) as compared to zero tilled rice- wheat crop rotation with residue mulching [23]. Sapkota et al. (2014) found the carbon dioxide efflux so also the global warming potential of wheat (through life cycle analysis) for its unit production under conventional tillage based practice is 10 times higher than no till-age based production. Introduction of legume in crop rotation and residue addition to the soil help reducing fertiliser requirement and energy need in arable systems. Considering the fact that, the annual global fertiliser leads to an annual release of 300 Tg of CO_2 into the atmosphere during fertiliser manufacturing process [24], any management practice that will reduce the chemical fertiliser requirement with optimised output is highly environment friendly. They also explained that the release of every 2.6–3.7 kg CO_2 per every 1 Kg of synthesised N, is produced from fossil fuel thus causing a net contribution to atmospheric amount of CO_2 [24].

While the carbon sequestration in soil will occur at a certain point of time (until saturation) depending upon the soil type, reduction in emission owing to less energy requirement, fossil fuel consumption and machinery use will continue until the practice is carried out [25]. Zero tillage cuts the fuel consumption for land preparation so also CO_2 emission. (Erenstein and Laxmi 2008) [26] found that adoption of ZT in wheat- maize system of the IGP could save an average of 36 L diesel ha–1 which is equivalent to a reduction in 93 kg CO_2 emission ha^{-1} yr.$^{-1}$ Sapkota et al. (2015) [20].

4.1.1 Mechanism of soil carbon sequestration in CA system

The carbon stock–enhancing effect of SOC management practice of conservation is possible due to reduced disturbance which is the prime factor in maintaining

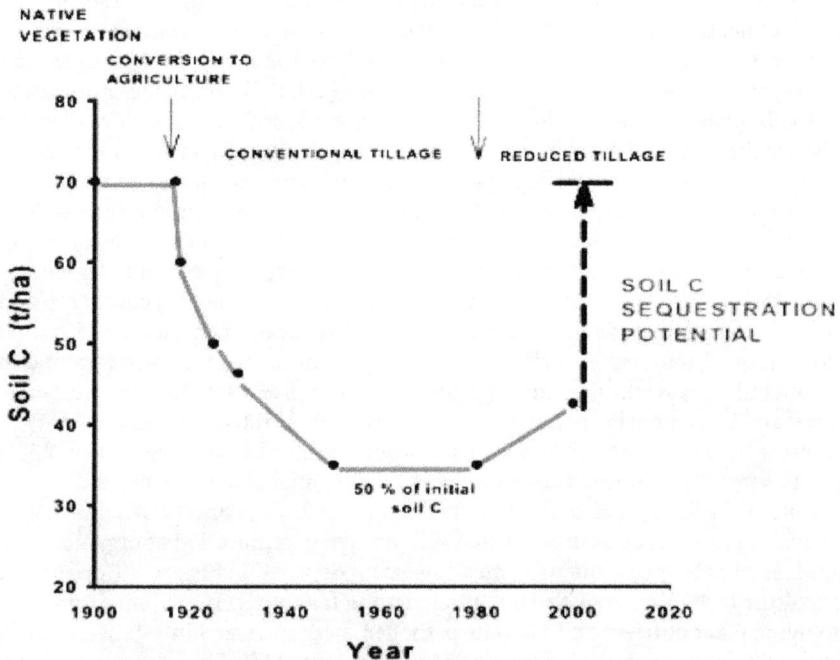

Figure 2.
Changes in SOC content in cultivated soil as a result of tillage pattern over years [4].

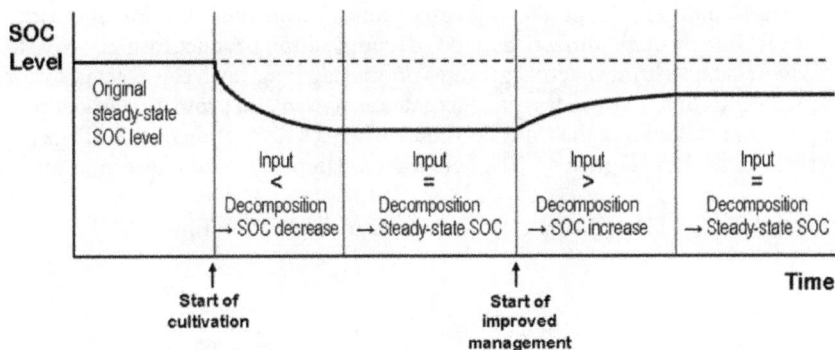

Figure 3.
Mechanism of achieving steady state SOC through input addition. Adopted from [22].

soils physical stability. This physical wellness of a soil system has positive effect on microbial habitat, their activities and the natural ecosystem functions of soil like nutrient cycling, buffering capacity, cation exchange etc.

The first principle is no tillage which is growing crops in soil without causing soil disturbance except for sowing or reduced tillage that is significant reduction of soil disturbance through less frequent passes of tillage, tillage in specific portion of the field which is in form of strip or ridge and shallower depth of tillage. Second principle aims at keeping a permanent organic cover on the soil surface in the form of residue mulch, growing cover crops both of which addresses many aspects of soil protection in the form of hindrance towards water, wind erosion, improved soil aggregation, enrichment of substate for microbial growth and functionality and many other chemical properties such as nitrogen fixation, carbon sequestration, etc. the third principle i.e. crop diversification is an essential tool for promoting better soil health as it has a role in allowing nutrient uptake of differently rooted crops from different depths, promoting microbial diversity, reducing disease and pest infestation there by allowing a better plant growth and biomass addition.

4.1.1.1 Aggregate formation and stabilisation

Soil particles are bound together by temporary (i.e., fungal hyphae and roots) and transient binding agents (i.e., microbial- and plant-derived polysaccharides through organic matter decomposition) [27]. In presence of these agents, aggregation is promoted and with time the microbially restructured carbohydrate molecules get attached with finer soil particles like clay and silt which is a stable form as compared to particulate organic matter (POM). With elimination of soil disturbance (tillage), soil organic matter gets strongly bound to clay particles in the form of macroaggregates and microaggregates within the macroaggregates. Again, microaggregates within the macroaggregates constitute a secure habitat soil microorganism, soil disturbance destroys the microbial habitat, affects its activity. In non-disturbed soil, the particulate organic matter present in macroaggregates get to be predominantly stabilised within microaggregates owing to the slow turnover rate [28]. On the other hand, a higher turnover of POM is seen due to tillage because they get exposed to rapid microbial attack preventing its incorporation into micro-aggregates as fine POM. In short, tillage leads to carbon loss through breakdown of C-rich macroaggregates and a decrease in microaggregate formation. Research has shown that 90% of total difference in SOC in soils of varying type and climate is due to the microaggregate-associated C fraction [29]. Thus, a slower turnover of this fraction in zero tillage allows greater protection and stabilisation of coarse POM over time through mineral-bound C decomposition product formation in the microaggregates-within-macroaggregates promoting long-term soil C sequestration in agricultural soils. The process of aggregate formation and protection under no tillage system is shown in the right flowchart whereas, disruption due to tillage is described in the left (**Figure 4**), The bold lines are implicative of higher amount.

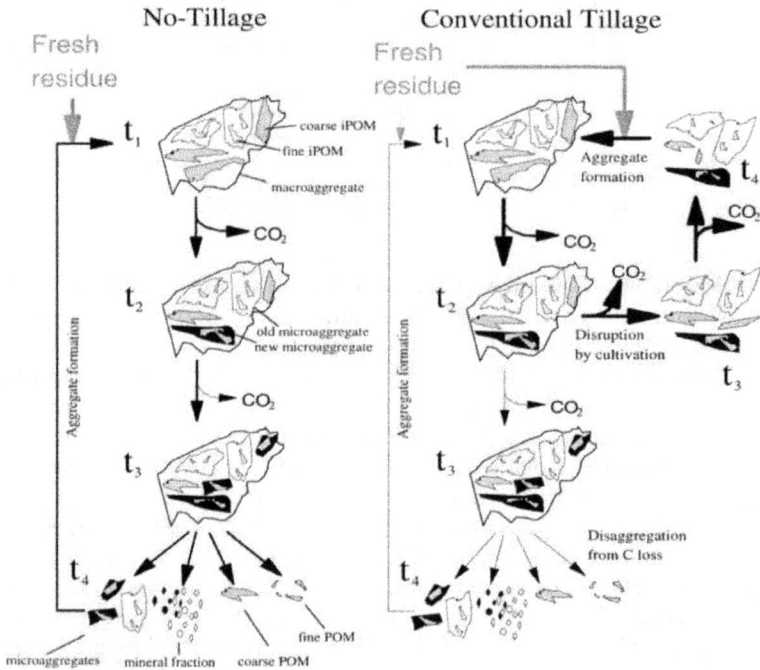

Figure 4.
Aggregate formation in a no-tillage as well as conventional tillage system. Adopted from [30].

4.1.1.2 Microbial population and diversity

Not only microbial habitat, but also macrofauna population is promoted under no tillage practices in absence of physical abrasion and habitat destruction as happens under conventional tillage practices.

Availability of protected habitat and higher C- input directly influence microbial population in a positive way. Generally, in tillage induced environment there is dominance of *r* strategists (with high reproduction rates and fast colonisation capacities) in soil biota shifting the ratio towards higher mesofauna vs. macrofauna or bacteria vs. fungi [31] and thus increased mineralisation versus humification [32] as well as low stability aggregate formation [27]. A fungal dominated system is considered to be a better carbon trapper because of higher metabolic growth efficiency of such class, which assimilate much of substate carbon in microbial biomass and by products but emit less CO_2. Higher the metabolic growth efficiency, lesser the loss of mineral associated carbon as CO_2 as the fungal products are more chemically resistant to decay [31]. The binding of microaggregates within macroaggregate by plant roots and microbial hyphae is described in **Figure 5**. The mechanism of higher microbial population (Fungi dominated) and aggregate stability are complementary to each other which is generally observed under high biomass input conservation tillage system. A higher amount of microbially derived carbohydrate C, acid hydrolysable C, amino acids, amino sugars and glomalin content is observed under no tillage soil than a tilled one [33]. The complex interlinking of carbon substrate addition, improved soil physical structure and physical & biological activity enables higher carbon capture under a conservation agriculture management system. More number of binding agents in an undisturbed agricultural soil

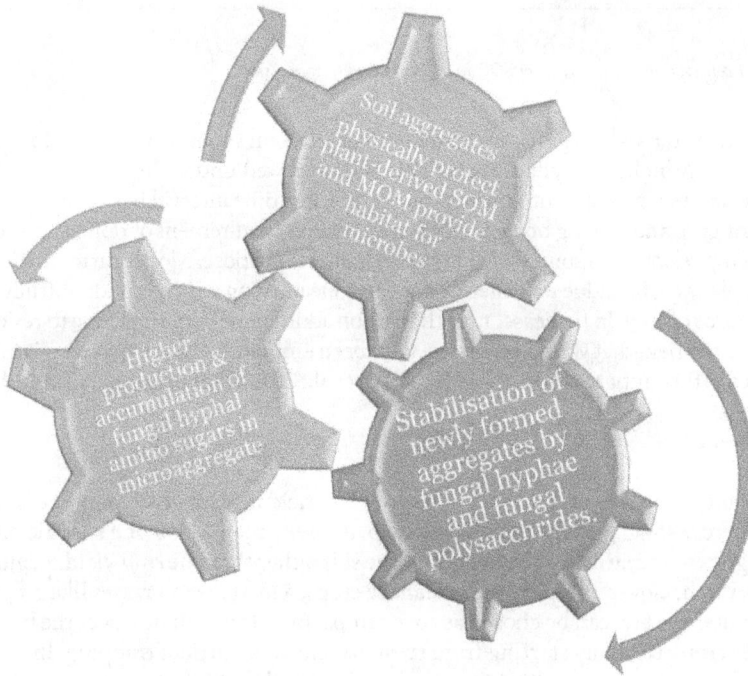

Figure 5.
Mechanism operated in soils under CA practice for enhancing C-pool size.

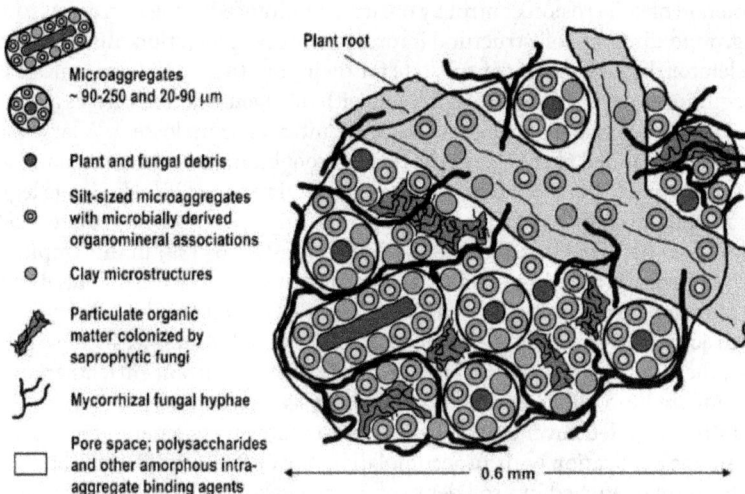

Figure 6.
Macro and micro aggregate formation in soil through binding agents. Adopted from [35].

Location	Cropping system	Depth	Years of adoption	SOC change Mg ha/yr	Reference
China	Maize, wheat, rice, soybean	0.2–1.0	Avg: 6.5	+0.25	[36]
Indo-Gangetic Plain	Rice- Wheat	0.05–1.05	2–26	+0.14	[34]

Table 4.
Impact of conservation agriculture on SOC in different countries of Asia.

promotes water stable aggregate formation and carbon sequestration within the structures. A higher enzymatic activity is also observed under CA.

The main social issue with farmers of IGP are, less time interval between harvesting of kharif crop and sowing of succeeding crop, fodder requirement of domestic animals, use of crop residue as a source of energy for domestic purpose. Mostly farmers adopt the simple way of residue management i.e., residue burning which is undoubtfully a huge source of CO_2. In that case, may the carbon addition be very small due to residue return to the field that would otherwise have been emitted to the atmosphere, is a sure shot CO_2 efflux mitigation principle (Powlson et al., 2016) [34] (**Figure 6**, **Table 4**).

4.2 Cover crop

The intercrops or catch crops can be grown in field instead of keeping the land fallow before sowing of the next fallow crop. A *cover crop* is a *crop* of a specific plant that is grown primarily for the benefit of the soil rather than the *crop* yield. Legumes such as vetch, clover, cowpea; green manure crops, a mixture of grasses like ryegrass, oats, winter rye etc. can be chosen as cover crops. In soils health prospect the benefits of cover crops are many starting from erosion control to nutrient trapping. In crops point of view they are excellent for reducing weed and pest infestation in the crop land resulting in a better crop stand. As a direct source of organic biomass to the land, growing cover crops is one of the most effective carbon management practices in Asia. The process of carbon management through cover crop is another interlinked phenomenon of soil erosion control by creating hindrance for the rain drops to splash on the ground directly, soil structural improvement and protection, microbial activity accelerator through supply of substate for their growth and carbon sequestration [37]. Legumes as cover crops enrich the soil with nitrogen whereas cereals and brassica are excellent nutrient scavengers (scavenge nitrogen from losses). A large part of the cover crop is added to the soil in the form of root biomass which was found to be a relatively stable carbon pool than the above ground residue [38]. No tillage legume can act as a potential sink of GHG with global warming potential of -971 to -2818 kg CO_2 equivalent ha^{-1} $year^{-1}$ as observed by Bayer et al. (2016) [39] in sub-tropical ultisols of Brazil. He also suggested that, these systems may act as a potential source of N_2O emission but the net effect is fully offset by CO_2 retention in soil organic matter which accounts for -2063 to -3940 kg CO_2 ha^{-1} $year^{-1}$. Along with below ground biomass, the cover crop is anyway an additional source of carbon enrichment to the soil as compared to a fallow period. A meta- analysis conducted by Poeplau and Don (2015) [40] concluded cover crop to be higher estimate management practice than sewage sludge application with an accumulation rate of 0.32 ± 0.08 Mg C ha^{-1} $yr.^{-1}$ until saturation is reached in a soil depth of 22 cm (mean) in 30 sites worldwide (in Asia sites under study are from India and Japan). This cumulative carbon sequestration through cover cropping has the potential to compensate for 8% of the annual direct greenhouse gas emissions from agriculture [41]. Dynamics of nitrogen is very essential for carbon stabilisation in soil. C: N ratio, quality of nitrogen is a major factor controlling nitrogen dynamics in soil. a low C:N ratio plant like legume, early killing of cereal crop can release nitrogen faster into the following crop whereas high C:N ratio cereal grains slow down N release rate. Nitrogen is very much needed in balancing soil organic carbon. Thus, reduced tillage system and high C:N ratio residue can temporarily increase optimum N requirement in crop field that will add to long term carbon storage in soils. Cover crops can contribute this N either by scavenging residual N or by N_2 fixation by legumes.

4.3 Crop diversification or rotation

Monoculture is a technique that favour strong outbreak of diseases and pests. Again, due to same root architecture in every season, plants access nutrient from a specific depth. These affect plant growth and production. On the other hand, the stratified root architecture associated with crop diversification allows plants to uptake nutrients from various depths of the soil. Rhizosphere provides suitable environment for microbial diversity and proliferation in different level of the soil. Crop diversification has been shown to reduce the emergence and damage of such pests and diseases. This promotes better above ground as well as below ground biomass production in crop plant by which crop diversification directly contributes to carbon enrichment in soil. Crop rotation or mixed cultivar use instead of single genotype are found to improve resilience towards climate change extremities, pest, disease occurrence, enhance yield stability and reduce fertiliser footprint which ultimately cuts contribution of crop production towards CO_2 emission. A study conducted by Hu et al. (2016) [42] showed that there is 46% less soil respiration and 10% less emission in wheat- maize intercropping as compared to maize monoculture in north-west China. In case of intensive cropping systems, minimum one legume crop is necessary for soil carbon stabilisation along with other soil quality benefits. Legume plants are characterised by deeper root system, high leaf shedding, higher root exudates accelerate rhizospheric activity [43]. The quality and quantity of both root exudates and microbial polysaccharides (rich in lignopolyphenol complexes) promote macro and meso aggregate associated carbon storage in "rotation with legume" system than "cereal- cereal" system which is a good indicator of carbon sequestration [44, 45]. A life cycle-assessment (LCA) review conducted by Clune et al. (2017) [46] from 2000 to 2015 around the world highlighted that pulses have a very low Global Warming Potential (GWP) values (0.50–0.51 kg CO_2 eq kg^{-1} which makes inclusion of a pulse crop in crop rotation, a win-win situation.

Pulse cultivation has other beneficial effect on soil environment viz.; pulses during summer can conserve moisture because soil covering through litterfall protects soil surface from atmospheric temperature. Not only the exudate or biomass quality but the management practices associated with crop rotation (irrigation, fertiliser dose, nitrogen fixation, amount of residue recycled for different crop rotations) cause variation on biomass input into a system. Legume crops acquire their N from biological nitrogen fixation (except for starter dose of nitrogen fertiliser) rather than from the soil as nitrate a slight decrease in pH of soil occurs. The reduction in soil pH in neutral and alkaline soil environments promote microbial activity in root zone and increase the nutrient availability [45]. Therefore, pulse in rotation enhances the macroaggregates rather than cereal- cereal system. Though the results of legume in rotation are strong for higher carbon management, a cereal- cereal rotation improve the passive carbon pool because higher carbon: nitrogen ratio of such crop residues [45]. Cereal in a rotation has also found to be important in environmental aspect as per a study conducted by Senbayram et al. (2016) [47] who found that monocropped faba beans lead to three times higher cumulative N2O emissions than that of unfertilized wheat whereas faba bean wheat intercropping could lower the cumulative N2O emissions by 31% as compared to N-fertilised wheat.

Proliferated root condition under diversified cropping system supports a hierarchy of aggregate formation (macroaggregates followed by microaggregates within macroaggregates). Plant roots are residues bind the individual soil particles together to form macroaggregates then fine root hairs grow into these aggregates. The organic acids, enzymes, and other C-rich compounds exuded by these roots support higher microbial populations and act as the nucleation centre for micro-aggregate formation [48, 49]. The microbially altered organic compounds get polymerised and are then strongly bound to finer particles (silt & clay) inside of the macroaggregates. These newly formed occluded microaggregates are C and N enriched [48, 49].

4.4 Integrated and balanced nutrient management

With increase in demand of food per capita per unit land area, farmers are adopting higher fertiliser application in hope of getting higher yield. But in contrast the expectation, over use of chemical fertiliser result in severe soil degradation which is a major contributor towards soil carbon loss and higher GHG emission. As a correction measure to such issue, many scient have looked for the role of integrated (chemical+ organic) and balanced fertilisation on GHG emission reduction and soil carbon enrichment. As per a study conducted in subtropical north-western states of India, application of organics along with chemical fertilisers reduces the gaseous N losses as compared to fertiliser nitrogen alone in rice-wheat system [50]. Addition of organics no doubt acted as the primary source of denitrification, but the carbon balance was still positive. The higher yielding cropping systems created a scenario of higher CO_2-C consumed by crops for photosynthesis than the total flux of CO_2-C from rice-wheat system even with the use of organics thus making it a sink of atmospheric CO_2-C [50].

Integrated nutrient management (INM) technology improves the physical, chemical and biological activity of the soil, which leads to a healthy plant population and higher yield. Organic treatments like FYM, sulphitation press mud (SPM), green gram residue (GR) and rice-wheat crop residues (CR) may consistently increase biomass yields and increase C inputs in soil. The strong influence from increasing C stock through long-term balanced fertilisation under rice–wheat cropping system was found by Nayak et al., 2012 [51]. Organic material incorporation improved soil aggregation and structural stability and resulted in higher C content in macroaggregates, thereby improved C sequestration potential in soils. However, the C accumulation in aggregates may determine by the kind and source of organic inputs. Thus, study by Das *et al.* (2014) [52] found that a combination of GR in rice and FYM in wheat significantly improved C content in macroaggregates, 100% N application through inorganic fertiliser. However, CR incorporation enhances coarse particulate organic matter (>0.25 mm) which substantially increase C content within macroaggregates. Intensive rice–wheat system through combination of inorganic and organic fertilisers and crop residues increases C content in microaggregates- within-macroaggregates [53] indicating higher potential of C stabilisation in soil.

Organic amendment like FYM, vermicompost, biochar etc. have higher humification rate constant but less decomposition rate thus, improve the amount and stability of SOC through their addition. An incubation study by Naher et al., 2020 [54] described that carbon mineralisation rate was 0.011 tonne year^{-1} for INM followed by balanced fertiliser and control which in turn enhance the scope for SOC sequestration in soil for sustainable rice production.

Treatments	Carbon efficiency ratio	Global warming potential (kg CO_2 ha^{-1})	% increase in yield over control	% increase in emission over control
NPK	13.82 ± 0.82	540.60 ± 21.25	—	—
NPK + Green manure	9.94 ± 0.24	887.40 ± 12.11	10.70	64.15
NPK+ Azolla compost	16.90 ± 0.25	625.20 ± 13.03	27.43	15.66
LSD (T)	0.634	21.068		

Table 5.
Effect of INM on emission and yield.

Soil attributes	Soil functions
Soil physical properties	Organic amendments support aggregate formation, aeration, higher water holding capacity
Nutrition supply	Release of nutrient over a long period of time, improved use efficiency, higher SOC, more aggregate SOC, reduce phosphate fixation, reduced the mining of K from the soil
Soil reactions	Increase CEC, buffering, rhizospheric elemental transformation
Soil biological property	Microbial species diversity, soil enzymes, microbial biomass C (MBC), slow establishment and persistence of pathogens
Agronomic properties	Better root establishment, higher grain, straw production

Table 6.
Effect of INM on various soil properties for better soil health and crop production.

A study conducted by Bharali et al. (2018) [55] in the north-eastern India showed that addition of organics (Azolla compost or green manure) along with chemical fertilisers resulted in higher emission worth of higher global warming potential however, the carbon efficiency ratio and amount of fixed carbon in terms of grain yield was found to be higher and lower in case of Azolla compost as compared to chemical fertiliser alone. Likewise, in case of NPK + green manure, there is 64% higher emission over the control, a lower carbon efficiency ratio but higher total C fixed in a form of grain carbon (**Table 5**). Though INM is not a direct solution for reducing C efflux, the extra organics added may result in more emission as compared to sole chemical fertiliser addition, it also contributes to sufficiently higher C fixation in the form of grain C which ultimately shows to have a positive carbon balance due to INM.

A review done by Wu and Ma (2015) [56] shows the effect on INM on different soil properties and crop growth in countries of Asia is summarised in **Table 6**.

A meta-analysis conducted by Waqas et al. (2020) [57] all over China to study the effect of balanced, imbalanced, integrated, sole fertilisation and their combinations on yield sustainability (YSI), yield variability index (YVI) suggest that balanced and integrated fertilisation has highest YSI and lowest YVI and balanced chemical fertilisation has less YVI as compared to sole organics addition or imbalanced chemical fertilisation. The result supports the fact that integrated and balanced fertilisation supports carbon addition through higher above ground and below ground biomass production. Even imbalanced+ organic fertilisation and organic fertilisation alone can increase SOC due to direct addition of stabilised carbon through organic amendments. Organic amendments are also supply additional nutrients (N, P, S, etc.) into the soil which are responsible for production of fine fraction of soil organic matter [58]. The direct and indirect carbon input through integrated fertiliser management is a great adoptive measure as carbon management practice. In general, cold temperature promotes carbon sequestration due to low rate of organic matter decomposition but in higher temperature region with higher productivity and consequently increased biomass carbon input into soil [59], SOC can be improved through stable aggregate formation.

Sole and continuous use of chemical fertilisers inhibit the micro-organisms and their biochemical compositions, which reduced the aggregate formation. But the fresh organic matter added through organic amendments supply promote microbial polysaccharide formation (water soluble and hydrolysable substrate) that also promote aggregate formation. In completely no fertiliser condition, higher root extraction causes shattering of macroaggregates and breaking up soil structure [60].

Biochar as an organic amendment is also a great choice because the carbon-rich material has many organic functional groups to which act as bridge to form strong complexes with soil and is also helpful to increase soil aggregation through charged surface, porous structure and high cation exchange capacity [61].

Biochar amendments has two mechanisms of improving SOC dynamics (1) promoting soil aggregation thereby physical protection of bound SOC (2) Negative priming by means of higher recalcitrant organic substrate pool having low decomposition rate [62] (**Tables 7** and **8**).

Treatment	Carbon mineralisation rate (t yr^{-1})	Carbon stock (t ha^{-1} year^{-1})	Carbon sequestration (kg ha^{-1}year^{-1})
Fertiliser control	0.009	10.95	−213
Balanced fertiliser	0.010	17.30	−72.15
INM	0.011	26.30	127.86

Table 7.
Carbon sequestration in soil with rice–rice–fallow cropping sequence for 10 years [54].

Country	Control	NPK	INM	Reference
New Delhi, India	7.53	8.50	**11.08**	[63]
Kanpur, India	3.73	4.59	**5.45**(50% RDF+ FYM+ residue+ biofertilisers)	[45]
Bangladesh	9.8	13.3	**17.1** (RDF + Gr. manure)	[64]
China	13.81	13.40 (Only N)	**15.12**(RDF + Straw/Biochar) **15.22** (RDF + Biochar)	[65]

Bold letters are to make the values distinctly visible from treatment details only.

Table 8.
Total organic carbon (TOC) content under INM and chemical fertilisation practice in various regions of Asia (TOC given in g/kg).

5. Conclusion

In the degraded land Soil organic carbon acts as the centre of soil health through positive regulation of soil physical, chemical, biological and ecological functions. The integrated management practice like conservation agriculture does not only add carbon to the soil directly but also reduce fossil fuel CO_2 emission, oxidation of SOC. Cover crops and crop diversity are beneficial for combating disease- pest occurrence, support healthy above ground and below ground biomass production. Legume in a crop rotation supports aggregate formation and stabilisation and ultimately protects the aggregate associated carbon through chemical polymerisation and physical occlusion. INM is beneficial over imbalanced chemical or sole chemical fertiliser application. Though biochar is another effective amendment for carbon sequestration in agricultural land, the higher carbon foot print associated with its production technique (CO_2 production during pyrolysis and more CO_2 emission from amended plots) can offset it as a climate change mitigative- adoptive practice. Soil C sequestration is not a permanent solution for all climate change related issues but is a holistic approach to restore degraded soil, reduce erosion, increase agronomic yields and reduce CO_2 emission into the atmosphere at the same time. Thus, careful selection of carbon management practice according to climatic and soil condition is necessary for making it agriculturally and environment friendly.

Author details

Biswabara Sahu*, Snigdha Chatterjee and Ruby Patel
Department of Agricultural Chemistry and Soil Science, Bidhan Chandra Krishi Vishwavidyalaya, Mohanpur, West Bengal

*Address all correspondence to: biswabara.kunu94@gmail.com

IntechOpen

References

[1] Our World in Data. Land Use [Internet]. 2019. Available from: https://ourworldindata.org/land-use#arable-agriculture-cropland.

[2] Intergovernmental Panel for Climate Change (IPCC). Summary for policymakers. Climate Change 2007: Synthesis Report. Fourth Assessment Report of the Intergovernmental Panel for Climate Change, 2007a, available at http://www.ipcc.ch/pdf/assessment-report/ar4/syr/ar4_syr_spm.pdf.

[3] Intergovernmental Panel for Climate Change (IPCC). Climate Change 2014: Synthesis Report. 2014, available at https://www.ipcc.ch/site/assets/uploads/2018/05/SYR_AR5_FINAL_full_wcover.pdf

[4] Chan Y. Increasing soil organic carbon of agricultural land. Primefact. 2008; 735, 1-5.

[5] Whitehead RJ. Asia produces nearly half of world's greenhouse gases from agriculture [Internet]. 2014. Available from: https://www.foodnavigator-asia.com/Article/2014/04/14/

[6] FAO, ITPS. *Status of the World's Soil Resources (SWSR) - Main Report.* (Food and Agriculture Organization of the United Nations and Intergovernmental Technical Panel on Soils, 2015; 1–648.

[7] Zhu T, Burton I, Huq S, Rosegrant MW, Yohe G, Ewing M, Valmonte-Santos R. Climate change and Asian griculture. Asian Journal of Agriculture and Development. 2010; 7(1362-2016-107686), 41-81. DOI: http://dx.doi.org/10.22004/ag.econ.199082

[8] Lal R. Soil carbon management and climate change. Carbon Management. 2013; 4(4), 439-462. DOI: https://doi.org/10.4155/cmt.13.31

[9] Lal R. Managing Soils and Ecosystems for Mitigating Anthropogenic Carbon Emissions and Advancing Global Food Security. BioScience. 2010; 60:708–721. DOI: 10.1525/bio.2010.60.9.8.

[10] Post WM, Kwon KC. Soil carbon sequestration and land-use change: processes and potential. Global Change Biol. 2000; 6:317–327. DOI: 10.1046/j.1365-2486.2000.00308.x

[11] Zomer RJ, Bossio DA, Sommer R, Verchot LV. Global sequestration potential of increased organic carbon in cropland soils. Scientific Reports. 2017; 7(1), 1-8. DOI: https://doi.org/10.1038/s41598-017-15794-8

[12] Lal R. Carbon Management in Agricultural Soils. Mitig Adapt Strat Glob Change. 2007; **12**, 303–322. DOI: https://doi.org/10.1007/s11027-006-9036-7

[13] Lützow M, Kögel-Knabner I, Ekschmitt K et al. Stabilization of organic matter in temperate soils: mechanisms and their relevance under different soil conditions: a review. Eur. J. Soil Sci. 2006; 57(4), 426–445. DOI: https://doi.org/10.1111/j.1365-2389.2006.00809.x

[14] Dexter AR, Richard G, Arrouays D, Czy_ z EA, Jolivet C, Duval O. Complexed organic matter controls soil physical properties. Geoderma. 2008; 144, 620–627. DOI: http://dx.doi.org/10.1016/j.geoderma.2008.01.022

[15] Minasny B, Malone BP, McBratney, AB, Angers DA, Arrouays D, Chambers A, ... Field DJ. Soil carbon 4 per mille. Geoderma. 2017; 292, 59-86. DOI: https://doi.org/10.1016/j.geoderma.2017a.01.002

[16] Minasny B, Mcbratney AB, Malone BP, Angers D, Arrouays D,

Chambers A, ... & Field DJ. 4 per 1000 soil carbon sequestration. In Global Symposium on Soil Organic Carbon; 21-23 March 2017b; Rome. Italy: 2007; p. 534.

[17] Dungait JAJ, Hopkins DW, Gregory AS, Whitmore AP. Soil organic matter turnover is governed by accessibility not recalcitrance. Glob. Change Biol. 2012; 18:1781–1796. DOI: https://doi.org/10.1111/j.1365-2486.2012.02665.x

[18] von Lützow M, Kögel-Knaber I. Temperature sensitivity of soil matter decomposition-what do we know? Biol. Fertil. Soils. 2009; 46:1–15. DOI: https://doi.org/10.1007/s00374-009-0413-8

[19] Murty D, Kirschbaum MUF, McMurtrie RE, McGilvray H. Does conversion of forest to agricultural land change soil carbon and nitrogen? A review of the literature. Glob. Change Biol. 2002; 8: 105–123. DOI: https://doi.org/10.1046/j.1354-1013.2001.00459.x

[20] West TO, Marland G. A synthesis of carbon sequestration, carbon emissions, and net carbon flux in agriculture: comparing tillage practices in the United States. Agr. Ecosyst. Environ. 2002; 91: 217–232. DOI: https://doi.org/10.1016/S0167-8809(01)00233-X

[21] Sapkota TB, Jat ML, Aryal JP, Jat RK, Khatri-Chhetri A. Climate change adaptation, greenhouse gas mitigation and economic profitability of conservation agriculture: Some examples from cereal systems of Indo-Gangetic Plains. Journal of Integrative Agriculture. 2015; 14(8), 1524-1533. https://doi.org/10.1016/S2095-3119(15)61093-0

[22] Govaerts B, Verhulst N, Castellanos-Navarrete A, Sayre KD, Dixon J, Dendooven L. Conservation Agriculture and Soil Carbon Sequestration: Between Myth and Farmer Reality, Critical Reviews in Plant Sciences. 2009; 28:3, 97-122, DOI: 10.1080/07352680902776358

[23] Sapkota TB, Majumdar K, Jat ML, Kumar A, Bishnoi DK, McDonald AJ, Pampolino M. Precision nutrient management in conservation agriculture based wheat production of Northwest India: Profitability, nutrient use efficiency and environmental footprint. Field Crops Research. 2014; 155, 233-244. https://doi.org/10.1016/j.fcr.2013.09.001

[24] Jensen ES, Peoples MB, Boddey RM, Gresshoff PM, Hauggaard-Nielsen H, Alves BJ, Morrison MJ. Legumes for mitigation of climate change and the provision of feedstock for biofuels and biorefineries. A review. Agronomy for Sustainable Development. 2012; 32:329–364.

[25] West TO, Post WM. Soil organic carbon sequestration rates by tillage and crop rotation. Soil Science Society of America Journal. 2002; 66, 1930–1946. https://doi.org/10.2136/sssaj2002.1930

[26] Erenstein O, Laxmi V. Zero tillage impacts in india's rice-wheat systems: A review. Soil and Tillage Research. 2008; 100 (1–2): 1–14. https://doi.org/10.1016/j.still.2008.05.001

[27] Six J, Ogle SM, Breidt FJ, Conant RT, Mosier AR, Paustian K. The potential to mitigate global warming with no-tillage management is only realized when practised in the long term. Glob. Change Biol. 2004b; 10: 155–160. DOI: https://doi.org/10.1111/j.1529-8817.2003.00730.x

[28] Six J, Paustian K, Elliott ET, Combrink C. Soil structure and soil organic matter: I. Distribution of aggregate size classes and aggregate associated carbon. Soil Sci. Soc. Am. J. 2000a; 64: 681–689. DOI: https://doi.org/10.2136/sssaj2000.642681x

[29] Denef K, Six J, Merckx R, Paustian K. Carbon sequestration in

microaggregates of no-tillage soils with different clay mineralogy. Soil Sci. Soc. Am. J. 2004; **68**:1935–1944. DOI: https://doi.org/10.2136/sssaj2004.1935

[30] Fiedler SR. Short-term biogeochemical effects of agricultural management measures in soils amended with anaerobic biogas digestate [thesis]. University of Rostock; 2017. https://www.researchgate.net/publication/320298638

[31] Six J, Frey SD, Thiet RK, Batten KM. Bacterial and fungal contributions to carbon sequestration in agroecosystems. Soil Science Society of America Journal, 2006; **70**(2): 555-569. DOI: https://doi.org/10.2136/sssaj2004.0347

[32] Wardle DA. 1995. Impacts of disturbance on detritus food webs in agroecosystems of contrasting tillage and weed management practices. **In:** Begon M. Fitter AH, Editors. Advances in Ecological Research. New York: Academic Press; 1995. Vol. 26, p. 105–185. DOI: https://doi.org/10.1016/S0065-2504(08)60065-3

[33] Arshad MA, Schnitzer M, Angers DA, Ripmeester JA. Effects of till versus no till on the quality of soil organic matter. Soil Biol. Biochem. 1990; 22:595–599. DOI: https://doi.org/10.1016/0038-0717(90)90003-I

[34] Powlson DS, Stirling CM, Thierfelder C, White RP, Jat ML. Does conservation agriculture deliver climate change mitigation through soil carbon sequestration in tropical agro-ecosystems? Agric. Ecosyst. Environ. 2016; 220, 164–174. DOI: 10.1016/j.agee.2016.01.005

[35] Jastrow JD, Amonette JE, Bailey VL. Mechanisms controlling soil carbon turnover and their potential application for enhancing carbon sequestration. Climatic Change. 2007; 80(1), 5-23. DOI: https://doi.org/10.1007/s10584-006-9178-3

[36] Du Z, Angers DA, Ren T, Zhang Q, Li G. The effect of no-till on organic C storage in Chinese soils should not be overemphasized: A metaanalysis. Agric. Ecosyst. Environ. 2017; 236, 1–11. DOI: 10.1016/j.agee.2016.11.007

[37] Dabney SM, Delgado JA, Reeves DW. Using winter cover crops to improve soil and water quality. Commun. Soil Sci. Plant Anal. 2001; 32, 1221–1250. DOI: https://doi.org/10.1081/CSS-100104110

[38] Kätterer T, Bolinder MA, Andrén O, Kirchmann H, Menichetti L. Roots contribute more to refractory soil organic matter than above-ground crop residues as revealed by a long-term field experiment. Agric. Ecosyst. Environ. 2011; 141, 184–192. DOI: https://doi.org/10.1016/j.agee.2011.02.029

[39] Bayer C, Gomes J, Zanatta JA, Vieira FCB, Dieckow J. Mitigating greenhouse gas emissions from a subtropical Ultisol by using long-term no-tillage in combination with legume cover crops. Soil Tillage Research. 2016; 161:86–94. https://doi.org/10.1016/j.still.2016.03.011

[40] Poeplau C, Don A. Carbon sequestration in agricultural soils via cultivation of cover crops–A meta-analysis. Agriculture, Ecosystems & Environment. 2015; 200, 33-41. https://doi.org/10.1016/j.agee.2014.10.024

[41] IPCC. In: Solomon S, Qin D, Manning M, Marquis M, Averyt K, Tingor MMB, ... Chen Z, editors. IPCC Climate Change: The Physical Science Basis. Cambridge: Cambridge University Press; 2007.

[42] Hu F, Gan Y, Cui H, Zhao C, Feng F, Yin W, Chai Q. Intercropping maize and wheat with conservation agriculture principles improves water harvesting and reduces carbon emissions in dry areas. European Journal of Agronomy. 2016; 74, 9-17. https://doi.org/10.1016/j.eja.2015.11.019

[43] Kou TJ, Zhu P, Huang S, Peng XX, Song ZW, Deng AX, Gao HJ, Peng C, Zhang WJ. Effects of long-term cropping regimes on soil carbon sequestration and aggregate composition in rainfed farmland of Northeast China. Soil Tillage Res. 2012; 118, 132–138. DOI: https://doi.org/10.1016/j.still.2011.10.018

[44] Six J, Bossuyt H, Degryze S, Denef K. A history of research on the link between (micro) aggregates, soil biota, and soil organic matter dynamics. Soil and Tillage Research. 2004a; 79(1), 7-31. DOI: https://doi.org/10.1016/j.still.2004.03.008

[45] Hazra KK, Nath CP, Singh U, Praharaj CS, Kumar N, Singh SS, Singh NP. Diversification of maize-wheat cropping system with legumes and integrated nutrient management increases soil aggregation and carbon sequestration. Geoderma. 2019; 353, 308-319. DOI: https://doi.org/10.1016/j.geoderma.2019.06.039

[46] Clune S, Crossin E, Verghese K. Systematic review of greenhouse gas emissions for different fresh food categories. Journal of Cleaner Production. 2017; 140:766–783. https://doi.org/10.1016/j.jclepro.2016.04.082

[47] Senbayram M, Wenthe C, Lingner A, Isselstein J, Steinmann H, Kaya C, Köbke S. Legume-based mixed intercropping systems may lower agricultural born N2O emissions. Energy, Sustainability and Society. 2016; 6:2. https://doi.org/10.1186/s13705-015-0067-3

[48] Six J, Elliott ET, Paustian K. Soil macroaggregate turnover and microaggregate formation: a mechanism for C sequestration under no-tillage agriculture. Soil Biology and Biochemistry. 2000b; 32, 2099–2103. DOI: https://doi.org/10.1016/S0038-0717(00)00179-6

[49] Fonte SJ, Nesper M, Hegglin D, Velásquez JE, Ramirez B, Rao IM, Bernasconi SM, Bünemann EK, Frossard E, Oberson A. Pasture degradation impacts soil phosphorus storage via changes to aggregate-associated soil organic matter in highly weathered tropical soils. Soil Biology and Biochemistry. 2014; 68, 150–157. DOI: https://doi.org/10.1016/j.soilbio.2013.09.025

[50] Aulakh MS. Integrated nutrient management for sustainable crop production, improving crop quality and soil health, and minimizing environmental pollution. In 19th world congress of soil science, soil solutions for a changing world; August, 2010; p. 1-6.

[51] Nayak AK, Gangwara B, Arvind K, Shukla S, Mazumdar P, Kumar A, ... Mohan, U. Long-term effect of different integrated nutrient management on soil organic carbon and its fractions and sustainability of rice–wheat system in Indo Gangetic Plains of India. Field Crop Research. 2012; 127, 129–139. https://doi.org/10.1016/j.fcr.2011.11.011

[52] Das B, Chakraborty D, Singh VK, Aggarwal P, Singh R, Dwivedi BS, Mishra RP. Effect of integrated nutrient management practice on soil aggregate properties, its stability and aggregate-associated carbon content in an intensive rice–wheat system. Soil and Tillage Research. 2014; 136, 9-18. https://doi.org/10.1016/j.still.2013.09.009

[53] Six J, Conant RT, Paul EA, Paustian K. Stabilization mechanisms of soil organic matter: implications for C-saturation of soils. Plant Soil. 2002; 241 (2) 155– 176. https://doi.org/10.1023/A:1016125726789

[54] Naher UA, Hossain MB, Haque MM, Maniruzzaman M, Choudhury AK, Biswas JC. Effect of long-term nutrient management on soil organic carbon

sequestration in rice-rice-fallow rotation. Current Science. 2020; (00113891), 118(4).

[55] Bharali A, Baruah KK, Baruah SG, Bhattacharyya P. Impacts of integrated nutrient management on methane emission, global warming potential and carbon storage capacity in rice grown in a northeast India soil. Environmental Science and Pollution Research. 2018; 25(6), 5889-5901. https://doi.org/10.1007/s11356-017-0879-0

[56] Wu W, Ma B. Integrated nutrient management (INM) for sustaining crop productivity and reducing environmental impact: A review. Science of the Total Environment. 2015; 512, 415-427. DOI: https://doi.org/10.1016/j.scitotenv.2014.12.101

[57] Waqas MA, Li YE, Lal R, Wang X, Shi S, Zhu Y, ... & Liu S. When does nutrient management sequester more carbon in soils and produce high and stable grain yields in China? Land Degradation & Development. 2020; 31(15), 1926-1941. DOI: https://doi.org/10.1002/ldr.3567

[58] Kögel-Knabner I. The macromolecular organic composition of plant and microbial residues as inputs to soil organic matter. Soil Biology and Biochemistry. 2002; 34(2), 139-162. DOI: https://doi.org/10.1016/S0038-0717(01)00158-4

[59] Godde CM, Thorburn PJ, Biggs JS, Meier EA. Understanding the impacts of soil, climate, and farming practices on soil organic carbon sequestration: a simulation study in Australia. Frontiers in plant science. 2016; 7, 661. DOI: https://doi.org/10.3389/fpls.2016.00661

[60] Lichter K, Govaerts B, Six J, Sayre KD, Deckers J, Dendooven L. Aggregation and C and N contents of soil organic matter fractions in a permanent raised-bed planting system

in the Highlands of Central Mexico. Plant Soil. 2008; 305, 237–252. DOI: https://doi.org/10.1007/s11104-008-9557-9

[61] Jien SH, Wang CS. Effects of biochar on soil properties and erosion potential in a highly weathered soil. Catena. 2013; 110(110), 225-233. https://doi.org/10.1016/j.catena.2013.06.021

[62] Zhang A, Bian R, Pan G, Cui L, Hussain Q, Li L, ... Han X. Effects of biochar amendment on soil quality, crop yield and greenhouse gas emission in a Chinese rice paddy: a field study of 2 consecutive rice growing cycles. Field Crops Research. 2012; 127, 153-160. https://doi.org/10.1016/j.fcr.2011.11.020

[63] Moharana PC, Sharma BM, Biswas DR, Dwivedi BS, Singh RV. Long-term effect of nutrient management on soil fertility and soil organic carbon pools under a 6-year-old pearl millet–wheat cropping system in an Inceptisol of subtropical India. Field Crops Research. 2012; 136, 32-41. https://doi.org/10.1016/j.fcr.2012.07.002

[64] Saha PK, Ishaque M, Saleque MA, Miah M, Panaullah GM, Bhuiyan NI. Longterm integrated nutrient management for rice-based cropping pattern: effect on growth, yield, nutrient uptake, nutrient balance sheet, and soil fertility. Commun. Soil Sci. Plan. 2007; 38, 579–610.https://doi.org/10.1080/00103620701215718

[65] Sui Y, Gao J, Liu C, Zhang W, Lan Y, Li S, ... Tang L. Interactive effects of straw-derived biochar and N fertilization on soil C storage and rice productivity in rice paddies of Northeast China. Science of the Total Environment. 2016; 544, 203-210. https://doi.org/10.1016/j.scitotenv.2015.11.079

Adapting to Climatic Extremes through Climate Resilient Industrial Landscapes: Building Capacities in the Southern Indian States of Telangana and Andhra Pradesh

Narendran Kodandapani

Abstract

There is now greater confidence and understanding of the consequences of anthropogenic caused climate change. One of the many impacts of climate change, has been the occurrence of extreme climatic events, recent studies indicate that the magnitude, frequency, and intensity of hydro-meteorological events such as heat waves, cyclones, droughts, wildfires, and floods are expected to increase several fold in the coming decades. These climatic extremes are likely to have social, economic, and environmental costs to nations across the globe. There is an urgent need to prepare various stakeholders to these disasters through capacity building and training measures. Here, we present an analysis of the capacity needs assessment of various stakeholders to climate change adaptation in industrial parks in two southern states of India. Adaptation to climate change in industrial areas is an understudied yet highly urgent requirement to build resilience among stakeholders in the Indian subcontinent. The capacity needs assessment was conducted in two stages, participatory rural appraisal (PRA) and focus group discussion (FGD) were conducted among various stakeholders to determine the current capacities for climate change adaptation (CCA) for both, stakeholders and functional groups. Our analysis indicates that in the states of Telangana and Andhra Pradesh, all stakeholder groups require low to high levels of retraining in infrastructure and engineering, planning, and financial aspects related to CCA. Our study broadly supports the need for capacity building and retraining of functionaries at local and state levels in various climate change adaptation measures; likewise industry managers need support to alleviate the impacts of climate change. Specific knowledge, skills, and abilities, with regard to land zoning, storm water management, developing building codes, green financing for CCA, early warning systems for climatic extremes, to name a few are required to enhance and build resilience to climate change in the industrial landscapes of the two states.

Keywords: climatic extremes, climate change adaptation, industrial parks, Telangana, Andhra Pradesh, capacity needs assessment

1. Introduction

A recent report indicates that national governments have until 2020 for the carbon emissions to peak in order to contain the global surface temperature below 2°C [1]. After staying flat for about three years, annual global carbon emissions have again begun to rise above 40 Gt and could rise further in the coming years due to the renewed economic growth across the globe. The present global pandemic has resulted in a small decline in global carbon emissions, however, with governments desperate to restore economic activities, this only appears temporary. The Paris Accord in 2017 was ratified and adopted with targets for countries to reduce emissions by 2030 to keep temperatures below 2°C and well below at 1.5° C by the end of the century. Annual carbon emissions will have to decline by greater than 50% compared to present levels by 2050 to keep the global surface temperatures below 2°C [2]. Already, global temperatures are above 1.1° C compared with pre-industrial levels, and there is warming occurring at the rate of 0.2° C/decade [3].

Anthropogenic greenhouse gas emissions has increased substantially since pre-industrial times and directly related to human population and economic growth. The concentrations of greenhouse gas emissions such as CO_2 have increased almost 40% since pre-industrial times [4]. Annual CO_2 emissions have increased from about 30 billion tons in 2000 to about 42 billion tons in 2016 [1]. However, in order to achieve the Paris Agreement, the remaining carbon budget, after accounting for past emissions, is in the range of 150 to 1050 gigatons CO_2.

Major contributors to the increased concentration of CO_2 in the atmosphere chiefly come from cement industries and thermal power plants through the burning of fossil fuels. Two main sinks for CO_2 are the ocean and land sinks. Countries such as China and India are industrializing at a rapid pace, emissions are increasing in these two countries. However, CO_2 emissions from India are much lower compared to China and the United States, CO_2 emissions of India in 2015 were a quarter, and half, of the emissions from China and the United States respectively [5].

Recent reports from the IPCC, such as the *fifth assessment report* has provided unequivocal evidence of the impacts on various earth system processes due to anthropogenic caused warming [4]. There is now greater confidence and understanding of the consequences of anthropogenic caused climate change [4]. The globally averaged land and sea surface temperature shows an increase of about 0.8° C over the period 1880–2012 and an increase of about 1° C over pre-industrial temperatures [6]. There has been an increase in decadal annual global mean surface temperature at an average rate of 0.07° C per decade during the 1880s, increasing to 0.17° C per decade during the 1970s, increasing to 0.3° C per decade during the 2000s. There has been a decline in the extent of ice-sheets, loss of glacial mass, and also northern hemisphere spring snow cover in several regions across the world. Likewise, the mean sea level rise between 1901 and 2010 has risen by about 20 cm [4].

1.1 Global impacts of extreme climatic events

Other impacts of climate change are the occurrence of extreme climatic events, recent studies indicate that the magnitude, frequency, and intensity of hydro-meteorological events such as heat waves, cyclones, droughts, wildfires, and floods are expected to increase several fold in the coming decades [7]. Globally, between 1980 and 2018, on average 400 disasters occurred annually, resulting in about 23000 fatalities, and leading to direct economic damage of about USD 100 billion each year [8]. Globally, the number of extreme events has increased from 200 during the 1980s to about 700 during 2016, therefore almost a three-fold increase in the number of extreme events in a thirty-five year period [9]. Globally, the number of extreme hydrological, climatological, and meteorological events has doubled in a 30 year period [5]. Worldwide, about 100 million people were affected by disasters in 2015, with climate being a factor in 92% of the events [10], about one-fifth of those affected were from India.

Not only are extreme events becoming more frequent, they are also becoming costlier, for example, Hurricane Harvey is the costliest disaster to strike the United States, estimated cost of damages is about $190 billion [11]. In 2016, insurance companies reported losses of about $175 billion around the world, three-fourth of these losses were caused by meteorological, hydrological or climatological events [12]. During the year 2015, the total economic loss from disasters globally was about $66.5 billion; the economic damage for India was about $ 3.3 billion [9]. The economic loss from the Chennai floods has been estimated to be about $ 0.3 billion of insured losses, and the total loss was about $ 2 billion. Similarly, the floods of Mumbai in 2005 resulted in economic losses of $1.7 billion. Likewise, the flooding in the southwestern state of Kerala in 2018 resulted in the death of 470 people.

The costs and risks of climate change are substantially high and all options to manage the impacts of climate change require urgent interventions. The *World Economic Forum Global Risks Perceptions Survey 2017–2018* [13] has ranked extreme events and natural disasters as top two risks in terms of likelihood. Similarly, extreme events, natural disasters, and failure on climate change mitigation and adaptation, are ranked second, third, and fourth in terms of impacts. These risks and impacts have been reported during successive years and there has been an increasing trend in environmental risks gaining prominence overtime over a ten-year horizon, commencing from 2008. Similarly, the recent, *Global Risks Perception Survey 2021* [14], indicates that extreme climate events are the most important risks to livelihoods. Infact, extreme climate events have emerged as the primary risk to humans since 2017.

2. Extreme climatic events in India

Recent studies in India, indicate increasing trends in the occurrence of hydro-meteorological extreme climatic events across the country. A systematic analysis of these events pan-India over the last five decades provides insights into the spatial and temporal patterns of these extreme events [15]. This same study indicates, that about 100 million people were affected by floods each year, similarly, 140 million people were affected by droughts each year, about 40 million people were affected by tropical storms and cyclones. Thereby a substantial number of people across the country were affected by these hydro-meteorological events each, almost a quarter of the population in India each year. An analysis of floods, droughts, and cyclones over the last five decades, at the district scale, reveals an increasing trend, especially during the last two decades. For example, the number of annual extreme flood events across India has increased from 2 to 16 during the last five decades (1970–2020). Likewise, the number of districts affected by floods has also increased from 10 to 150 districts. The number of districts affected by tropical storms and cyclones has increased from 5 to 90. Overall, there is an acceleration in the frequency of floods, increasing at the rate of 25%/decade, likewise, tropical cyclones are increasing at the rate of 6.5%/decade.

While the decadal number of droughts has not revealed any substantial changes, the number of districts affected by droughts has increased from 10 to 250. Prolonged droughts have occurred in different regions of India, for example in the early 2000s, the Western Ghats entered a period of drought, possibly as a result of global climatic change [16]. Also, the Indian sub-continent and the Western Ghats receive a significant proportion of their annual rainfall from the south-west monsoon. Rainfall data from long-term (1951–2003) observations suggest decreasing trends in both early-monsoon and late-monsoon rainfall and the number of rainy days [17]. Frequent droughts has been a recurrent feature of climate variability in India, between 1901 and 2015, India has experienced 23 drought events [18].

2.1 Climatic extremes in AP and Telangana

Climatic extremes such as cyclone/storms, droughts, floods, and heatwaves are prevalent in the two Indian states of Andhra Pradesh and Telangana. An analysis of cyclones on the western and eastern coasts of India, over a 100 year period, indicated that close to 92 of the 262 cyclones occurred in a 50 km stretch on the east coast of India [19]. Andhra Pradesh is particularly vulnerable to cyclones formed in the Bay of Bengal. About four cyclones have occurred per decade in Andhra Pradesh since the 1980s, with one out of the four classified as severe.

Precipitation in AP and Telangana are largely due to the Indian monsoon system. The south-west monsoon is the main source of rainfall for the two states [20]. The monsoon advances from the southern tip of peninsular India at the end of May and spreads across the entire country within 10 to 15 days [21]. The monsoon gradually withdraws at the end of September commencing from northern India and reaching the tip of southern peninsular India by early December. Thus the advancing phase (South-West Monsoon) and, the withdrawal phase (North-East Monsoon) contributes to the rainfall pattern in the two states. Apart from the seasonality of rainfall, the distribution is variable, for example, the coastal areas of AP receive higher annual rainfall (750–1500 mm), whereas in the interior areas of AP and Telangana rainfall is much lower (300–500 mm), thus rainfall variability is high, 20 to 30%, particularly high in coastal areas. Thus, heavy rainfall occurs in cyclone prone areas along the coast in AP, besides, the two major river systems in AP and Telangana, the Krishna and Godavari rivers, which originate in the Western Ghats, can cause flooding due to the accumulated discharges of water from upstream areas. The flooding in delta flood plains is exacerbated by the combination of cyclonic rainfall and storm surges.

Droughts are a recurrent feature of the two states, 12.5 million ha in the two states are drought prone [22], defined here as areas that receive annual rainfall less than 75% of the normal (30 year average) in 20% of the years examined and where less than 30% of the cultivated area is irrigated.

Heat waves are another climatic extreme prevalent in the two states, during the months of April and May, maximum temperatures of about 49°C can be recorded. Not only are maximum temperatures high, the duration of maximum temperatures has increased from 7 days during the 1980s to 19 days during the 2000s.

3. Climate Change adaptation (CCA)

Adaptation has been defined by the IPCC as the adjustment in natural or human systems in response to actual or expected climatic stimuli or their effects, which moderates harm or exploits beneficial opportunities. Increasingly, there has been an emphasis on mechanisms that can anticipate and prepare communities to effectively deal with both gradual climate change and also extreme climatic events, commonly referred to as climate change adaptation strategies. The annual average adaptation costs for India in 2100 under various C-C scenarios can range from 0.36% of GDP to 1.32% of GDP [23]. CCA can result in win-win outcomes, by moderating the impacts of climate change, enable sustainable development, and also useful for disaster risk reduction [23]. Hence it will be important to understand how management actions can be developed, refined, and employed in the context of well-developed and flexible management systems in order to enhance our ability to cope with climate change [24].

3.1 CCA and Industries in India

India has embarked on an ambitious industrial program, wherein it plans to increase industrial growth rate from 10 to 14%, increasing the contribution of the industrial sector to 25% of GDP, adding 100 million jobs, by 2022. However, these milestones could be impacted, if industries are not resilient to emergent climate change threats, such as extreme weather. Industries are vulnerable to losses, stemming from extreme climate, due to flooding, tropical storms, and heatwaves. Adaptation, to these emerging threats need to be incorporated into master planning while setting up industrial parks, industrial estates, and special economic zones. There is an urgent need to enhance climate resilient industrial parks throughout the country, resilience could be incorporated at the planning stage for extreme climatic events, it could be made mandatory to include CCA in the EIA. For example adaptation measures such as, managing drainage for run-off of excess water; mitigation of heat islands; increase provision of intermediate water storage facilities; provision for water recycling; creation of green spaces/blue spaces; storm reduction measures; separate storm water/sewage; maintenance of drainage networks. It is important to locate critical infrastructure at higher elevations to prevent flooding. Industries could shift to renewable energy, and also have provision for sustainable backup of power. With increase in the frequency of heat waves, there is need for cooling for ICT facilities; develop shade and cool storage facilities; design green buildings and enhance thermal regulation within elevated buildings; design roofs of industry to cope with tropical storms/hurricanes. There is also need to improve circularity and resource efficiency, through increased water efficiency in production, reuse of water, including recycled materials into the production process, and also reduce dependency on climate impacted raw materials [25].

The efforts to decarbonise and adopt climate resilient development would require substantial knowledge up grading, skill development, and awareness among various stakeholders of the industrial areas in these two states. Capacity development for all stakeholders connected with industrial development in the two states will be critical to meet the needs of a climate resilient industrial development future. Decision makers in departments such as the environment, industries, planning, will have to integrate climate resilience into planning processes. Capacity of technicians and engineers to develop and design novel and innovative technologies, for example the ability to install technologies in renewal energy, will be important. Currently, organizational structures and management skills are lacking to make decisions under climate uncertainty.

Capacity development is central to the improvement of societies, organizations, and individuals and is essential to strengthen and maintain the abilities of these structures at all levels of management. In fact development can encompass a wide variety of measures ranging from, the creation of enabling policies that accelerate the capacity development of organizations/individuals, to changes in individual behaviors through knowledge and skills.

4. Objectives

The main objective of our chapter was to systematically examine learning and skill development needs for the stakeholders in industrial parks of two southern states of India, Telangana and Andhra Pradesh (AP) in the wake of climate change and extreme climatic events. APIIC (Andhra Pradesh Industrial and Infrastructure Corporation) and TSIIC (Telangana State Industrial and Infrastructure Corporation) are two umbrella organizations responsible for catalyzing industrial development in the two states, in a planned manner. There are about 257 industrial parks in AP spread over 13 districts, similarly in Telangana there are about 118 industrial parks (IPs) (**Figure 1**). A stakeholder analysis was conducted to assess the current capacity and gaps in existing knowledge, skills, and abilities to meet the needs for CCA.

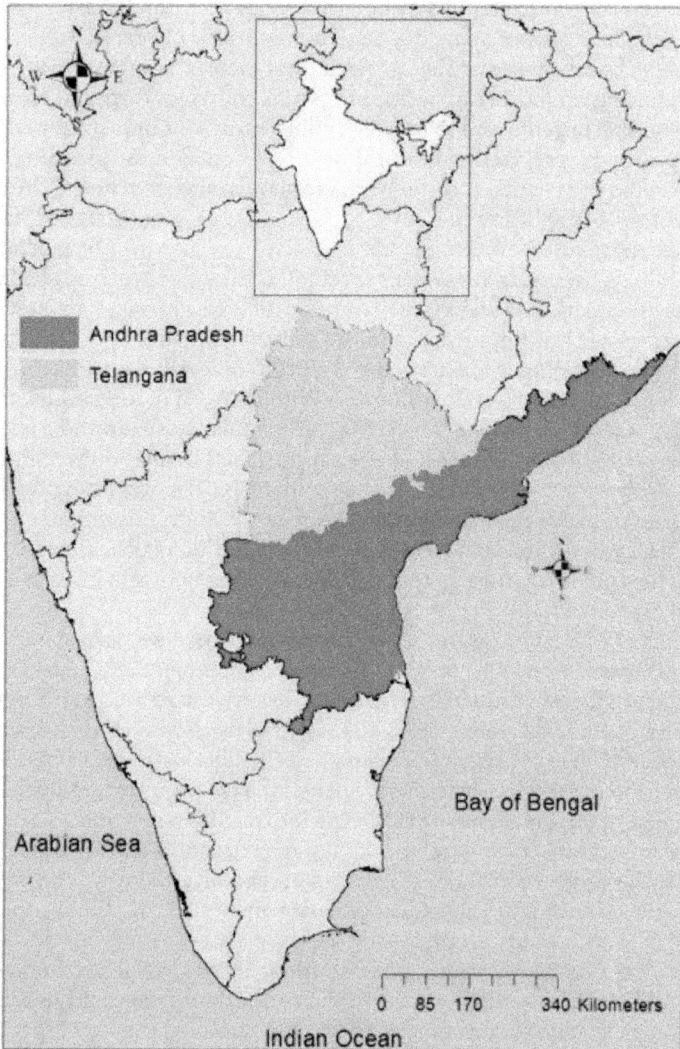

Figure 1.
Location of the two study states in India.

5. Methods

Stakeholder consultations were conducted in three IPs in Telangana State, including Cherlapally (I to IV), Hitech City, Madhapur, and Jeedimetla. Similarly, stakeholder consultations were conducted in four IPs in the state of Andhra Pradesh, including, Kakinada, Gajulamandhyam, Gajuwaka, and Ongole. Details regarding the type of industries in these IPs, the number of plots in each, and the extent of these IPs can be found elsewhere [25]. The criteria used for selecting the IPs, was based on the exposure of existing IPs to climatic changes, especially extreme climatic conditions in recent times such as severe cyclones, droughts, heat waves, and floods. Likewise, we included a variety of industries ranging from information technology to pharmaceuticals.

The capacity needs assessment was conducted in two stages, during the first stage, a brief presentation was made on climate change, and distinctions between gradual and extreme climatic events and its impacts. During the second stage, a focus group discussion (FGD) was conducted among the various stakeholders by dividing them into smaller groups to elicit the current level of capacity with regard to various functions and tasks with regard to CCA for both stakeholders and functional groups [25].

Information on current capacities and future needs for CCA were assessed, based on a numerical scale, scores were computed for current capacities for stakeholder groups (industry, local, and state functionaries), and simultaneously for functional groups (financial tasks and functions, planning tasks and functions, and engineering and infrastructure tasks and functions), and likewise scores were computed for target capacities.

$$AS = \sum_{i=1}^{n} \text{Current tasks} \tag{1}$$

$$TS = \sum_{i=1}^{n} \text{Expected tasks under CCA} \tag{2}$$

Where AS = Actual score; TS = Total score

$$\text{Current Capacity} = \sum_{i=1}^{n} \text{Current tasks} / \sum_{i=1}^{n} \text{Expected tasks under CCA} \tag{3}$$

$$\text{Capacity Gap} = \left(1 - \sum_{i=1}^{n} \text{Current tasks} / \sum_{i=1}^{n} \text{Expected tasks under CCA}\right) \tag{4}$$

Eq. (4) provides the capacity gap, high value indicates greater training needs.

6. Results

6.1 Analysis of capacity needs for CCA in Telangana

Our analysis indicates that in the state of Telangana, all stakeholder groups and functional groups require significant training on various issues related to CCA.

The stakeholder consultations and analysis revealed that for the stakeholder group industry, capacity building needs are moderate, likewise, for the stakeholder group, local functionaries; capacity building needs varies from low to moderate (**Table 1**). Similarly, for the stakeholder group state level functionaries, capacity building needs varies from moderate to high (**Table 1**). Currently, about one-third to half of the task and functions with respect to CCA is performed by various stakeholders, for about half of the remaining tasks and functions required under CCA, it is required to be developed through training and other mechanisms.

Capacity gaps with regard to functional aspects, such as finance, engineering and infrastructure, and planning revealed moderate to high needs. While, capacity needs with respect to engineering and infrastructure and planning are moderate especially among the officials from the head office, such as TSIIC. Capacity needs were high with regard to tasks and functions related to finance knowledge, skills, and abilities (**Table 2**).

6.2 Analysis of capacity needs for CCA in AP

Our analysis indicates that in the state of Andhra Pradesh, all stakeholder groups and functional groups require significant training on various issues related to CCA. For the stakeholder group industry, capacity building needs varies from low to moderate. For the stakeholder group local functionaries, capacity building needs varies from low to high, probably indicating higher training needs. Similarly, for the stakeholder group state functionaries, capacity building needs varies from moderate to high (**Table 1**). Thus in Andhra Pradesh from the perceptions of the various stakeholders in the consultations it appears that approximately 25 to 50% of the tasks and functions with respect to climate change adaptation is performed by various stakeholders. There is a substantial need for capacity building required under CCA to be developed through training and other mechanism.

When it comes to specific tasks and functions related to CCA with reference to finance, engineering and infrastructure, and planning, the stakeholder consultations indicated moderate needs among the officials from the head office, such as APIIC (**Table 2**).

Sl. No.	Stakeholder	Telangana	AP
1	Industry	Moderate	Low to Moderate
2	Local functionaries	Low to Moderate	Low to High
3	State functionaries	Moderate to High	Moderate to High

Low: <50%; Moderate: > 50 < 75%; High >75%.

Table 1.
Capacity needs gap for stakeholder group industry in Telangana state and Andhra Pradesh.

Sl. No.	Functional group	Telangana	AP
1	Finance	High	Moderate
2	Engineering & Infrastructure	Moderate	Moderate
3	Planning	Moderate	Moderate

Low: <50%; Moderate: > 50 < 75%; High >75%.

Table 2.
Capacity needs gap for specific tasks and functions required for CCA in Telangana state and Andhra Pradesh.

7. Discussion

Our analysis from the stakeholder consultations provides support that various stakeholder groups require capacity building for CCA and climate resilient development outcomes. Individual industries in both states appear to have higher capacity with regard to CCA, nevertheless, for the overall success of CCA, there is a need to simultaneously develop the capacities of decision makers at the scale of individual IPs as well as state level organizations. In particular, officials working in various agencies such the environment agency, the district planning agencies, the local municipalities require assistance to improve resilience of industries for managing climate change impacts. There is an urgent need to develop capacity of decision makers to identify solutions based on the impending climate impacts, with emphasis on local conditions, for example, capacity of officials could be developed in such a manner that they identify and implement local interventions, such as land zoning, storm water management, and developing building codes.

The ability of local institutions and functionaries to intervene is a critical component in moderating the harmful effects of climatic extremes [4, 26]. At the local scale, in IPs, local municipalities need capacity building to identify key climate related hazards, the spatial and temporal pattern of these hazards, the susceptibility, and the resilience of IPs, climate change adaptation measures, and locally viable climate resilient strategies need to be developed. Likewise, the industry stakeholder group requires substantial training for developing climate resilient measures and climate proofing industries to extreme climatic events. For example, industries could identify alternate sources of supplies and markets for products. Innovations in identifying new markets could lead to numerous new opportunities and spinoffs provided by CCA. Similarly, factoring the long-term risks of C-C and the early adoption of C-C adaptation measures could result in resilient and profitable industrial landscape. An interesting finding from the stakeholder analysis was the higher capacity with regard to CCA among certain industries, such as the ICT (Information and Communication Technologies). This could be explained by the global nature of this industry, hence it is better prepared to deal with extreme climatic events, both in terms of financing as well as the knowledge base.

Better planning for mainstreaming of climate change adaptation into plans of the government would be important. National and state level plans for climate change adaptation in Andhra Pradesh and Telangana could provide overall policy and goals to develop climate resilient pathways. Some of these measures include providing policy framework to guide decisions at the state level; providing legal framework; directing actions in key sectors. Similarly, multi-sectoral and multi-spatial planning across sectors such as environment, agriculture, industries, urban, irrigation, and other sectors are crucial for CCA. The stakeholder analysis identified specific needs with regard to development of master plans, including climate resilient measures at inception stage of project, mock drills, risk analysis for C-C, development of early warning systems for floods, cyclones, heat waves, and droughts in both states. A key aspect lies in the dissemination of the CCA planning strategies, it is critical that ICT (Information and Communication Technologies) are leveraged by officials in both states to ensure effective outcomes. A reassuring finding of the stakeholder analysis was the existing capacity among various officials, especially among engineers, with regard to engineering and infrastructure. However, specific capacity development with regard to mainstreaming climate change adaptation into their existing capacities would enhance CCA outcomes [27]. In line with Nationally Determined Contributions (NDC) of the Paris Agreement, development of green infrastructure, climate resilient infrastructure and nature based solutions to address climate change would lead to win-win outcomes.

During the stakeholder consultations a frequently occurring need was regarding capacity to leverage financial resources to meet the demands of CCA. Specific gaps in skills included, sustainable financing, green financing, and green budgeting. In this regard, specific training to attract funding from the Green Climate Fund (GCF) would be important [28]. Specific training programs garnered towards financial incentives, including taxes and subsidies; insurance, including weather based insurance schemes; creation of catastrophe bonds; payments for ecosystem service; differential water tariffs; microfinance; and disaster contingency funds were identified during the analysis. Financing that includes diversity of portfolios, such as public and private funding mechanisms; debt and equity for climate financing; export credits and foreign direct invests for climate change adaptation could be important information to be included in capacity development for CCA [29].

8. Conclusions

Climate change and the occurrence of extreme climatic events could have deleterious effects on industrial activities. The occurrence of disastrous events such as floods, droughts, storms, and heatwaves could have implications for human lives and economic losses for organizations. Climate change adaptation to these impending threats to industrial activities could moderate the impacts of these extreme climatic events, thereby reducing loses to industries. The study indicates that, there is an urgent need to build capacity among various stakeholders involved in the industrial development landscape. Functionaries at local and state levels need critical training in various climate change adaptation measures, similarly, industry owners, need support to alleviate the impacts of climate change. Specific knowledge, skills, and abilities, with regard to infrastructure and engineering capacities, planning capacities, and financial aspects, especially green financing of CCA activities are required. The chapter provides important information on assessing capacity gaps with regard to CCA in the industrial sector. Further, the chapter also identifies critical gaps in capacity among various stakeholders.

Acknowledgements

I would like to thank the funding agency GIZ for a grant to successfully carry out the research in the two states. I would also like to thank the various officials and industry representatives in the two states for their active participation and feedback in preparing the capacity needs assessment. I would especially like to thank Dr. Dieter Brulez, Dr. Hrishikesh Mahadev, and Dr. Rajani Ganta for their useful comments.

Author details

Narendran Kodandapani
Center for Advanced Spatial and Environmental Research (CASER),
Bangalore, India

*Address all correspondence to: svknaren@gmail.com

IntechOpen

References

[1] Figueres C, Schellnhuber HJ, Whiteman G, Rockstrom J, Hobley A, Rahmstorf S. Three years to safeguard our climate. Nature. 2017:546:593-595.

[2] IPCC, 2018: Summary for Policymakers. In: Global Warming of 1.5°C. An IPCC Special Report on the impacts of global warming of 1.5°C above pre-industrial levels and related global greenhouse gas emission pathways, in the context of strengthening the global response to the threat of climate change, sustainable development, and efforts to eradicate poverty [Masson-Delmotte, V., P. Zhai, H.-O. Pörtner, D. Roberts, J. Skea, P.R. Shukla, A. Pirani, W. Moufouma-Okia, C. Péan, R. Pidcock, S. Connors, J.B.R. Matthews, Y. Chen, X. Zhou, M.I. Gomis, E. Lonnoy, T. Maycock, M. Tignor, and T. Waterfield (eds.)].

[3] Jolly, W. M., Cochrane, M. A., Freeborn, P. H., Holden, Z. A., Brown, T. J., Williamson, G. J., & Bowman, D. M. J. S.. Climate induced variations in global wildfire danger from 1979 -2013. Nature Communication. 2015:|6:7537|DOI: 10.1038/ncomms8537.

[4] IPCC, 2014. Impacts, Adaptation, and Vulnerability. Part B: Regional Aspects. Contribution of Work- ing Group II to the Fifth Assessment Report of the Intergovernmental Panel on Climate Change. Cambridge University Press, Cambridge, United Kingdom and New York, NY, USA. 2014:1327-1370.

[5] GEO-6. Healthy planet healthy people. Cambridge University Press, Cambridge, UK. 2019.

[6] Hughes TP, Barnes ML, Bellwood DR, Cinner JE, Cumming GS, Jackson JBC, Kleypas J, van de Leemput IA, Lough JM, Morrison TH, Palumbi SR, van Nes EH, Scheffer M. Coral reefs in the Anthropocene. Nature. 2017:546, 82-90.

[7] IPCC, 2012. Managing the risks of extreme events and disasters to advance climate change adaptation. A special report of working groups I and II of the Intergovernmental Panel on Climate Change. IPCC, Cambridge University Press, Cambridge.

[8] Munich RE (2020). NatCatSERVICE analysis tool. Retrieved from https://natcatservice.munichre.com

[9] The Economist. A rising tide: Natural disaster loss events by cause. September 2017.

[10] Economic Times, 2016. Disaster Count.

[11] Tubiana L. The people's pledge to fight climate change. THE HINDU|The New York Times (30th December, 2017 issue).

[12] Hov O, Terblanche D, Carmichael G, Jones S, Ruti PM, Tarasova O. Five priorities for weather and climate research. Nature. 2017:552:168-170.

[13] WEF 2018. The global risks report 2018.

[14] WEF 2021. The global risks report 2021.

[15] Mohanty, A. 2020. Preparing India for Extreme Climate Events: Mapping Hotspots and Response Mechanisms. New Delhi: Council on Energy, Environment and Water.

[16] Kale MP, Ramachandran RM, Pardeshi SN, Chavan M, Joshi PK, Pai DS, Bhavani P, Ashok K, Roy PS. Are climate extremities changing fire regimes in India? An Analysis using MODIS fire locations during 2003-2013 and gridded climate data of Indian Meteorological Department. Proc. Natl. Acad. Sci., India, Sect. A Phys. Sci. 2017: 87(4), 827-843.

[17] Ramesh KV, Goswami P.. Reduction in temporal and spatial extent of the Indian summer monsoon. Geophysical Letters. 2007: 34: L23704, doi: 10.1029/2007GL031613.

[18] Borah PJ, Venugopal V, Sukhatme J, Muddebihal P, Goswami BN. Indian monsoon derailed a North Atlantic wavetrain. Science. 2020:370:1335-1338.

[19] EPTRI 2012. State Action Plan on Climate Change for Andhra Pradesh. Submitted to Ministry of Environment and Forests, Government of India, New Delhi. Available online at www.moef. nic.in/sites/default/files/sapcc/Andhra-pradesh.pdf

[20] Pai DS, Sridhar L, Rajeevan M, Sreejith OP, Sathbai NS, Mukhopadhyay B (2014) Development of a new high spatial resolution (0.25° × 0.25°) Long Period (1901-2010) daily gridded rainfall data set over India and its comparison with existing data sets over the region. Mausam, 65, 1-18.

[21] Pascal J-P (1986) Explanatory booklet on the Forest map of South India. (French Institute, Pondicherry).

[22] World Bank 2005. Drought in Andhra Pradesh: Long term impacts and adaptation strategies. South East Asia Environment and Social Development Department.

[23] Kabisch S, Bollwein T, Bank P, Brulez D, Varaprasad SS, Mahadev HR, Ganta, R. Climate change adaptation for sustainable industrial development: A strategy outline for the implementation of the "Climate Change Adaptation Project (CCA)" in industrial areas of AP and Telangana, India. 2015.

[24] Julius SH, West JM, Nover D, Hauser R, Schimel DS, Janetos AC, Walsh MK, Backlund P. Climate Change and US Natural Resources: Advancing the Nation's capability to adapt. Issues in Ecology. 2013.

[25] Kodandapani N, Bala Subramanyam G. Understanding capacity needs requirements for different stakeholders of climate change adaptation for industrial areas of Andhra Pradesh & Telangana. Funding agency, Integration and GIZ. 2016.

[26] Denton, F., T.J. Wilbanks, A.C. Abeysinghe, I. Burton, Q. Gao, M.C. Lemos, T. Masui, K.L. O'Brien, and K. Warner, 2014: Climate-resilient pathways: adaptation, mitigation, and sustainable development. In: Climate Change 2014: Impacts, Adaptation, and Vulnerability. Part A: Global and Sectoral Aspects. Contribution of Working Group II to the Fifth Assessment Report of the Intergovernmental Panel on Climate Change [Field, C.B., V.R. Barros, D.J. Dokken, K.J. Mach, M.D. Mastrandrea, T.E. Bilir, M. Chatterjee, K.L. Ebi, Y.O. Estrada, R.C. Genova, B. Girma, E.S. Kissel, A.N. Levy, S. MacCracken, P.R. Mastrandrea, and L.L. White (eds.)]. Cambridge University Press, Cambridge, United Kingdom and New York, NY, USA, pp. 1101-1131.

[27] Klein, R.J.T., G.F. Midgley, B.L. Preston, M. Alam, F.G.H. Berkhout, K. Dow, and M.R. Shaw, 2014: Adaptation opportunities, constraints, and limits. In: Climate Change 2014: Impacts, Adaptation, and Vulnerability. Part A: Global and Sectoral Aspects. Contribution of Working Group II to the Fifth Assessment Report of the Intergovernmental Panel on Climate Change [Field, C.B., V.R. Barros, D.J. Dokken, K.J. Mach, M.D. Mastrandrea, T.E. Bilir, M. Chatterjee, K.L. Ebi, Y.O. Estrada, R.C. Genova, B. Girma, E.S. Kissel, A.N. Levy, S. MacCracken, P.R. Mastrandrea, and L.L. White (eds.)]. Cambridge University Press, Cambridge, United Kingdom and New York, NY, USA, pp. 899-943.

[28] UNFCCC, 2007: Investment and Financial Flows to Address Climate Change. The United Nations Framework

Convention on Climate Change
(UNFCCC), UNFCCC Secretariat,
Bonn, Germany, 272pp.

[29] Chambwera, M., G. Heal, C.
Dubeux, S. Hallegatte, L. Leclerc, A.
Markandya, B.A. McCarl, R. Mechler,
and J.E. Neumann, 2014: Economics of
adaptation. In: Climate Change 2014:
Impacts, Adaptation, and Vulnerability.
Part A: Global and Sectoral Aspects.
Contribution of Working Group II to the
Fifth Assessment Report of the
Intergovernmental Panel on Climate
Change [Field, C.B., V.R. Barros, D.J.
Dokken, K.J. Mach, M.D. Mastrandrea,
T.E. Bilir, M. Chatterjee, K.L. Ebi, Y.O.
Estrada, R.C. Genova, B. Girma, E.S.
Kissel, A.N. Levy, S. MacCracken, P.R.
Mastrandrea, and L.L. White (eds.)].
Cambridge University Press,
Cambridge, United Kingdom and New
York, NY, USA, pp. 945-977.

Chapter 13

Climate-Driven Temporary Displacement of Women and Children in Anambra State, Nigeria: The Causes and Consequences

Akanwa Angela Oyilieze, Ngozi N. Joe-Ikechebelu,
Ijeoma N. Okedo-Alex, Kenebechukwu J. Okafor,
Fred A. Omoruyi, Jennifer Okeke, Sophia N. Amobi,
Angela C. Enweruzor, Chinonye E. Obioma,
Princess I. Izunobi, Theresa O. Nwakacha, Chinenye B. Oranu,
Nora I. Anazodo, Chiamaka A. Okeke,
Uwa-Abasi E. Ugwuoke, Uche M. Umeh,
Emmanuel O. Ogbuefi and Sylvia T. Echendu

Abstract

With increasing periods of extreme wet seasons, low lying geographic position, with socioeconomic, and political factors; some communities in Anambra State, Nigeria experience heightened floods annually resulting in loss of shelter, displacement of people with breakdown of livelihoods, particularly in rural communities worsening their risks and vulnerabilities. In 2012, a major flood event in the state temporarily displaced about 2 million people. In this chapter, we used a community-based adaptation approach to investigate the causes and consequences of climate-related temporary displacement on community members in Ogbaru LGA, Anambra State following flood events. We used global positioning system to obtain the community's ground control points and gathered our data *via* field observation, transects walks, focus group discussions, photography, and in-depth interviews. Our findings reveal a heightened magnitude of flood related disasters with decreased socio-economic activities, affecting their health and well-being. Also, the community members have a practice of returning to their land, after flood events, as a local mitigating risk management strategy. For multilevel humanitarian responses at the temporary shelter camps, it becomes imperative to meaningfully engage the community members on the challenging risks and vulnerabilities they experience following climate-driven temporary displacement to inform adaptation and resilience research, policy change and advocacy.

Keywords: climate-driven temporary displacement, women, children, flood reoccurrence, CBA and Anambra State

1. Introduction

Over the years in Africa, changes in climate worsened by disasters and fragility are becoming increasingly common, particularly in some vulnerable Nigerian weather-impacted communities threatening disprivileged populations. With accompanying periods of extreme wet seasons and their effects on heightening flood events, there have been *climate-related temporary displacements* from excess water coursing through the lands [1]. Climate-related temporary displacement is events impacted the anthropomorphic aspect of climate change, such as loss of shelter, movement of people, and breakdown of livelihoods of rural and urban communities with added risks and vulnerabilities, besides other socio-economic factors that are affected [2]. Due to the need to continue to protect community biological diversity, health, and natural resources [3], the community members resort to community-based climate risk management options, such as climate-related temporary displacement. This temporary displacement involves going back to the affected community after flooding events to manage and reduce impacts of climate change and local pressures [4]. As defined by [4], *community-based adaptation* is a local mitigating risk management strategy that involves the process of migrating back to their community periodically after an environmental stressor, such as flooding. Community-based adaptation (CBA), a process led by the community, is aimed to meet and prioritize the needs of communities, leveraging on their built capacities and the knowledge they hold, and empowering members with the required information and activities for favorable climate change outcomes.

Community-based approach (CBA) utilizes a solution-oriented lens to confront challenging complex issues associated with human displacement, internal dislocation, and relocation, as well as addressing the present climate-fueled flood reoccurrence events in affected communities. With its beneficial approach, CBA has been shown to increase and respond to comprehensive climate change threat(s) and its impacts on disprivileged populations [5]. Notably, CBA exhibits a partnering relational procedure between groups, such as community and institutional stakeholders, and not activities that were decided on, and imposed on the locals [6]. This procedure can build, strengthen, and bridge on existing adaptive capacities, relational values, and skills, while holding on prevailing local knowledge and technologies to encourage the communities on their community-led goals [7, 8].

For our study area: Anambra State is made of 21 Local Government Areas (LGA), and Ogbaru LGA is one of them, with Akili Ogidi as one of its main towns in this LGA. Ogbaru LGA is in the Southwest of Anambra State in Nigeria (see **Figures 1** and **2**). And with a population of about 221,879 [9], Ogbaru LGA occupies an area of 388 km² and density of about 762.3/km² [10] (see **Figure 2**). To the North, Akili Ogidi is bounded by Onitsha South Local Government Area, on the South by Rivers state and Imo state, while on the Western side, it is bounded by Delta state and on the East by Idemili South, Ekwusigo and Ihiala Local Government Areas. (https://en.wikipedia.org/wiki/Ogbaru?msclkid=737721afcf3911ec90c 5d2cabc0b2eb2).

Lying on the latitudes 5°42′N to 6°10′N and longitudes 6°41′E to 6°50′E, respectively, Akili Ogidi is noted for its agricultural activities [11]. The flood/alluvial plains of the Niger River are shown to form the major parts of our study site (see **Figures 2** and **3**). The vegetation is characterized by modified green areas and grasslands in remote areas [12]. With an elevation of 25 m above sea level, the area is dominated by shallow aquifers, while the climate is tropically characterized by high precipitation averaging between about 1800 and 2300 mm [13]. For the relative humidity in our study area, which is within an average of about 60–70% in July, the average daily annual temperature of this area is in the range of 24–28°C with a

Figure 1.
Map of Nigeria showing Anambra State.Source: Department of Environment Management, COOU (2021).

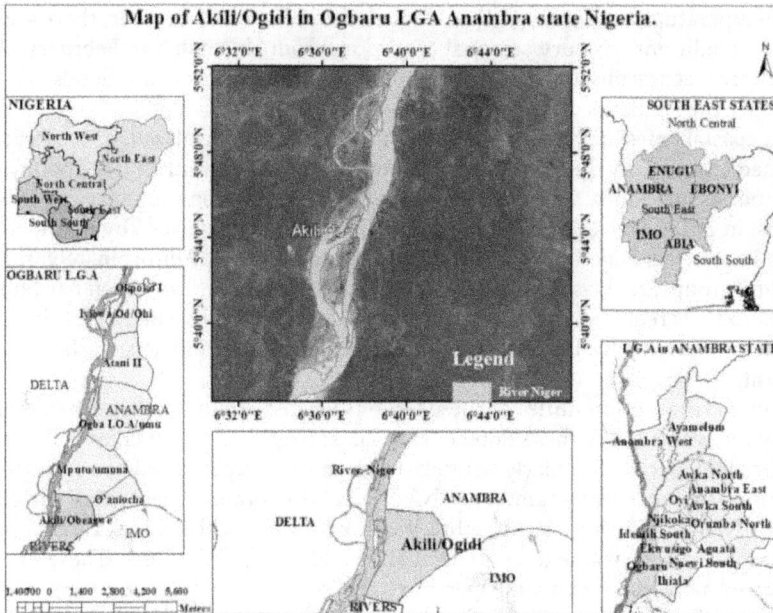

Figure 2.
Maps showing Southeast States, Anambra State and Ogbaru Local Government Area and Akili Ogidi—study area surrounded by River Niger.Source: Department of Environment Management, COOU (2021).

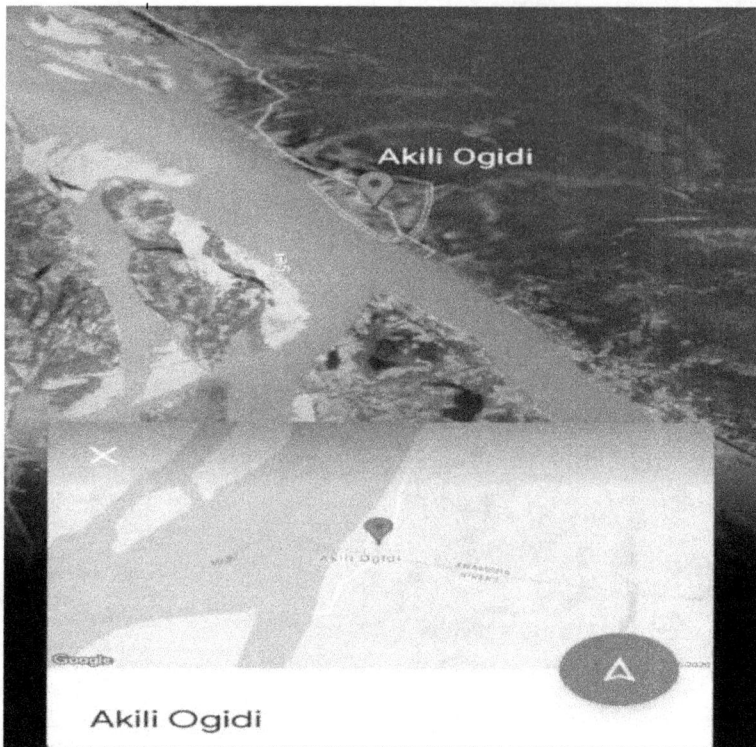

Figure 3.
Satellite image showing Akili Ogidi.Source: Google Map (2021).

night temperature of 16–18°C [14]. Averagely from March to October, there are wet climatic conditions with dry seasonal conditions from November to February [15]. The longer wet seasons are prevalent with intense storms that cause floods displacing the residents during the entire wet season.

For coastal cities in Nigeria, they are continually being displaced, dislocated, and relocated temporarily due to continual events of floods resulting in large population of temporary migrants, further threatening the life of the communities and individuals. In 2018, over 1.9 million persons were internally displaced by the floods across 12 States in Nigeria and over 500,000 were homeless. Unfortunately, marginalized groups, such as rural women and children majorly, have been hard hit leading to their reduction in income levels, pressure on food security, health, poor nutrition worsening health status of children, besides other impacts on the environment.

Though rural communities of these regions located around the river banks are usually the worse hit by these flood events, most Nigerian coastal cities and communities in Nigeria particularly Southeast region have experienced continual sea-level rise, from increasing rainfall. In Nigeria, all the communities within the bank of River Niger and Benue, in Anambra, Kogi, Edo, Delta, Nassarawa, Taraba, and Adamawa states, continue to be flooded almost like an annual event. These states are located along the two main Nigerian rivers: the Niger and Benue rivers. These flooding events further implicate the rising water levels of these two rivers as the primary causes behind the flooding experienced in Nigeria. Further, the ecosystem-dependent livelihoods, such as rainfed agriculture, are breaking down due to the changing temperatures and associated erratic high-volume patterns of rains impacting agricultural procedures [16, 17]. Of note, these floods threaten agricultural production levels and food security [18] aggravated by increased inter-annual variability of precipitation, which may heighten temporary displacement out of the lower rural land production areas to the urban regions.

The 2063 agenda for the African Union's (AU) agenda is a call to strengthen humanitarian activities and to transform the continent within 10 years. IPCC Fourth Assessment Report [19] in its summary of climate change projections and impacts for Africa reported that Africa is a vulnerable continent. Africa's vulnerability to climate change may be hinged on the wicked problems, such as extreme poverty, unregulated emissions from fossil fuel combustion, poor physical planning, governance failures, corruption, violence extremism, inequalities, food insecurity, unemployment [20], and uncontrolled population rise with an imbalanced demographics that characterize the continent. These multiple intractable problems are worsened by lifestyles, such as indiscriminate waste disposal, uncensored felling of trees, and deforestation that aggravate anthropogenic emissions leading to extensive influence on population and natural ecosystems.

Globally, World Health Organization (WHO) has estimated that in the next 30 years (by the year 2030 and 2050), climate variability may result to increasing environmental disaster-related annual deaths (about 250,000 additional deaths). In these contexts, declaring 2019 as the "Year of Refugees, Returnees and IDPs" by AU is aimed to promote lasting solutions, particularly to forced displacement witnessed in Africa. Of note, the launching of the African Climate Mobility Initiative (ACMI) will continue to assist the works of the African Union member states in addressing the complex issues of climate-related forced internal and external migration and leverage on possibilities that come with the African climate-induced migration. Notably, climate changes will escalate diverse impacts, such as rising sea level and extreme weather in the West African region, particularly Nigeria [21] in addition to the worsening socioeconomic milieu. However, in recent times, modern sophisticated modeling technologies coupled with improved research studies have provided atmospheric scientists the ability to determine and comprehend the causes of the majority of weather events, particularly the events influenced and heightened by climate change.

In light of the aforesaid discourse, the goal of our chapter is to investigate the causes and consequences of climate-driven floods and temporary displacement prevalent among community members, particularly women and children in Ogbaru LGA, Anambra State. This study provides information needs shared by affected community to reduce the risks of climate variability by involving them in planning of adaptational practices, activities, and planning.

2. Materials and methods

A qualitative case study approach was employed in this study to provide a holistic and in-depth explanation of the community-based flood problem and its implications for temporary displacement on the people of the Ogbaru community, particularly women and children. Our primary data were key informant interviews, observations, semi-structured interviews, photographs, focus group discussions, and informal interactive dialogs with indigenes that were recorded. We also used secondary sources of information. For our reviews, we also included an analysis of our operational environment, livelihood sources, challenging vulnerabilities and hazards, and the dynamics of the socioeconomic milieu. Also, we used the purposive sampling technique (convenient and snowball methods) to select our informants to ensure that their core views and voices of women and the participating male informants of problems under study were appropriately represented. Forty community members were randomly interviewed using unstructured questions.

Our interview and discussion groups were carried out using the Igbo and English languages. Effective interpretation back to the English language was done by our research assistants. We included different variables from our target relative to educational background, marital status, occupation, and socioeconomic data of the perceptions of the interviewees on the effects of river flooding. We also employed the focus group discussion (FGD) to interact with 60 women using unstructured questions. There were separated into six women groups (10 in each group) namely single, married, and aged/with disabilities. Forty persons were randomly interviewed using semi-structured interviews, while eight (8) persons (5 women and 3 men) were recruited to be involved in our key in-depth interviews (see **Figures 4–7**).

Figure 4.
The community chairman Mr. Uche Ijomah (red arrow) who came to receive the researchers at bank of River Niger to the study area (blue arrow).

Figure 5.
Different FGD sessions of Akili Ogidi women.

Figure 6.
A photo session of the researchers with the community women after the focus group discussion.

Figure 7.
A photo session of some of the women researchers with the community chairman (red), community chief (blue), elders, and women leader (red).

Each session of the three women groups lasted for about an hour. The FGD provided the researchers with the primary information about the community women's daily activities and how their roles intersected with their coping mechanisms and pressure from flooding. We, manually, analyzed our qualitative data, and for our photographs and the field observations, the images gave a vivid description of flood activity. The GPS obtained ground control points. We used other sources of data to confirm the results reflecting unique findings from our study area.

3. Results and discussion

3.1 Sample characteristics

In Nigeria, temporary-displaced migrants from communities are majorly affected by climate-related flooding, and they move out of their indigenous lands to places of refuge making them high-risk and a vulnerable group [22]. For a qualitative study, our sample size (n = 100) was a large one, ensuring that a greater number of community members were included. Moreover, most of the participants were eager to be part of the study. Demographic variables, such as sex, age, income, and occupation, were asked, with a summary of some social data (economic, environment, gender roles, and health data).

Table 1 shows the summary of core characteristics of our participants. The highest age group was in the 30 to 40 years category, while those in the 50 to 60 years had the lowest proportion of 18%. Most of our respondents were between the ages of 18–60 years, with lived experience of flood activities. A total of 72% of the participants were females and 18% were males. For their educational levels, all the participants had completed their basic primary education, but there were no

Variables	Frequency
Age	
18–30 years	22
30–40 years	36
40–50 years	24
50–60 years	18
Gender	
Male	27
Female	73
Education level	
FSLC	53
WAEC/WASC	27
NCE/OND	00
BSC/HND	00
NONE	20
Marital status	
Single	28
Married	35
Widowed	17
Divorced	08
Separated	12
Household size	
1–4	20
5–9	42
10–14	38
Years of living in the community	
1–2	00
2–5	08
5–10	20
10 years and above	72
Income level (monthly)	
₦1000–10,000	32
₦11,000–₦20,000	28
₦21,000–₦30,000	15
₦31,000–₦40,000	10
₦41,000–50,000	07

Variables	Frequency
₦50,000 and above	08
Internal migration of respondents	
Gender	
Male	33
Female	67
Total respondent	100

Source: Researchers' Analysis (2020).

Table 1.
The socio-economic characteristics of participants in Akili Ogidi.

participants with complete or some level of completion of graduate-level education. More than 30% of our participants were married. A large number of the interviewed men had married several wives with large households ranging from 1 to 12 persons. For the household size, 38% of the participants had more than 10 persons within their household, while 20% had between one and four persons. Majority of the community have lived in the affected community for over 10 years or more. For their income level, more than 90% had very low income $1USD/day. From our discussions and interviews, a large number of community members were internal migrants, and women were in the majority (67%), and they migrated with their children particularly the young ones. Although we were targeting women, men wanted to be part of the study and we had to randomly select the men who consented to be part of the key in-depth interviews.

The results from qualitative case study approach provided comprehensive discussions on reoccurring flood activity, social consequences of flooding, as well as the effect of climate-related internal migration on women and children from the study area.

4. Causes of persistent flood activity in Akili Ogidi community

Unarguably, flooding events are aggravated by changing climate and it requires emergency rescue and rehabilitation of affected population and communities, because of the high level of accompanying morbidity. For example, the Nigeria Displacement Report of 2013 reported that about 4,189,650 people suffered from food insecurity, 3,883,215 people were targeted for malnutrition challenges, and 194, 859 were harbored in IDPS from floods. Clearly, climate change is fast becoming in Nigeria a major driver of human displacement. According to International Red Cross [23], more people will be displaced from weather-related and climatic events than war (http://www.icrc.org/en/document/internally-displaced-people). Climate-driven displacement, environmental migrants, or climate migrants come with adverse effects from climate persecution and violence to temporary displacements sequel to flooding, which has been prevalent for over 10 years in Anambra state. Climate-driven temporary displacement is a previous situation in Anambra-riverine communities annually as people flee from flood disasters to become displaced and can be stranded inside their own communities. Here, the floods have been extreme, particularly the flood events of 2012, 2013, 2018, 2019, and 2020 [24–29].

Key informant interviews reveal important insights on the major causes that contribute to the (re)occurring floods in the study area, and our findings are consistent with the past flooding events. Firstly, the community occupies a vast wet area that lies in the SouthWest part of Anambra state, at the bank of River Niger, which is a fragile and flood prone area with heavy rainfalls during the wet season. The flooding leads to a rise in water levels of River Niger, adjoining creeks, and ponds overflowing into the community. The leader for the women group described the anxiety and despair they face as the wet season ensues:

> ... The rainy season come with heavy storms and floods that threaten our lives, we are left with no choice than to face the harsh conditions it bring. You know that our community is facing the River Niger. We have accepted our fate that our problems are locational. This is worse because we have little help and assistance.

One of the male interviewees who emphasized that, apart from the floods of 2012, it was majorly influenced by the mismanagement of dams and water reservoirs in Nigeria and our neighboring countries. Undoubtedly, subsequent floods have been sourced from the overflow of the banks of River Niger. According to the elder (male):

> ... this makes our community vulnerable to flood occurrences annually during the wet season, and to flood and disaster. our people will always migrate within the state in search of safe locations and the experience brings diverse devastating consequences.

Ajaero and Mozie [30] reported that Ogbaru LGA is continually impacted by the floods due to the low and flat nature of the area. Ogbaru LGA slopes at angle of 1°–3°. The relative low disposition of Ogbaru land makes the area to be continually flooded for longer than 6 months, and our study site is situated where the River Niger has its highest discharge rates. The overflowing river bank affects the superimposed plain land, which is one of the main geographic characteristics of Ogbaru, particularly Akili Ogidi. The people of Akili Ogidi live in the "hotspot" flood region facing limited resources and support to adapt to an increasingly hostile environment. Notably, in 2012 and 2018, NEMA [31] declared Anambra state a flood disaster region twice [32].

Additionally, the community Chief informed us that there have been elongated noticeable changes in the weather pattern indicating rising rainfall intensity, frequency, and increased destruction occurring in the past 10 years in the community:

> ... there are clear changes in the weather. We have noticed this over time, especially as the onset as our wet season now come early, stays longer. sometimes it is irregular and the rains are heavy making us to be inactive as our community is flooded.

Clearly, the location of the state and its changing climate have added to the reoccurring and increased flood status of the state. Mgbenu and Egbueri [13] confirmed in their study that Ogbaru LGA is dominated by shallow aquifers of about 25 m above sea level with a tropical climate that is characterized by high precipitation averaging between about 1800 and 2300 mm. Notably, Akanwa and Ezeomedo [17] affirmed that additionally unpredictable weather conditions have aggravated the high precipitation levels in the area resulting in increased flooding and erosional problems traceable to variable timing and intensity of rainfall from the changing climate in Anambra State. Further, Nzoiwu et al. [33] also reported about noticeable changes in climate causing a gradual shift in the rainy seasons such that the rainfall season has extended beyond June to September to the months of October and November. For the dry seasons, the months have extended from December to March. Unfortunately, these changes were proven by the report provided by NEMA [31] that Anambra state is affected by a vast range of hydro-meteorological and climatological hazardous changes, such as storms and temperatures as the State is situated by the river Niger making it flood prone.

5. Consequences of climate-related floods

5.1 Destruction of houses and limited movement/access to residents

Results indicated that about 70% of the Akili Ogidi residents' economy, social life, and health were affected by the 2020 floods. Residential houses were totally abandoned and streets deserted as the floods made movement impossible except with the use of local canoe (see **Figures 8** and **9**). During the discussion session, we found that the majority of the displaced community members have nowhere to go, and within the onset of dry seasons that begin in late November and December periods when the flood has receded, they usually return back to their homes and lands but to a devastated community. In 2012, all 16 communities in Ogbaru LGA were submerged by the flood destroying more than 100 buildings and displacing hundreds of people [24]. Similarly, in 2013, Ogbaru LGA was flooded between the months of June and August causing a displacement of about 124, 859 people [25]. In 2018, about 9000 people from 1500 household were internally displaced from their homes in Ogabru LGA, and in 2020, about 5000 persons are displaced (homeless) in September during heavy storms and flooding in Ogabru LGA [33]. People could only access their homes using canoe, though movement to closer bigger cities, such as Onitsha town, can also be accessed *via* speed boats. However, using speed boats is expensive, costing about 1500 naira per person ($3.64 based on http://exchangerate. guru on January 3, 2022). The residents complained that it was expensive, and their income is affected by poor source of livelihood negatively affected by the flood. The situation was worsened by the fact that it was the only source of transportation during the 8-month flood season.

Figure 8.
Flooded streets and abandoned houses where residents have migrated.

Figure 9.
More flooded houses and streets that have become unhabitable and temporarily abandoned.

The community members fled their homes to find sanctuary with relatives and friends living on higher grounds. Some of them had to stay in public facilities such as schools and churches, designated as IDP camps though majority were stranded in their houses having nowhere to go, because of COVID-19 restrictions. The youth (female) leader reported that ...

> ... *our situation was pathetic since majority of our people were forced to leave our homes and our properties to find shelter in public buildings used as IDP camps to stay safe from the floods.*

A woman elder said this:

> ... *the effect of flood was worse, because of the COVID-19. most of our relatives were nonchalant about receiving flood victims into their home, so many of our people were forced to stay and face the harsh conditions of the flood.*

Additionally, the secretary of the women group reported that for other flood episodes she relocated with her three children to nearby town—Onitsha to stay with relatives until the floods were over. However, the case was different because of the heightened awareness of spread of COVID-19 virus. She said:

> ... *we have experienced heavy rainfall every year, and this year, 2020 was intense. The water filled my house unlike other times the flood hardly gets into my house. I decided to remain in my house because I can't leave the only place, I have called my home. I, husband and three children battled the floods due to the COVID restrictions. We had nowhere to go".*

For our observations during field surveys, the flooding impacts were aggravated by the poor-quality houses constructed with low materials that have become weak over time (see **Figures 10** and **11**). Also, inadequate infrastructure and the absence of physical planning aided the rapid destruction of buildings during floods (see **Figure 10**). Further, the situation in the study area was worsened by the absence of drainage systems to check the flood problem.

Although the government of Anambra State provided communal shelters that were recognized as IDP camps to accommodate most of the stranded people, most

Figure 10.
A collapsed fence and an abandoned house with the roof destroyed by the heavy storms.

Figure 11.
Poorly developed houses, inadequate facilities, and poor living conditions of the residents.

of them refused to relocate to these (IDPs) camps. One of the interviewees admitted that the IDP camps were over populated having inadequate mattresses, insufficient mosquito nets and food items for their upkeep, and the affected residents to remain in their flooded houses. The temporary shelters were characterized with discomfort and pain worsening their vulnerability. This was confirmed by one of community women ...

> ... I and most of the people had to leave the camp because there was little space, too many mosquitoes, inadequate sleeping mats and the food provisions were too little to take care of the large number of stranded people. More people keep arriving daily, many were looking tired, disappointed and exhausted from the flooded condition of their homes'.

According to Punch [29], it confirmed that those who were accommodated in IDP camps experienced shortages of essential items, such as food, limited medical personnel, insecticide-treated bed nets needed for their daily survival, despite the increased numbers of women, pregnant mothers, and injured children.

6. Destruction of infrastructural resources with human dislocation and loss of lives

Findings from discussion groups and interviewees reported that the floods submerged infrastructures, such as shops, roads and footpaths, market, schools, churches, healthcare center, electricity poles were destroyed by the flood, although the use of low-building materials has not helped the impact of the consistent flooding in the community. Following the flooding, the schools were shut down once again after the initial three (3) months COVID-19 pandemic lockdown. This was a source of worry to community members, who were not only grieved by the disruption of their academic pursuit, but by the deaths of their infants during the floods. Notably, the low standards of the learning environment and the schools built with poor-quality structures can easily give way to intense storms and floods (see **Figure 12**).

The pupils and teachers have been studying under a collapsing wooden building because there are no classrooms to learn as past floods have submerged the buildings. The learning environment is not conducive and may affect the teaching and learning process, in addition to the long periods of flood intervention where the schools are closed annually. Due to the absence of electricity (floods submerged poles), there are no technological structures to facilitate Internet learning.

Obviously, the early childhood educational development in the area has been consistently affected. Early childhood education is a critical time of learning and foundation to leverage other levels of education. One of the elders during the course of interview on the schools added that ...

> ... our people suffer from flood problem all the time. The state government have tried to help the situation, but it is still persistent. Our children education is suffering because we have shut down the schools during the floods to avoid our children from death, drowning or sustaining injuries. On several occasions the school have lost all their chairs, desks, office furniture to the flood. We are begging the state government to bring a permanent solution to the flood.

Figure 12.
Poorly developed school structures and learning environment in the study area.

Also, the community Chief added

> ... we found that our building were going under the water and our people were
> trapped, displaced, many injured and cases of deaths, especially children drowning
> during the flood.

The closure of all socioeconomic, religious and educational activities in the
community was for over 6 months for the floods to recede. Also, the flood affected
infrastructure, the quality of water polluting the rivers in Ogbaru LGA, and affecting economic activities [34]. Again, the flood caused the influx of snakes and other
dangerous reptiles into homes-seeking refuge from the flood placing the life of
people at greater risk of attack [35].

7. Destruction of family bonds and social/community ties

The menace from the flooding continues to tear families and individuals apart as
they seek for temporary spots or relief camps to stay alive from the flood. One of the
representatives of the women group reported thus:

> ... majority of the women were separated from their families for 8 months adding up
> to the entire flood season and this affected our family relationship.

Generally, the close knitted and family culture of Akili Ogidi people is
heavily threatened by the annual flooding [36]. Indigenous communities in
Anambra State are characterized by shared learning, beliefs, and shared bonds of
fellowship that set their standards or patterns of behavior [37]. A common culture
such as language, livelihoods, and administration with close bond to biodiverse
natural resources is threatened yearly. It is unfortunate that these underlying
commitments within families, social groups and customs are threatened
annually. Worse still, most families experience the fragmentation of their social
family unit and ties. Similar published studies affirmed that climate-induced
floods affect fragile communities forcing internal dislocation of the marginalized
populations [38–44].

According to the report of the women leader during an interview session, she
indicated that the severity of the floods keeps increasing and has become an annual
event that the community has to deal with. She further informed us thus:

> ... families face the pain of not only leaving the comfort of their homes, but also, the
> disintegration of the members of the family to other locations due to the harsh
> consequences of the floods that sweeps through our homes. Women, children and
> husbands are separated from one another. The social disconnection is usually
> unbearable that most mothers risk being with their young children in IDP camps or
> other locations in order to avoid social disconnection.

Clearly, flood in the study area was responsible for unexpected deaths, inflicting
injuries on family members and also the separation of families creating societal
vulnerabilities for vulnerable persons. Evidently, the displacement from the normal
patterns of life creates unrest, tension, and violence (www.absradiotv.com). The
flooding events have been shown to create strains on family knitting and commitment heightening intimate partner- and gender-related violence when marginalized
groups, such as women, children, and the disabled, are left in relief camps exacer-
bating their vulnerability levels. According to a Pacific study involving six

Pacifician island nations, they reported that for all the women surveyed, more than 75% of participants experienced some types of violence, either the physical or the sexual form from their close partners and significant others around them (UN Women, n.d).

The findings further showed that these women experienced major forms of violence, such as assault and violence, implicating intimate partners and family members as perpetrators. Also, according to Wan [45] separating children from their parents come with complex devastating issues. This may be as a result of effect on critical bonds in human life. Floods can traumatize its victims for a long time, and often times, survivors are left with diminishing social determinants of health. Worse still, on such occasions of flood crisis, minors are given new roles as principal care givers, earners, and heads of households that can destabilize their mental and emotional health [46].

One of the affected community women reported:

> ... we have these floods every year, but the magnitude of this year's flood, 2020 was high almost similar to 2012.
>
> The flood damaged our house, crops such as rice, maize and cassava. I had to separate my four children in different places, so I and my husband can gather our lives and little belongings left from the flood. I miss my children all the time. I can hardly see them because we have little money left with me to fix our damaged house.

It is forecasted that there may be greater flood impacts on in Ogabru LGA, threatening community ideology, values, and belief systems as climate changes has remained unpredictable (www.channelstv.com//ag). Indeed, climatic extremes and uncertainties have become the norm where the community have to face prolonged flood crisis.

8. Destruction of farmlands: a threat to local income and livelihoods of women

As reported by Ugwu and Ugwu [47] one of the sectors impacted by flood is the agriculture, particularly in the global South nations. The Nigerian Hydrological Service Agency (NIHSA) and Nigerian Meteorological Agency [48] predicted the year 2020 will be characterized by a wetter season with thunderstorms and heavy rainfall, worsening the rising of water levels of rivers, intensifying longer lasting and more intense flash floods in Nigerian coastal cities than the previous year. Undoubtedly, agriculture in Nigeria is a huge industry that contributes to a substantial amount of about 26.09% of the national GDP, besides employing about 36% of the total Nigerian labor force and the largest direct employer of labor [49]. The Akili Ogidi community operates an agrarian economy that is dependent on rain-fed agriculture and the (in)direct impact on sustaining livelihoods. Notably, agriculture has played prominent roles in supplying food products, giving employment to community members. Akili Ogidi supplies food items and serves as food baskets to other urban markets and locations. Egbueri [14] mentioned that the community is majorly involved in fishing and farming activities due to the presence of River Niger. Clearly, the location of the study area by the River Niger, its high dependency levels on rain-fed agriculture and its high rurality index, climate change, and poor accessibility during floods are factors that place the community at high risk of flooding annually, where the poor conditions of the inhabitants are even a much bigger challenge to deal with providing no other option than for them to migrate internally.

Flooding negatively affects plants by uprooting their stem and roots thereby limiting its growth, when runoffs from heavy storms sweep the entire farm surface. It has led to widespread damage to crops, reducing quality and quantity of livestock and farm products being transported to larger markets. NEMA [25] reported in 2013, that about 2, 217 farmlands were destroyed by flood in Ogbaru LGA. The impact of floods was also felt on soil deposits and quality from heightened erosion challenges and associated fluvial deposits worsening the deep-layered mottled Ogbaru soils making the soils unfavorable to plant growth (see **Figure 11**). Several households have experienced financial loss, hardship resulting from loss of farm-lands, food insecurity, and hunger. Generally, this is a setback on local innovation, productivity, and development in of community.

Findings from this study confirmed the predictions of intense and last-longing floods of 2020 in Anambra state caused havoc on local farmlands, crops, and seedlings, bringing all farming activities to a halt. The inferences from interviews and discussions revealed that the community is hugely involved in the farming products such as cassava, citrus fruits, rice, and oil palm products. Also, rice cultivation is prevalent in the area, and particularly, the swamp rice cultivation and cassava farming were affected [50]. Drawing from the results of this study, about 72% of women report that their household income was from agriculture-related activities especially subsistence farming, fishing, and animal rearing. Findings from discussions revealed that women farmers were usually the worse hit during flood disaster where their farmlands are totally submerged destroying their farms and food products, economic/fruit trees, and farm animals (see **Figures 9** and **10**).

Generally, men and women continue to suffer from persistent floods-associated setbacks, yet women are impacted differentially as their livelihoods and general well-being are grossly depleted as they get more involved in gender roles, such as family care and nursing sick children. Women are found majorly in the agricultural sector, and invariably, the changing climate continues to impact women relative to accessibility to the resource opportunities they need to be more efficient. If the acclaimed unequal access to the resources and opportunities persists, then it will be difficult to combat food insecurity, hunger, malnutrition, and poverty that have become dire consequences. There is need for emphasis and actions that would promote gender equality while empowering women in agriculture to maximize their capacities in mitigating food insecurity, hunger, and extreme poverty. In the interview session, one of the community women leaders complained that their crops were either washed away by the floods or get rotten in the affected farmlands. She further informed that...

> ... we are usually afraid at the onset of the wet season because we experience huge destruction of our farmlands and our root crops are destroyed and over the years the situation has worsened knowing that we are riverine and our occupation is agriculture."

Majorly in Akili Ogidi, women are involved in the entire agricultural value chain, such as farm managers, suppliers of labor, harvesters, processors, and selling of farm products, making them vital and strategic in food security and agricultural production. Also, most women-headed households are all disproportionately affected by the reoccurring flood in the community. As confirmed by one of the women leaders in our group discussion, who has lost her husband, she said,

> "Women suffer the consequences of the flood especially the huge loss of income, major source of employment, livelihoods and food security for their family members. The

> prices of food products are high and transportation is equally high since the floods
> block the only access route in the community.

However, the women interviewees informed us that the community farmers have improvised local adaptation strategies where they farm twice in the year. The first planting season comes early in the year about late January/early February before the onset of wet season in March-April. This enables the farmers to harvest their farm products before their farms are submerged in the flood by May/June. The second planting season starts after the floods recede in late October/November, so that harvest can take place on or before January to prepare for another planting season to commence (see **Figure 11**). However, the floods of 2020 shattered their strategy, because the wet seasons extended longer than usual causing the floods to last longer as well. It is unfortunate that climate variability has remained unstable in recent times. One of the women interviewees responded saying ...

> *"We need government to support our farming activities so we can survive the floods*
> *and have food to give to our children since the flood takes away everything from us.*
> *We need farm inputs like fertilizer, tractors, loan, good roads to assist us in accessing*
> *our farmlands".*

Additionally, findings showed that the women and men were heavily dependent on the River Niger for their supply of large fish species during the floods. The community is traditionally surrounded by two Rivers Niger and Ulasi that provide various species of fishes and aquatic life. The participants informed during survey that they experience huge sales from large species of fishes caught and sold (see **Figure 12**). The floods provide an advantage for their fishing business because the majority of the fishes are brought to the river surface and even to their doorsteps during the flood events. During an interview session, one of the fisher women added that the supply of fishes is the only advantage of the floods ...

> *"The floods bring to us large amounts of fish from the river. It drops the fishes for us*
> *at the bank of the river. It makes it easy for us to collect the fishes. This makes so*
> *many women to go into fishing business in the wet season to increase our income. We*
> *have work to do this time apart from house jobs.*

It is on record that women face high levels of discrimination such as poverty, inadequate knowledge, and low execution of human rights among others, because women are overwhelmingly burdened with the huge roles of caring for children, the elderly, and people with disabilities and this places them at higher risks during flood disasters. A study carried out by Ihaji and Aondoaver [51] confirmed that women and children were the main casualties in flood relief camps in Cross River, Nigeria. There is need to support Akili Ogidi women to prepare for climate-driven floods. Hence, women should be given more considerations in terms of food, healthcare accommodation, and social facilities.

9. Effect of flooding on health of women and children

Findings from the study showed that flooding had extensive and significant effects on the health of community members—men, women, children, and people with disability. However, women and children were extensively affected by the harsh conditions provided by the flood. Alderman et al. [52] categorized health effects into short and long terms. Moreover, short-term health effects observed during the floods include infections, drowning, mental health challenges, physical injuries, and water-borne diseases. The longer-term health effects result from physical dislocation, shortages of safe water, injuries, depression over the loss of personal property, and death of a family member though most of the deaths in Akili Ogidi are from drowning of children and women. However, globally, it has been shown that mortality rates after major flooding increase by 50%, while with a prevalence of 8.6–53%, mental and psychological distress continue beyond 2 years after the flooding events [52].

Studies have confirmed that men and women are not affected in an equal magnitude because women have disproportionate vulnerability to flood disasters, compared with men. Considering, the flood disaster in Akili Ogidi, the women and children were more vulnerable because women are often concerned over their children's safety. During the focus group, discussion session of the married women, a nursing mother related her experience during the flood. She reported that ...

"During the 2020 floods I was pregnant and I had four other children aged 6, 5, 4 and 2years which I had to cater for. Worse still, our community health center was shut down because of the flood. I and the children suffered from exhaustion, stress, body pains, fever and physical injuries such as cuts, sprain from falls and insect bites from mosquitoes worsening my health condition".

Understandably with the physiology of the pregnant state, such injuries may indirectly contribute to complications such as abortions, abruption placentae leading to vaginal bleeding in pregnancy (in the case of trauma or fall on the gravid abdomen), and even death from being carried away by the flood. Also, access to health facilities were further compounded by the destruction of such properties by the floods. It was noted by observation and during interactions with the women that they had restricted access to contraception, which would have been obtained as part of family planning services during postnatal care where the health facilities to function as required.

Report showed that women and children were badly affected as there were reported cases of deaths. According to Duncan [53] more women die during flooding disasters than men, which may be related partly due to less physical ability to run, and most of these women died trying to save their children. It was noted that the Anambra flood in 2019 killed four persons—a nine (9) year old boy, a (7) year girl, an (18) year boy, and a pregnant woman [28].

Flood-related or flood-prone injuries occurred as people tried to save themselves, their family, and their valuable possessions from the flood and theft. There were also reported cases of sexual violence (women and children being raped and abused), concern over missing children, and properties. Other health features seen were loss of appetite, lack of sleep, nightmares, tiredness, and irritability. All these can affect the psychological and mental health of the women and even the children whom flood has made orphans and homeless. Makwana [54] affirmed that mental and emotional/psychological trauma experienced by women and children contribute to intense impacts psychologically such as anxiety, low mood, and frustration. Even the children's mental health was badly affected by hunger and food (in) security from intense poverty in our study area. Although the devastating mental health flood impacts on women may not be comparable to men, women will need more professional assistance to recover and stabilize their mental/emotional state due to challenging patriarchal gender biases.

Further, the flood damages the conditions of the environment (water, land and air) making it vulnerable to pollutants that sponsor increased fecal oral transmission of disease [55]. Generally, there is poor state of hygiene maintenance increasing the risk of waterborne illnesses, such as hepatitis A and cholera. With the high levels of poverty, negligence and poor access to clean water supply, and sanitation strategies, women are more susceptible to infections [56]. The major source of water supply was surrounding rivers that were polluted by the flood. Also, studies have reported post-flooding events bring an increase in cholera, nonspecific diarrhea, poliomyelitis, rotavirus, and typhoid fevers [57]. Intense precipitation influences waterborne disease outbreaks such as cholera, cryptosporidiosis, non-specific diarrhea, rotavirus, typhoid, and paratyphoid [58–60].

This is because floods can easily move (in)animate objects such as debris carrying parasites, bacteria, and viruses rapidly into the moving water systems and spread waterborne and related diseases. Also, studies showed that during the September 2012 flooding in Lagos, the flooded water came with lots of fecal pathogens and pollutants affecting major parts of Lagos. This is a challenge that pollutes drinking water, the associated poor sanitation [61], can lead to skin irritation and diseases as people wade through the polluted muddy-infested water to reach their homes.

10. Effect of climate-driven temporary displacement on women and children

It is recorded that over 50 million people will be uprooted in their countries when they are affected by disasters that deteriorate living conditions [62]. Achieving the sustainable development goals will not be realized, when disprivilged persons, such as women, disabled, and LGBTQ+ continue to experience one form of violence due to climate-related stressors [63], such as flooding and consequent temporary displacement. Although sexual and global based violence (SGBV) are reported in the global North, they are worse in the global South, particularly in the rural regions. These regions have a higher propensity and impact of effect of natural disasters, which have been shown to increase the risk of domestic violence.

This violence can be linked to psychosocial tensions relating to income loss from climate change impact on agricultural sector. UNFCCC [63] further reported that one in five women who are displaced from climate-related disasters has been violated sexually. With rising temperature and heightening of other related climate stressors, vulnerable groups and persons are often impacted disproportionately. Flooding exacerbated by climate changes put women and children at higher risks as they are displaced from their comfort zones worsening the link between gender-based violence (GBV) and climate-related flooding. Although any gender can experience any form of violence, such as domestic violence, sexual assault, and forced prostitution, the violence is worsened by deep-rooted sociocultural norms heightening the vulnerability of women and gender minorities, when they are not protected by laws. But these laws and policy can be gender unfriendly, particularly when they limit access to resource control of women due to scarce resources for the family. For example, early marriage and sexual exploitation are some examples of sociocultural tensions that women may go through to necessitate their survival. Also, women and children experience higher levels of vulnerability and natural weakness during flood emergencies and disasters. Findings from this study on 2020 flood disaster in Akili Ogidi provided the adverse effects on women and children. Some of the children were left in pitiable conditions without love and attention (see **Figure 13** and **14**). Women are forced to endure or go through issues that are not acceptable to them. During interview, one of the women reported that ...

> *... our women faced difficult situations during the flood unlike men who could easily run and escape, we could not do it. Even in the camp we had no privacy to the bathroom. We had to use behind the buildings and bushes to urinate, defecate, bath and change our clothes.*

Figure 13.
A flooded cassava farm in the study area where the farmer's children tried to harvest some of the farm products.

Figure 14.
Some of the hurridedly harvested Cassava tubers by the children from the flood.

Figure 15.
Soil layer washed away by the flood leaving it bare and dry and the onset of the second planting season after the floods have receded to enable early harvest.

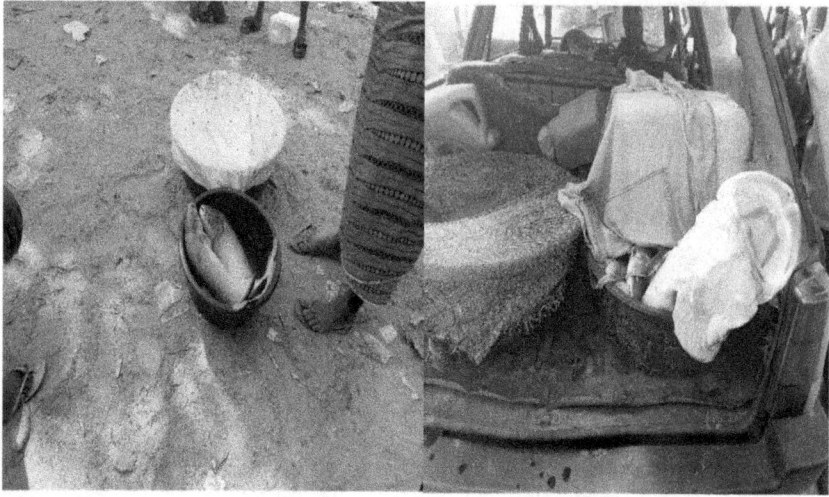

Figure 16.
The large fish species caught and sold during the floods at the bank of the river Niger.

Figure 17.
The conditions of some of the children displaced by flood at the government provided IDP camp most of them were left without a care giver, guardian, or parents.

Drawing from the study, limited access to basic infrastructres worsened the events that may have led to loss of human lives (**Figures 15–17**). Worsening family bonds and community's social ties left people lonely, anxious and mentally oppressed, and diminishing source of income, employment, and livelihoods left community members more improvised, injured, hungry, sick, dead, and infected. Worse still, a large population were displaced, homeless, and helpless in their own town.

11. Conclusion

The study confirmed that flood negatively affects the many aspects of the lives of marginalized people dislocating them from their safe home to become temporary displaced. It resulted in the loss of lives and damage of infrastructure, farmlands, agricultural products, properties, and personal belongings that were worth of billions of naira. It is responsible for accidents, congestion, hunger, food insecurity, and loss of beautifying values of the environment. Also, it causes overcrowding, spreads communicable diseases, and waterborne diseases, which has become prevalent in the area. Unfortunately, predictions indicate that climate change will influence frequent intense floods across world regions [64].

Although science has proven that there is no doubt on climate change and whether it is real or a hoax. The time is now to find out what necessary actions can be taken at the community levels for a more resilient globe, and the way forward to tackle climate events and changes. Climate change is here, and we all have to do our part to lessen its detrimental impact on our communities. We must consider options for more efficient energy, renewable energy options such as solar energy and wind power, which will contribute immensely in carbon neutralization in the atmosphere and adoption of eco-relatable friendly lifestyle that is safe for the next generation.

Consequently, the behavior changes in climate protection, as an ongoing process, need not only to give tools to let people know what to do but also to provide an enabling environment and incentives. Akili Ogidi community is already an environmental hazard, and strategies must be agreed upon and with the community members on the way forward and out of these complex-wicked flooding challenges. To check and alleviate the problems of flood occurrence in Akili Ogidi, it is critical that the Federal and state governments of Nigeria set up financial aid strategies to help rural women in recovery of farmlands after the flood disaster. Further, to alleviate the situation, there should be more funding allocated to critical agencies related to disaster management and organization that support women development to enable them perform and execute their duties optimally. Also, involving the community in local environmental management has been reported as a solution for environmental problems. Hence, researchers should employ participatory methods to encourage people to change their lifestyles.

Acknowledgements

We thank the Akili Ogidi Community in Ogbaru LGA and the participants who contributed to this community-based research, particularly the women for their courage in sharing difficult and painful experiences on internal migration due to annual flooding in their native land. We also acknowledge Nigerian Coalition for Eco-health Research (NCEHR) for their research contribution to this study. The support of Intimate Vessels Church (IVC), Awka, Anambra State for their support toward this research.

Funding acknowledgement

This study has been possible with the generous support of Spirit Filled Women International (SFWI) through their charity Fund sponsored by Intimate Vessels Church (IVC). SFWI is a non-governmental organization (NGO) and community-based research group driven toward advocacy for women's welfare and the girl child.

Author's contributions and statement

This is a community-based research project and all the authors collaborated in the development, writing, data analysis, review, and finalization of this manuscript.

Several authors took the lead for certain sections of the manuscript; AAO, NN, IN, and OKJ created the outline of the manuscript and developed ideas presented with the ongoing engagement and dialog with co-authors for input and feedback. AAO, NN, and EAC took the lead in writing and editing sections 4, 5, and 6. OCB AND ANI took the lead in sections 7and 8. AAO, OCE, NTO, OCA, and UUE took the lead for sections 10, 11, and 12. OJ, EAC, and ANS focused on Section 13. OKJ, AAO, and ANI took the lead for Section 14. IIP, ANA, NN, IN, OJ, UUM, OEO, and EDT took the lead in Section 15. AAO, OFA, NN, IN, OJ, and EAC took the lead in data analysis, discussion, synthesis of findings, policy development with ongoing engagement, and dialog with co-authors for input and feedback. The content and views expressed in this study are those of the authors and do not necessarily reflect those of the Government of Nigeria.

Conflicts of interests

There are no financial or other conflict of interest. None of the authors benefitted as a result of this study.

Abbreviations

ACMI	African Climate Mobility Initiative
AUC	African Union Commission
GPS	Global Positioning System
IDI	Internal Displacement Index
IDMC	Internal Displacement Monitoring Centre
IFRC	International Federation of the Red Cross
IPCC	Intergovernmental Panel on Climate Change
LGA	Local Government Area
LGTBQ	Lesbian, Gay, Transgender Bisexual, Queer and Plus Others
NCEHR	Nigerian Coalition for Eco-Health Research
NAN	News Agency of Nigeria
NEEDS	National Environmental, Economic and Development Study for Climate Change
NEMA	National Emergency Management System
NIHSA	Nigerian Hydrological Service Agency
NIMET	Nigerian Meteorological Agency
SEMA	State Emergency Management Agency
UNFCCC	United Nations Framework Convention on Climate Change
WHO	World Health Organization
SFWI	Spirit Filled Women International
SGBV	Sexual and Gender Based Violence
GBV	Gender Based Violence

Author details

Akanwa Angela Oyilieze[1,2,3*], Ngozi N. Joe-Ikechebelu[3,4,5], Ijeoma N. Okedo-Alex[6], Kenebechukwu J. Okafor[2,7], Fred A. Omoruyi[8], Jennifer Okeke[2,9], Sophia N. Amobi[2,10], Angela C. Enweruzor[2,11], Chinonye E. Obioma[2,12], Princess I. Izunobi[2,13], Theresa O. Nwakacha[2,14], Chinenye B. Oranu[2,15], Nora I. Anazodo[2,16], Chiamaka A. Okeke[2,17], Uwa-Abasi E. Ugwuoke[2,18], Uche M. Umeh[3,19], Emmanuel O. Ogbuefi[3,20] and Sylvia T. Echendu[21]

1 Faculty of Environmental Sciences, Department of Environmental Management, Chukwuemeka Odumegwu Ojukwu University (COOU), Uli, Anambra State, Nigeria

2 Spirit Filled Women International (SFWI), A Non-Governmental Organization (NGO) for Women Development, Awka, Anambra State, Nigeria

3 Nigeria Coalition on EcoSocial Health Research (NCEHR), Uli, Anambra State, Nigeria

4 Department of Community Medicine and Primary Healthcare, Chukwuemeka Odumegwu Ojukwu University (COOU) Teaching Hospital (COOUTH), Awka, Anambra State, Nigeria

5 Social Dimensions of Health, School of Public Health and Social Policy, University of Victoria, British Columbia, Canada

6 Department of Community Medicine, Alex Ekwueme Federal University Teaching Hospital Abakaliki, Abakaliki, Ebonyi State, Nigeria

7 Faculty of Management Science, Accountancy Department, Nnamdi Azikiwe University, Awka, Anambra State, Nigeria

8 Faculty of Physical Sciences, Department of Statistics, Nnamdi Azikiwe University, Awka, Anambra State, Nigeria

9 Department of Physiology, Chukwuemeka Odumegwu Ojukwu University Teaching Hospital (COOUTH), Awka, Anambra State, Nigeria

10 Faculty of Social Sciences, Department of Psychology, Nnamdi Azikiwe University, Anambra, Nigeria

11 Faculty of Social Sciences, Department of Sociology and Anthropology, Nnamdi Azikiwe University, Awka, Anambra State, Nigeria

12 Faculty of Education, Department of Science Education, Nnamdi Azikiwe University, Awka, Anambra State, Nigeria

13 Faculty of Health Science and Technology, Department of Medical Laboratory Science, Nnamdi Azikiwe University, Awka, Anambra State, Nigeria

14 Faculty of Management Science, Department of Co-operative Economics and Management, Nnamdi Azikiwe University, Awka, Anambra, Nigeria

15 Faculty of Physical Sciences, Department of Geology, University of Benin, Benin City, Edo State, Nigeria

16 Faculty of Agricultural Science, Department of Soil Science and Land Resources Management, Nnamdi Azikiwe University, Awka, Anambra State, Nigeria

17 Faculty of Education, Department of Early Childhood and Primary Education, Nnamdi Azikiwe University, Awka, Anambra State, Nigeria

18 Faculty of Management Sciences, Department of Business Administration, Nnamdi Azikiwe University, Awka, Anambra State, Nigeria

19 Department of Community Medicine and Primary Healthcare, College of Health Sciences, Chukwuemeka Odumegwu University, Awka, Anambra State, Nigeria

20 Faculty of Bioscience, Department of Parasitology and Entomology, Nnamdi Azikiwe University, Awka, Anambra State, Nigeria

21 Life International Hospital, Agu-Awka, Anambra State, Nigeria

*Address all correspondence to: angela.akanwa1@gmail.com

References

[1] Djimesah IE, Okine AND, Mireku KK. Influential factors in creating warning systems towards flood disaster management in Ghana: An analysis of 2007 Northern flood. International Journal of Disaster Risk Reduction. 2018;**28**:318-326

[2] Shalal, A. Climate Change Could Trigger Internal Migration of 216 Million People – World Bank. 2021. Available from: https://www.reuters.com/business/environment/climate-change-could-trigger-internal-migration-216-mln-people-world-bank-2021-09-13/

[3] Eric AS, Wishnie M. Conservation and subsistence in small-scale societies. Annual Review of Anthropology. 2000;**26**(29):493-524

[4] UNDP. Community-Based Adaptation. 2012. Available from: https://sgp.undp.org/areas-of-work-151/climate-change/community-based-adaptation-177.html

[5] Huq S, Reid H. Community-Based Adaptation: A Vital Approach to the Threat Climate Change Poses to the Poor. IIED Briefing Paper. International Institute for Environment and Development (IIED); 2007

[6] Ayers J, Huq S. Adaptation, development and the community. In: Palutikof J, Boulter SL, Ash AJ, Smith MS, Parry M, Waschka M, Guitart D, editors. Climate Adaptation Futures. 1st ed., John Wiley & Sons, Ltd.; 2013. pp. 203-214

[7] Kirkby P, Willaims C, Huq S. Community based adaptation (CBA): Addressing conceptual clarity to the approach and establishing its principles and challenges. Climate and Development. 2018;**10**(295):1-13

[8] Ensor J, Berger R. Understanding Climate Change Adaptation: Lessons from Community-Based Approaches. Rugby, UK: Practical Action Publishing; 2009

[9] National Population Commission (NPC). Population Census Figures for 2006. Abuja: Official Gazette; 2006

[10] Media Nigeria. List of Towns and villages in Ogbaru LGA, Anambra State accessed April 18th, 2018. 2020. http://www.medianigeria.com/list-of-towns

[11] Onwuka SU, Ikekpeazu FO, Onuoha DC. Assessment of the environmental effects of 2012 floods in Umuleri, Anambra East local government area of Anambra State, Nigeria. International Resource Journal. 2015;**3**(1):1-15

[12] Ezenwaji EE, Orji MU, Enete CI, Ahiadu HO. The effect of climate change on the communities of Ogbaru Wetland of South West Anambra State, Nigeria. New York Science Journal. 2014;**7**(10):68-74. Available from: http://www.sciencepub.net/newyork

[13] Mgbenu CN, Egbueri JC. The hydrogeochemical signatures, quality indices and health risk assessment of water resources in Umunya district, Southeast Nigeria. Applied Water Science. 2019. DOI: 10.1007/s1320 1-019-0900-5

[14] Egbueri JC. Evaluation and characterization of the groundwater quality and hydrogeochemistry of Ogbaru farming district in southeastern Nigeria. SN Applied Sciences. 2019;**1**(8): 1-16. DOI: 10.1007/s42452-019-0853-1

[15] Ejikeme JO, Ojiako JC, Onwuzuligbo CU, Ezeh FC. Enhancing food security in Anambra State, Nigeria using remote sensing data. Environmental Review. 2017;**6**(1): 27-44

[16] Ching L, Edwards S, El-Hage SN. Climate Change and Food Systems Resilience in Sub-Saharan Africa. Rome: Food and Agriculture Organization of the United Nations (FAO); 2011

[17] Akanwa A, Ezeomedo IC. Changing climate and the effect of gully erosion on Akpo Community Farmers. Journal of Ecology and Natural Resources. 2018; **2**(6):2-12

[18] Williams PA, Crespo O, Abu M, Simpson NP. A systematic review of how vulnerability of smallholder agricultural systems to changing climate is assessed in Africa. Environmental Research Letters. 2018;**13**:103004

[19] IPCC Fourth Assessment Report. Climate Change (2007) the Physical Science Basis, Contribution from Working Group 1 to the Fourth Assessment Report, Policy Maker Summary. Cambridge, UK: Intergovernmental Panel on Climate Change. Cambridge University Press; 2007

[20] Akanwa AO. Covid-19 confinement in Nigeria: A consequence for increased violent extremism among youths. John Foundation Journal of EduSpark. 2020; **2**(3):1-24

[21] Akanwa AO, Ngozi Joe-Ikechebelu (2019) The developing World's contribution to global warming and the resulting consequences of climate change in these regions: A Nigerian case, In: Global Warming and Climate Change. Edited by John Tiefenbacher, Published by Intech Open, London, United Kingdom, 2020. DOI: 10.5772/intechopen.85052

[22] UNHCR. Nigeria Emergency. 2019. Available from: https://www.unhcr.org/Nigeria-emergency.html

[23] International Red Cross. Red Cross: Nigeria Flood victims a 'major emergency'. 2018. http://wwwvoanews.com

[24] Premium Times. Anambra Flood Victims Stranded in open fields, October 3rd, 2012. 2012

[25] NEMA. 90 LGA Likely to be Affected by Floods in 2013. 2013. http://www.premuimtimesng.com. [Accessed: August 10

[26] NEMA. Measures to Mitigate Impact of Flood in 2013. Leadership 20th March, 2013. 2018

[27] Reliefweb. Nigeria Floods 2018. Work Report. 2018. Relief.int/report/nigeria

[28] Sun News. Anambra State flood Archives-The Sun Nigeria. 2019 Available from: http://wwwsunne wsonline.com. [Accessed September 18, 2018]

[29] Punch. Anambra Displaces 5000, Submerge Houses, Schools, Churches. 2020 punch.com. [Accessed: October 2020]

[30] Ajaero CK, Mozie AT. Socio-demographic differentials in vulnerability to flood disasters in rural Southeastern Nigeria. In: International Seminar on Demographic Differential Vulnerability to Natural Disasters in the Context of Climate Change Adaptation, Kao Lak, Thailand. Southeastern Nigeria; 2014. pp. 23-25

[31] NEMA. Preventing Flood Disaster in Nigeria. 2012Available from: http//.MEMANigeria.com. [Accessed: December 19, 2012]

[32] Vanguard Flood Kills Girl, 9 in Ogbaru, Anambra. Vanguard Nigeria. 2018. [Accessed: September 26, 2018]

[33] Nzoiwu PC, Ezenwaji EE, Enete IC, Igu NI. Analysis of trends in rainfall and water balance characteristics of Awka, Nigeria. Journal of Geography Regional Planning. 2017;**10**(7):186-196

[34] Augustina OU, Rita OU. Disaster vulnerability, severity of flood losses and information dissemination in Ogbaru Local Government Area of Anambra State, Nigeria. International Journal of Advances in Agricultural & Environmental Engineering. 2017;**4**(1): 102-106

[35] Guardian. Making of a natural disaster, repeat of the 2012 great flood. 2018. Available from: guardain.ng

[36] Ozah M. Can we dance together? Gender and performance space discourse in Égwú Àmàlà of the Ogbaru of Nigeria. Yearbook for Traditional Music. 2010;**42**:21-40

[37] Uzoka AF. Ogbaru: Our people, our dreams. Ogbaru Association official website; 1999. Available from: http://wwwogabru.net/news

[38] Douglas I, Alam K, Maghenda M, Mcdonnell Y, Mclean L, Campbell J. Unjust waters: Climate change, flooding and the urban poor in Africa. Urbanization. 2008;**20**:187-205

[39] Black R, Arnell NW, Adger WN, Thomas D, Geddes A. Migration, immobility and displacement outcomes following extreme events. Science & Policy. 2013;**27**:S32-S43

[40] Dube E, Mtapuri O, Matunhu J. Flooding and poverty: Two interrelated social problems impacting rural development in Tsholotsho district of Matabeleland North province in Zimbabwe. J'amb'a: Journal of Disaster Risk Studies. 2018;**10**:a455

[41] Eilander D, Couasnon A, Ikeuchi H, Muis S, Yamazaki D, Winsemius H, et al. The effect of surge on riverine flood hazard and impact in deltas globally. Research Letters. 2020;**15**: 104007

[42] EM-DAT. International Disaster Database, 2019. Natural Disaster 2018.

2019. Available from: www.emdat.be/publications

[43] Shultz JM, Ceballos AMG, Espinel Z, Oliveros SR, Fonseca MF, Florez LJH. Internal displacement in Colombia: Fifteen distinguishing features. Disaster Health. 2014;**2**:13-24

[44] Tafere M. Forced displacements and the environment: Its place in national and international climate agenda. Journal of Environmental Management. 2018;**224**:191-201

[45] Wan W. What Separation from Parents Does to Children: The Effect is Catastrophic. Washington Post. 2018. Available from: washingtonpost.com/nation. [Accessed: June 18, 2018]

[46] Ugwu LI, Ugwu DI. Gender, flood and mental health: The way forward. International Journal of Asian Social Science. 2003;**3**(4):1030-1042

[47] FAO. Closing the knowledge gap on gender in agriculture. Gender in Agriculture. 2011;**2010-2011**:3-27

[48] Nigerian Meteorological Agency (NIMET). 2018 seasonal rainfall predictions. In: A Report by NIMET. Nigeria: Nigerian Meterological Agency (NIMET); 2018. p. 47

[49] Eljuwairiya ICCDI Africa. Effects of flooding on Agricultural production. teet-chat or article 1(2)page1. 2020.

[50] Okenmuo FC, Anochie CO, Ukabiala ME, Asadu CLA, Kefas PK, Akamigbo FOR. Agro-science discrete assessment of agricultural potentialof floodplain soils in Southeastern Nigeria. Journal of Tropical Agriculture, Food, Environment and Extension. 2020; **19**(3):51-61

[51] Ihaji EO, Aondoaver U. Gender and children registered at flood camps in Makurdi in the year 2012. International

Journal of Humanities and Education (IJHSSE). 2014;**1**(6):29-33

[52] Alderman K, Turner LR, Tong S. Foods and human health: A systematic review. Environment International. 2012;**47**:37-47. DOI: 10.1016/j.envint.2012.06.003

[53] Duncan K. Global Climate Change and Women's Health. Women and Environment. Internal Magazine Issue. 2007

[54] Makwana N. Disaster and its impact on mental health: A narrative review. Journal of Family Medicine and Primary Care. 2019;**8**(10):3090-3095

[55] Peate I, Wild K, Nar M. Nursing Practice: Knowledge and Care. New Jersey: John Wiley & Sons; 2013

[56] Yamin A. Why Are the Poor the most Vulnerable to Climatic Hazards (E.G. Floods)? A Case Study of Pakistan. Term Paper. Germany: University of Potsdam; 2014. pp. 1-19

[57] Sur D, Dutta P, Nair GB. Severe cholera outbreak following floods in India. Indian Journal of Medical Research. 2000;**112**(1):78-82

[58] Chen MJ, Lin CY, Wu YT, Lung SC, Su HT. Effect of extreme precipitation to the distribution of infectious diseases in Taiwan, 1994–2008. PLoS One. 2012;7(6):e34651

[59] Cann KF, Thomas DR, Salman RL, Wyn-Jones AP, Kay D. Systemic review: Extreme water-related weather events and water-borne diseases. Epidemiology Infections. 2013;**141**:671-686

[60] Brown L, Murray V. Examining the relationship between infectious diseases and flooding in Europe: A systemic literature review and summary of possible health interventions. Disaster Health. 2013;**1**:1-11

[61] Sessou E. Flood Takes Over Lagos, Destroys Properties. Vanguard. 28 June. 2012. Available from: https://www.vanguardngr.com/2012/06/flood-takes-over-lagos-road-destroys-properties/. [Accessed: June 14, 2014]

[62] IDI Report. Internal Displacement Index. IDMC (Internal Displacement Monitoring Centre Publications); December 2021. Available from: 2021-displacement.org/p

[63] UNFCCC. Climate Change Increases the Risk of Violence against Women. Article. 2019. Available from: unfccc.int. [Accessed: November 25, 2019]

[64] IPCC. Climate Change Impacts, Adaptation, and Vulnerability. Cambridge, United Kingdom: Cambridge University Press; 2001

Consequences of Climate Change Impacts and Implications on Ecosystem and Biodiversity; Impacts of Developmental Projects and Mitigation Strategy in Nepal

Ramesh Prasad Bhatt

Abstract

Climate change impacts and implications towards ecosystems and biodiversity, water resources, food production, and infrastructures can be mitigated through adapting, reducing or avoiding adverse impacts and maximizing positive consequences. It can have numerous effects on the world's natural ecosystems and their functions. IPCC projections showed approximately 10% of species to be at an increasing high risk of extinction for every 1 °C rise in global mean temperature and recommended to limit global temperatures below 1.5 °C. To identify consequences of climate change, impacts, and implications, data collected from different sources, reviewed, assessed and analyzed, discussing dimensional impacts and mitigation strategies adopted. Nepal's 118 major ecosystems and 75 vegetation types with 44.74% forestland comprising 0.1% of global landmass harboring 3.2% flora and 1.1% fauna of the world's biodiversity critically influenced by the regional climate change and intervention of developmental projects. Since 2000, Nepal lost forest area by 2.1% including several endangered and threatened species. Nepal is highly vulnerable towards natural disasters like GLOF, Glacier retreat, flooding, landslide and global warming. Therefore, it is crucial to plan climate resilience infrastructures adopting effective environmental management tools, formulation of strong plan, policy and strategy, mitigation of greenhouse gases, climate resilient adaptation and restoration of degraded ecosystems.

Keywords: climate change, ecosystem and biodiversity, development projects, impacts and implications

1. Introduction

The overwhelming global effects of climate change and its implications subtle changes in the earth's orbit belongs long term and immediate impacts. The impacts on atmosphere, ecosystem and sea as a consequences of human activities, warming globe, raising air and sea temperature dropping by 0.01 °C every 100 years for the last 7,000 years and since 1970 the temperature has been rising at an alarming rate of 1.7 °C per century. Consequently, IPCC efforts recommended limit global temperatures below 1.5 °C to avoid the more severe impacts of global warming [1].

The warmer temperature due to climate change increases frequency of ground level ozone, lethal air pollutants, and smog components [2]. Both the immediate impacts of climate change regards extreme weather, heat waves, storms, and flooding, forest fire, compromised safety, economic challenges and long-term effects likewise on human health, ecosystem, threats on water and food resources, altitudinal and tree line shifting are the consequences of climatic change issues today. The pragmatic climate change effects sensed on environment globally comprises shrinking glaciers, shifted plants and animals ranges and sooner tree flowering, loss of sea ice, enhanced sea level rise, penetrating heat waves affecting our dependency upon water, energy, terrestrial and aquatic ecosystems, agriculture, and human health. The future prediction of climate change has great challenge to preserve existing ecosystem and biodiversity. The projections showed 10% approximate species to be an increasingly high risk of extinction for every 1 °C rise in global mean temperature, within the range of future scenarios modelled in impacts assessments (typically <5 °C global temperature rise) as per IPCC AR4. The vulnerable ecosystems are freshwater habitat and wetlands, arctic and alpine ecosystems, cloud forests, mangroves, coral reefs to the impacts of climate change.

The impacts of climate change has extremely affected human societies and the natural environment. The most vulnerable to climate change impacts on an ecosystems area associated with the shifting alpine ecosystem to higher elevations and shrink in area, modification on tropical and subtropical rainforests, affected coastal wetlands by sea-level rise and saline intrusion, affected inland ecosystems by changed rainfall patterns and affected tropical savannahs by changes in the frequency and severity of bushfires. According to Dolling et al. [3]; Giambelluca et al. [4]; Frazier et al. [5]; increased drought stress of native plants and fire occurrence is due to decreasing rainfall and increasing temperatures. According to the IPCC 5th Assessment (IPCC 2013) [6], ocean surface temperature and combined land and warming globally was around 0.85 [0.65 to 1.06] °C, over the period 1880 to 2012. The issue of climate change have created by human activities is about 1 °C of global warming above pre-industrial levels with range of 0.8° to 1.2 °C [7]. The large influence on forest and agricultural systems due to increases in temperature, changes in precipitation patterns, and changes in the occurrence of floods and droughts has predicted by the Álvaro-Fuentes et al. [8]; Anaya-Romero et al. [9]; Conant et al. [10]; Lal [11]; Muñoz-Rojas et al. [12].

As an important environmental factor, climate influence ecosystems and biodiversity several ways whereas climate change affects habitats of species and existing food chain interacting other human stressors like development. The stressors cause minor and major affects including dramatic ecological changes some of their cumulative impacts [13]. Due to changes in the timing of seasonal life cycle, many species influences their migration, blooming and reproduction including reduced their growth and survival [14, 15]. The habitat range of the species both terrestrial and aquatic environment shift to higher elevations due to increased temperatures including changes in vegetative biomes [16] and expanding into areas of river and streams previously inhabited by Coldwater species [17]. Similarly, particular species can ripple through a food web and affect a wide range of other organisms as the loss of sea ice affected entire food web, from algae and plankton to fish to mammals [18]. The extreme events like wildfires, flooding, and drought serves ecosystems as natural buffers and human modification restrict ecosystems' aptitude to temper the impacts of extreme conditions which increases vulnerability to damage. Likewise, the main stressors of climate change like habitat destruction and pollution subsidize to species extinction and spread of pathogens, parasites, and diseases, with potentially serious effects on human health, agriculture, and fisheries. Thus, the species extinction rate globally is exceeding the observed natural rate of extinction [19].

In Nepal, the major driving force to degrade biodiversity and ecosystem is development of infrastructures such as construction of road, hydropower, irrigation canal, railway, transmission line, tourism industry and airport. The accelerated impacts of these activities allies with deforestation and degradation of natural forests, habitat fragmentation, infrastructure development on protected areas, encroachment of forest and forest land, destruction of natural habitats both terrestrial and aquatic species. The urbanization and population growth leads to unplanned infrastructure development ultimately has increased demand of natural products, and pressure on biological resources threatening natural ecosystem and biodiversity.

The rainfall patterns and temperature regimes altered by the climate change and its impacts on water, agriculture and biological resources are crucial issues in Nepal. The policy of the government in biodiversity conservation and economic development is ineffective for implementation integrating the development actions. An effective implementation of regulatory framework and strategy to protect biological resources including protected, endangered, threatened, rare and endemic species can support to conserve the biodiversity and ecosystem. Thus, the chapter assisted overall climate change scenario describing impacts and implications, focusing on ecosystem and biodiversity, including impacts of developmental projects, adoption of national mitigation strategy and use of assessment tools with Nepal's perspective.

2. Material and methods

2.1 Study area

A landlocked country Nepal situated between 80°and 88° E longitude and from 26° to 31° N latitude which covers an area of 147,181 Sq.km measuring 880 kilometers (547 mi) along its Himalayan axis by 150 to 250 kilometers (93 to 155 mi) across. The new

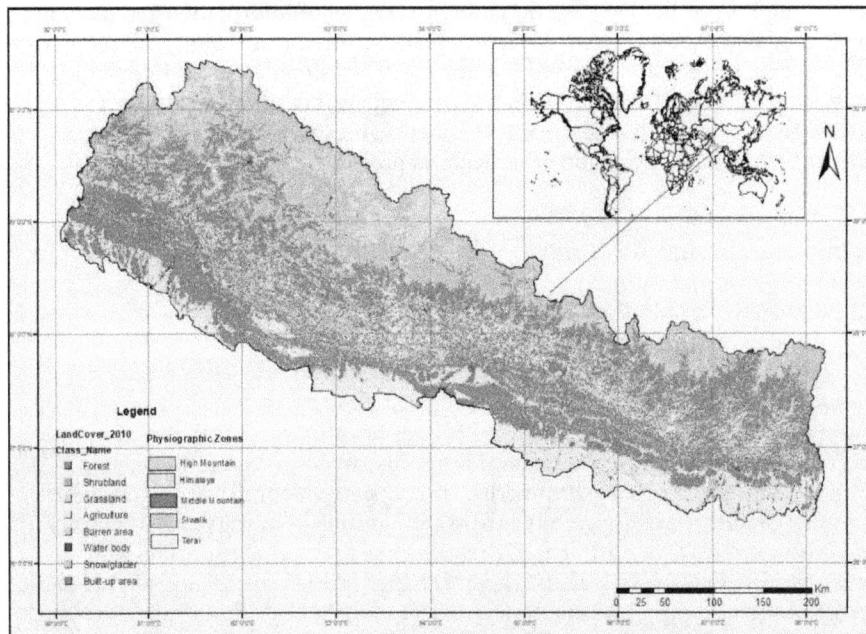

Figure 1.
Landuse pattern of Nepal.

map showing Kalapani, Limpiyadhura and Lipulekh as Nepali territories, Government of Nepal has further declared as an extended area of 147, 516km^2 (56,956 sq. mi) [20]. Nepal resembles five physiographic zones and seven bio-climatic zones identified by Dobremez [21] extending from East to West, including the High Himal, High Mountains, Middle Mountains (or Middle Hills), Siwalik (or Chure), and Terai [22] (**Figure 1**). The genetic, species and ecosystem level biodiversity of Nepal spreads over tropical forests from Terai (67-1000 masl), mid hill (1000-2000 masl) subtropical and temperate regions (2000–3000 masl) to Himalayas sub-alpine and alpine pastures and snow-covered peaks of the Himalayas (> 3000 masl). As shown in **Figure 1**, land cover in Nepal includes different types of forests (broadleaf, needle leaf, mixed), agriculture, shrub lands, grasslands, bare lands, river and lakes, and glacier/snow [23, 24].

3. Methods

Relevant literatures reviewed, summarized current estimates of the impacts of climate change as per IPCC and The World Bank data, and explained how those estimates assembled in order to identify the main sources of uncertainty and approximately affecting them. In addition, discussed scenarios of development actions and impacts, uncertainty influences and mitigation measures adopted through legal instruments and policymaker's decisions. The reviewed national and international policies, plan, strategies, Act and regulations were incorporated with the relevancy and implications in ecosystem and biodiversity and other thematic areas. AcrGIS_10.4.1 was used to delineate land use pattern and forest cover together with identification of Physio geographic zones of Nepal with reference of GoN/MoFSC, 2018 [25] and national scale forest resource assessment carried out over the period of 2010–2014 [26]. In the decision making process, EIA legislation and its process subjected to an environmental assessment during preparation, and before adoption. Similarly, integration of EbA approach into policies and plans for facilitating EbA technology, synthesis and packaging of information and planning tools was carried out to found alternative scenario for climate change adaption with resilience ecosystem. Impacts and implication of developmental projects, legal provisions to conduct environmental assessments and mitigation strategy were identified through review pertinent documents and study of infrastructure projects such as hydropower, roads and other sectoral projects.

4. Impacts and implications

4.1 Ecosystem and biodiversity

The increasing impacts of climate change on biodiversity become a progressively more significant threat in the coming decades as projected. The pressure of ocean acidification, resulting from higher concentrations atmospheric CO_2 and loss of Arctic sea ice threatens biodiversity across an entire biome and beyond. The frequent extreme weather events with changing pattern of rainfall and drought, warming temperatures as projected there are significant impacts on biodiversity and ecosystem [27]. According to IPBES the three major challenges like land degradation, biodiversity loss and climate change increasingly dangerous impact on the health of our natural environment degrading land, pollution and overexploitation to land-use change and habitat loss threat to wildlife globally. As per the IPCC [28] AR4 projection 10% of species assessed approximately are in high risk of extinction for every 1 °C rise in global mean temperature, within the range of future scenarios modelled in impacts assessments (typically <5 °C global temperature rise). In particular, aquatic freshwater habitats and wetlands, mangroves, coral reefs, arctic and alpine ecosystems, and cloud forests are vulnerable to the impacts of climate change.

Nepal is rich for globally significant biodiversity with 118 major ecosystems and 75 vegetation types harboring important flora and fauna occupying 39.6% (5,830,360 ha) forest area. Currently, out of total area of the country, 44.74% (6.61 million ha) is covered by the forests land including 17.32% Protected Areas (PAs) and 82.68% by other forest management regimes [29]. Nepal has established 20 protected areas with the goal of nature conservation covering 23.39% (34419.75 sq.km.) land area with2 12 National Parks, 1 Wildlife Reserve, 1 Hunting Reserve, 6 Conservation Areas, and 13 Buffer Zones extending from lowlands of Terai to high mountains.

Globally, Nepal occupies about 0.1% of global landmass harboring 3.2% flora and 1.1% fauna of the world's biodiversity including 5.2% of mammals, 9.5% birds, 5.1% gymnosperms and 8.2% bryophytes [30]. Similarly, Nepal government has declared 27 mammals, nine birds, 3 reptiles [31] under protection category. Among 208 mammal species recorded in Nepal, regionally 8 of them critically Endangered, 26 Endangered, 14 Vulnerable and 7 species nearly Threatened. Out of 886 bird species found in Nepal, 42 of them are globally threatened [32]. The species of flora and flora of Nepal listed in CITIES includes, 50 mammals, 108 birds, 29 Reptiles, 2 amphibians and 476 species of plants [33]. The updated checklist of CITES flora after CoP 17 includes, 154 species with 1 species in Appendix I, 149 species in Appendix II and 4 species in Appendix III [34].

The use of resources like grazing, fodder and fuelwood collection, timber extraction, collection of herbs, medicinal and aromatic plants, poaching, hunting and fishing are major threats for conserve biodiversity conservation. Among the 700 species medicinal plants in Nepal, more than hundred species exploited for commercial purposes. Loss and fragmentation of natural terrestrial and aquatic habitats and restricted mobility of migrant species due to habitat fragmentation is crucial problem. Poaching of rare species such as the tiger, rhino, bear (*Selenarctos thibatenus*), musk deer, snow leopard, gharial, and others is also critical problems. Nepal were to lose its remaining humid tropical forests, 10 species of highly valuable timber, six species of fiber, six species of edible fruit trees, four species of traditional medicinal herbs, and some 50 species of litde known trees and shrubs would be lost forever. Likewise, severely affected wildlife habitats are 200 species of birds, 40 species of mammals, and 20 species reptiles and amphibians [35].

The loss of biodiversity and vegetation leads to the changing pattern of natural environmental conditions occurring from numerous fundamental systems and decreasing number of species biodiversity due to modified environment and increased pressure on forest and vegetation [36]. The major impacts on biodiversity and ecosystems are loss of habitat due to encroachment of forest areas, expansion of agriculture and settlement in forest area, development of infrastructure within the forest area, planned priority projects within forestland to uplift economic development. Other major problem also leads to degradation of habitat due to overharvesting of biological resources, overgrazing and uncontrolled forest fire. Similarly, poaching and illegal wildlife trade, human-wildlife conflict, invasion by alien plant species, stone, gravel and sand mining threats ecosystem and biodiversity of the country. Nepal lost forest area by 2.1 percent and 1.4 percent during 1990–2000 and 2000–2005, respectively [37]. Nepal's forest area heavily degraded with loss of important biodiversity imposing landslides, and soil erosion, felling of trees for building materials and over lopping for fodder and fuel wood [38]. The unplanned infrastructure development such as schools, hospitals, temples, water storage tanks and other infrastructure within the forest area, particularly in the Tarai and Siwalik, 82,934 hectares forestland was reported under illegal occupation [39].

The major identified climate change impacts on forest and biodiversity in Nepal resembles; (a) increased temperature and rainfall variability, (b) changes in phenological cycles of tree species and shifting of tree line in the Himalaya, (c) shifts in agro-ecological zones, prolonged dry spells, and higher incidences of pests and diseases, (d) increased emergence and quickened spread of invasive alien plant species, (e) increased incidence of forest fire in recent years, (f) depletion of wetlands [40, 41]. The large scale development is a hallmark of the modern world, providing society with things humans' value but at an environmental cost [42, 43]. The major impacts on terrestrial or aquatic ecosystem associated with the developmental projects are:

- habitat degradation through overgrazing practices wetland drainage for agricultural, industrial or urban development practices;

- habitat loss, with attendant consequences on fish and wildlife because of excessive deforestation practices;

- changes in habitat and associated fish and wildlife species due to the construction and operation of hydropower projects;

- loss of critical habitat for endangered or threatened species as result or timber harvesting, recreational developments, and or military training activities;

- multiple aquatic and terrestrial ecosystem effects from acid rain formed as a consequence of SO_2 emissions from coal-fired power plants; and

- potential toxic effects to plants and or animals as a result of air-or water pollutant discharges or of waste disposal activities of industries and municipalities.

4.2 Infrastructures and development

The major infrastructures like road, hydropower, irrigation, water supply, housing, tourism and communication are associated with socioeconomic development of the developing country. Monsoon floods, landslides, and siltation-damaged infrastructures impede the developmental activities and affects transportation facility, electricity supply, industry, business, markets, and other allied activities. Furthermore, more extreme events associated with climate change forebodes great threat on infrastructures with increased instances of drainage congestion, scouring, inundation, slope instability, land subsidence, erosion, and collapse of structures. The existing roads, bridges, water supply and sanitation, settlement, hydropower and irrigation are affecting seasonally due to flooding, landslides, and debris deposits cause large impacts on socioeconomic development of the country. The degradation of key natural resources land, water, and forests and impacts on agriculture and livestock farming, agricultural production, transportation, infrastructure, forest-based industries, and hydropower, are associated with intensifies disasters such as landslides, floods, soil erosion drought and storms.

4.3 Hydropower development

Nepal is rich for water resources comprising 6,000 rivers with 220 billion cum annually run off spreading length of over 45,000 km. [44]. An estimated potential for hydropower to be around 83,000 MW whereas 4300 MW to be technically feasible [45]. The major river basins in Nepal are Sapta Koshi, Karnali, Sapta Gandaki, Mahakali, and the Southern rivers [46]. In Nepal, with total installed capacity of 1,332,858 KW, 14 major hydropower stations, 17 small hydropower plants, 23 Small Hydropower Plant (isolated), Two Thermal Plants and Solar plants are currently in operation generating 56315KW, 577394KW, 4536KW, 53410 KW, and 1350KW energy respectively. Similarly, eight Hydroelectricity Project (HEP) with installed capacity of 943100KW are under construction and nine with installed capacity of 3219200KW are planned and proposed. Hydropower construction is growing development in Nepal due to rapid increment demand of electricity by about 10% every year (NEA) and projected demand for year 2020/2021 to be around 2,203 MW [47]. The hydropower development considerably cause high impacts on construction phase rather than its operation along with cumulative and long-term impacts can harass to sustain ecosystem and biodiversity. As location of the project, size and type of the project, socioeconomic condition the environmental situation, beneficial and harmful impacts of hydropower are considered manifold [48, 49].

Nepal in progress of installation of renewable source of energy such as mini and micro hydropower, solar energy, biomass energy and biogas targeting access public to provide clean, reliable and affordable renewable energy solutions by 2030. As different Nations agreed in Paris in 2015, limit the global average temperature rise to as close as possible to a maximum 2 °C reducing energy-related CO_2 emissions by more than 70% by 2050 which can only be achieved with the massive deployment of renewable forms of energy such as wind, solar and hydro, combined with energy efficiency [50]. NEA is responsible to generate electricity, transmission, and distribution but inadequate planning, policy, regulation and inadequate capacity with poor governance leading to underinvestment and difficulty in sustaining the growth of the Hydropower and Economic Growth in Nepal [51].

To institutionalize hydropower sector, Government of Nepal endorsed Water Resources Act, 1992, Water Resources Regulation, 1993, Electricity Act, 1992, Electricity Regulation, 1993 and Hydropower Development Policy, 1992, Hydropower Development Policy, 2001, Water Resources Strategy, 2002 and National Water Plan, 2005. GoN has also prioritized development of hydropower including contemplation of Environment Protection Act, 2019 and Environment Protection Rules, 2020. The perennial nature of Nepalese rivers and the steep gradient flows from Himalayas and high mountains towards the plain of the Terai has largest hydropower potential for hydroelectric projects in Nepal whereas challenges remaining to protection towards ongoing potential impacts on natural environment. Although, hydropower plants provide cost efficient and environment friendly power supply to improve energy services a displacing imported fossil fuels but most of the hydropower constructing in Mountains, Mid hills and Siwalik regions of the country are great threats to biodiversity and ecosystems of the region. Landslide dam outburst floods (LDOFs) is one of the major challenges for hydropower development in Nepal due to its rugged topography, susceptible to landslides, very high relief, and intense precipitation during the monsoon period. Thus, there is a high need to protect the ecosystem together with the hydropower development by improving resilient hydropower infrastructure through good planning, design and sitting, construction, operation and maintenance, contingency planning, and restoring ecofriendly environment [52]. As per nature, extent, magnitude and duration of all observed adverse environmental issues/impacts on cultural and physical, chemical, biological and socio-economic during construction and operation phases of hydropower projects are highlighted in the **Table 1** below.

The climate change impacts significantly higher to water resources and hydropower sector like rising temperatures retreat glacier that in turn causes greater

Construction phase	Operation phase
a. *Physical environment*	
• Changes in site geomorphology, topography and land use patterns due to project structures	• Changes in hydrology and sedimentation
• Construction access road and labour camps	• Drainage disruption
• Stockpiling of Construction Materials	• Noise and vibration in powerhouse area
• Operation of quarry sites, crusher plant (noise and vibration, spoils from crusher, land instability)	• Change in water quality due to reduced flow and reservoir flushing
• Change in river morphology and drainage pattern	• Leakage of oil, grease and other chemicals
• Landslide and soil erosion, and sedimentation due to excavation works	• Change in microclimate
	• Cumulative impacts
• Generation of solid waste and spoil disposal	• Landslide, flooding, LDOF
• Leakage of oil, grease and other chemicals	• Generation of greenhouse gas
• Increase level of noise and vibration, change in air and water quality	• Change in river morphology and hydrological flow regime
• Pollution to surface water due to siltation, inadvertent disposal of waste oils, wastewater, and solid wastes into the river	• Sediment flushing on dewatered section or reduced in-stream
• Contamination of groundwater springs, aquifers and hydrogeological conditions due leakage of chemicals	• Alternation of geomorphology upstream and downstream of the dam and the powerhouse
• Slope destabilization and loss of top soil	• Generation of greenhouse gas
• Landslide Outburst Flood (LDOF)	• Generation of waste oil and used chemicals
	• Reduction downstream flow on riparian zone
b. *Biological Environment*	
• Loss of forest land, loss of forest and vegetation cover due to site clearance	• Disturbance to fish migration, loss of spawning area, loss of habitat
• Increase collection of firewood, timber, NTFPs and medicinal Plants	• Decrease aquatic flora and fauna in dewatered zone
• Habitat fragmentation and wildlife disturbance	• Destruction wildlife habitat and wildlife movement
• Illegal hunting and poaching	• Forest encroachment and easy access to forest
• Loss of aquatic species due to diversion of river	• Illegal hunting and poaching
• Increase fishing activity	• Impact on terrestrial and aquatic ecology
• Loss of biodiversity	• Sudden release of water in downstream and reduction in river flow in the dewatered zone
• Possible forest fire	• loss of biodiversity and important species and introduction of invasive species
• Loss of or encroachment on critical habitats, protected areas, wetlands, and forestlands	• Inundation of flora and fauna
• Reduction of terrestrial and aquatic biodiversity	• Forest fire inducement
• Impact on endangered, threatened species, protected and their habitats	• Electrocution effects to wildlife
• Disturbance to aquatic habitat	• Impacts on protected species
• Disturbance and/or removal of riparian vegetation	

Construction phase	Operation phase
c. Socioeconomic and cultural Environment	
• Land acquisition and private property	• Decrease or withdrawal in economic activities
• Impacts on livelihoods of affected families	• Occupational, health and safety hazards, navigation risks
• Loss of standing crops	
• Physical and economic displacement of PAF	• Reduction of agricultural land and food security
• Occupational health and safety hazard	• Sudden release of water downstream in the dewatered stretch impacts people and livestock
• Increase in pressure on local health and sanitation facilities	
• Economic flux	• Headrace alignment and penstock crossing and project structures on springs and water sources
• Pressure on community infrastructures, social service facilities and resources	• Dislocation/disturbance to foot trails along penstock alignment
• Impacts on social, cultural and religious rights	• increase in tourism activity and its associated pressure on local resources
• Impacts on gender and disadvantage groups	• Water use right and loss of cultural rights
• Impacts on springs and drinking water sources	• Generation of solid and Liquid waste
• Generation of solid and liquid waste	• Impacts of visitor and tourists
• Increase demand on energy sources, firewood & timber	• Impacts on use of community service facilities
• Demographic changes to the area and communities	• Contamination of water sources HHs liquid and solid waste, sewage disposal
• Loss of livelihoods and loss of access to natural resources and cultural heritage	

Table 1.
Potential Adverse Impacts of Hydropower.

variability (and eventual reduction) in streamflow, and glacial lake outburst floods posing significant risk to hydropower facilities, infrastructure and human settlements. The climate induced risks to water resources and hydropower facilities related to flooding, landslides, and sedimentation, intense precipitation events, flow variation in dry season. Resiliency of hydropower assets is essential to face of increased frequency of extreme weather events and rapid changes in hydrological patterns to reduce the risk of climate-related disruptions as hydropower contributes significant reduction of GHG emissions. Hydropower plants prevents the emission of about 3 GT CO_2 per year (9% of global emissions) compared with conventional coal power plants [53].

4.4 Roads and transportation

The majority of the population in Nepal does not have reliable and adequate access on transportation services (**Figure 2**). Thus, the development of physical infrastructure services like roadways, railways, waterways, subways, flyovers and ropeways, transport (Air Transport) and transit management and its operation and implementation are rapidly growing. As per data, about 29031 km of roads (53 percent paved roads) and 1952 bridges in the country are in operation [54].

The availability of road infrastructure as per road density, Nepal stands at 139 km per 1000 km^2 [55], where 60% or more road network are concentrated in the lowland (Terai) areas. However, Nepal's 20% population residing in urban areas but Nepal considered fastest urbanizing country with annual growth rate of 5% on an average since 1970s [56]. The Global Competitiveness Report 2016 ranks, Nepal

Figure 2.
Temporary Means for Access on Mahakali River, Darchula Nepal.

130 of 138 in infrastructure [57]. Government of Nepal step-up to capital expenditure in infrastructure, in particular, sectors like water, communication, transportation and electricity from 2009 to 2016 received greater priorities [58].

An immense facility and services provides by the infrastructure development accessing agriculture, market, commerce, industry, and social sectors including education, health, communication, livelihood and quality of life but construction and maintenance works adversely affects natural environment. Nepal's young and fragile geology, poverty, vulnerable communities, construction of infrastructures brings significant impacts to the local and regional environmental settings. The major impacts associated with such developmental activities comprises; slop instability and slope failures, landslides and soil erosion, impacts on national parks, conservation areas and wildlife reserves, national forests and agricultural lands and interference with water courses, loss of terrestrial and aquatic biodiversity, river regime change, extraction of sand/gravel, irrigation facilities, run off and sedimentation, occupational health and safety, land acquisition, damages on cultural properties and effects on the unique life of ethnic and minorities communities.

4.5 Climate change and its consequences

The water, food, health, land, environment, infrastructures are vulnerable affecting by severe consequences of global climate change with irreversible loss of many species around the world. The increasing in extreme weather events damages infrastructures by natural disasters, storms and flooding. According to IPCC 2007, climate has been getting warmer since 1960 and this will continue. Global temperature will be increased at the end of the 21st Century in relative to the end of the 20th Century ranges from 0.6 to 4 °C and 3.3 °C in South Asia with the min-max range as 2.7–4.7 °C. Climate change projections for 1961–1990, in East Asia shows relative to the average for mean temperatures will be 1.9°–2.6 °C higher across the region in 2050, and 3.8°–5.2 °C higher in 2090 [59]. The increasing pressure on natural resources in Asia associated with the rapid urbanization, industrialization, and economic development bringing challenges to protect the degrading environment. Asia comprises 51 countries/regions with land and territories, divided into six sub regions based on geographical position and coastal peripheries. Forest carbon pools affected by the Climate change in some countries of the region and observed annual mean temperatures over South Asia in the past is increasing significantly about 0.75 °C per century. The physiology, phenology and distribution of plant and animal affecting by the climate change and also increase the risk of mortality and injury from wind storms, flash floods, coastal flooding and expected numbers of vector-borne diseases in the near future.

According to the 2020 edition of Germanwatch's Climate Risk Index, Nepal is ninth hardest-hit nation by climate calamities during the period 1999 to 2018, as one of the most vulnerable countries to the climate change effects [60]. The average annual maximum temperature has been increasing by 0.056 °C per year from 1971 to 2014 [61] and extremely increasing precipitation [62]. Similarly, more than 80% property had lost due to water disasters like floods, landslides and glacial lake outburst floods (GLOFs) [63].

Biodiversity is essential for Earths functioning ecosystem but growing human population, habitat loss and over exploitation of resources is the main factors to loss of biodiversity. According to IPBES, more than 1 million IUCN red list threatened species are risk of disappearing which includes 41% of amphibians, 25% of mammals, 34% of conifers, 13% of birds, 31% of sharks and rays, 33% of reef-building corals, and 27% of crustaceans [64]. The issue of global warming is concerned with the raising temperature by trapping earth's emissions in the atmosphere due to exposure of greenhouse gas emissions like carbon dioxide, nitrous oxide, methane, ozone, and water vapor and burning of fossil fuels. Increased lake and stream temperatures have significant implications affecting frequencies of disease in fish species and their altered growth, increased energy expenditures, thermal barriers on adult and juvenile migration, delayed and reduced spawner survival, altered egg and juvenile development, changes in biological productivity and altered species distribution [65].

The living components of ecosystems like plants, herbivores, carnivores, and soil organisms influences by the climate change in their functional ecosystem and characteristics in regards of energy and chemicals flow altering ecosystems properties and species distributions. The major resilience biodiversity in its ecosystems with variation of life on the earth interdepending webs of living organisms and physical environment provides us clean air, fresh water, food, resources and medicine. The humankind activities like air and water pollution, habitat destruction and fragmentation, and the introduction of invasive species likely to be exacerbated ecosystems. The valuable goods and services provided to the human societies by the ecosystem threaten to jeopardize the numerous economic and social values due to climate change effects [66]. The detrimental effects can be mitigated through preserving and maintaining habitat and species to maintaining overall ecosystem structure and species composition together with adapting biodiversity conservation strategy by reducing fragmentation and degradation of habitats, increasing connectivity among habitat blocks and fragments, and reducing external anthropogenic environmental stresses.

Nepal's water resources, public health and terrestrial ecosystems are most vulnerable sectors [67] with associated issues of food security, poverty reduction and environmental degradation. Adaptation of appropriate technology for changing cropping patterns, enhancing mitigation for emerging pests and disease and protection of changing landscapes of Nepal can cope with changing climatic pattern. Nepal's changing climatic pattern as experienced in temperature and mean precipitation, data on temperature trends from 1975 to 2005 showed 0.06 °C rise in temperature annually with significantly decrease in mean rainfall on an average of 3.7 mm (−3.2%) per month per decade. The mean annual temperatures under various climate change scenarios for Nepal has projected to increase between 1.3–3.8 °C by the 2060s and 1.8–5.8 °C by the 2090s with reduction of annual precipitation to be in a range of 10 to 20% across the country [68]. The increased number of glacial lakes in Nepal exceeds by 11% on an average by 38 km² per year and 29% about 129 km³ ice reserve has estimated between the period of 1977 and 2010. Nepal is one of the most vulnerable countries to disasters and warming trends of Nepal is increasing as country's averaged mean temperature increases of 1.2 °C and 3 °C projected by 2050 and 2100 [69]. Glacier retreat and significant increases in the size and volume of glacial lakes of Nepal Himalayas making them more prone to Glacial Lake Outburst Flooding (GLOF). Climate change enhance more disaster's like landslides and soil erosion on excavation slopes, drainage disruption, flooding in roads, bridges and airport runways, drinking water sources and infrastructures of lowland areas so that roads, bridges, tunnels and other infrastructures are vulnerable to increased precipitation, groundwater levels, temperatures and winds. Transportation sector is affected by climate change and it contributes to greenhouse gas emissions. The mountain ecosystem and topography is highly affected by the climate events like high rainfall, accelerate surface run-off, increase flows in gullies, drainage channels, streams and rivers expose landslides and flooding, which instable road sections, bridges, and other infrastructures as well as inundations in lowland areas. Design and construction of climate resilient and environmental friendly infrastructures can support to reduce climate change impacts. Local microclimate can adapt through rod site plantation, bioengineering and climate resilient transportation system to reduce GHG level in atmosphere. The GHG emission of Nepal is around 0.027% of total global emissions and increasing GHG trend is from energy sector [70]. As Nepal is a party of UNFCCC supporting to limit temperature rise to well below 2 °C leading to 1.5 °C above pre-industrial levels in order to reduce the risks and adverse impacts of climate change.

Climate change and risks associated towards adverse impacts on ecosystems and biodiversity, water resources, food production, and infrastructure with global warming correlates to adapt reducing or avoiding adverse impacts and maximize positive consequences towards the risks. The adaptation process is complex towards risks posed by climate change and variability as Moser & Ekstrom 2010, stated adaptation process is constant of awareness and understanding, planning, implementation and monitoring and review [71]. To mitigate or avoid the projected impacts of climate change, adaptation urgently needed towards extreme changes and impacts that may occur [72]. According to Article of UNFCCC, evaluation of risk associate with climate change refers "dangerous anthropogenic interference with the climate system".

Although, Climate change affects biodiversity and ecosystem services that are not all negative, with some species either thriving or adapting. Climate change and associate impacts is an integrated and integral portion of the major developmental sectors of the nation that can be mitigated through effective implementation of mitigation tools. Environmental assessment a planning application for the development activities mainly propose mitigation measures as obligations within the legal framework for implementation but not fulfilled legal requirement seriously for monitor and implementation. Hence, the assessment strategy to address the likely environmental impacts of the planned sectoral projects is limited on fulfillment of the legal requirement.

5. Mitigation of environmental risks and climate change adaptation

5.1 Policy and legal instruments

Since 1980, Nepal has integrated environmental management related policies and strategies to address major consequences of developmental projects enunciating environment conservation related policies in the seventh plan [73]. In order to enforce policy and legal instruments for environmental protection from the project formulation stages for development strategy has considered avoiding or minimizing adverse effects on the ecological system. For this, EIA as a tool has adopted emphasizing emerging environmental and ecological issues envisioned from developmental projects such as industry, tourism, water resources, transportation, agriculture, forest and other developmental projects. To institutionalize EIA system in Nepal, eighth plan period [74] has remarkable contribution adopting and implementation of National EIA Guidelines of 1993 [75] and enforcement of Environment Protection Act, 1997 and the Environment Protection Rules, 1997 [76]. Nepal has enacted a number of regulatory measures for the consideration of the environmental aspects in the development project and programme, however still inadequate mainstreaming of biodiversity into national development plan and programmes. Generally, EIA implementation in developing countries appears to work best if legal and institutional arrangements have evolved gradually through an "organic" process, rather than one "imposed" from the outside [77]. The main Plan, Policy, Act, and Regulations related to sectoral projects, those addressed major environmental issues are outlined below:

- National Environment Policy 2019 aims to control pollution, manage wastes and promote greenery so as to ensure citizens" right to live in a fair and healthy environment. The policy was framed to guide the implementation of environment related laws and other thematic laws, realize international commitment and enable collaboration between all concerned government agencies and other sectors on environmental management actions. This policy aims to lessen and prevent all types of environment pollutions, manage wastes emanated from all sectors including home, industry and service, expand parks and greenery in urban area and ensure environment justice to the pollution-affected population. In order to meet the policy goals and objectives, the policy has specified special measures, including setup of effective systems for checking and reducing pollution of all types, encouragement for the use of environment-friendly technology in industry, hospital and vehicles, regulation of harmful pesticides in production and protection of human health from unauthorized food intake.

- Government of Nepal enacted Environmental Protection Act 2019 replacing previous EPA 1997 new amendments. The EPA 2019 empowers environmental protections; preserve right to clean and healthy environment of the society, maintain adaptation between environment and development, reduce adverse impacts on natural environment and biodiversity and climate change adaptation. According to the act, permission will be granted to such projects to use part of national forest if there are no other alternatives and it does not cause significant adverse impact on the environment. As such, it calls for such projects to go through environmental assessment procedure adhering with the prevailing act to ensure that it has clear provisions for compensating the forestland acquired and trees felled by such projects. Provisions relating to forest conservation area and its management, ecosystem service and payment for ecosystem service and establishment of forest development fund are some new highlights of this act. This act provides for establishment of a forest development fund to implement the objectives of this act, conserve and enhance forest. The provisions relating to management of forest as government managed, community, leasehold and religious forest are retained in this act.

Considering previous Environment Protection Regulation (EPR), 1997 and its amendments, EPR endorsed EPR 2020 under the provisions of the EPA 2020. The EPR adopts the environmental assessment criteria mentioned in the National Impact Assessment Guideline 1993. However, the EPR establishes the administrative framework for assessing, exhibition and determination of EIA/IEE, in terms of issues needing to be addressed and the format/layout of the EIA/IEE document. Schedule 1, 2 and 3 listed proposal to be conducted

Brief Environmental Study, Initial Environmental Examination (IEE) and Environmental Impact Assessment (EIA) studies of sectoral projects respectively. The criteria for screening sectoral projects to conduct environmental studies before commencement of the projects has included Forest Sector, Industrial, Mine and Minerals, Road, Residential, Buildings, settlement development and Urban development areas, Water Resource and industry, Tourism, Drinking Water, Solid Waste Management, Agriculture, Health and Education Sectors. Besides these provisions, Section 4 Climate Change, Rule 25 has defined preparation of National Report by the Ministry regarding climate change situation, impacts and risks in every 5 years period. Rule 26 of same section provisioned Ministry should prepare National Adaptation Plan for each 10 years period and Rule 27 provisioned implementation of Mitigation Plan. The detailed framework of EPR 2020 shown in the **Figure 3**.

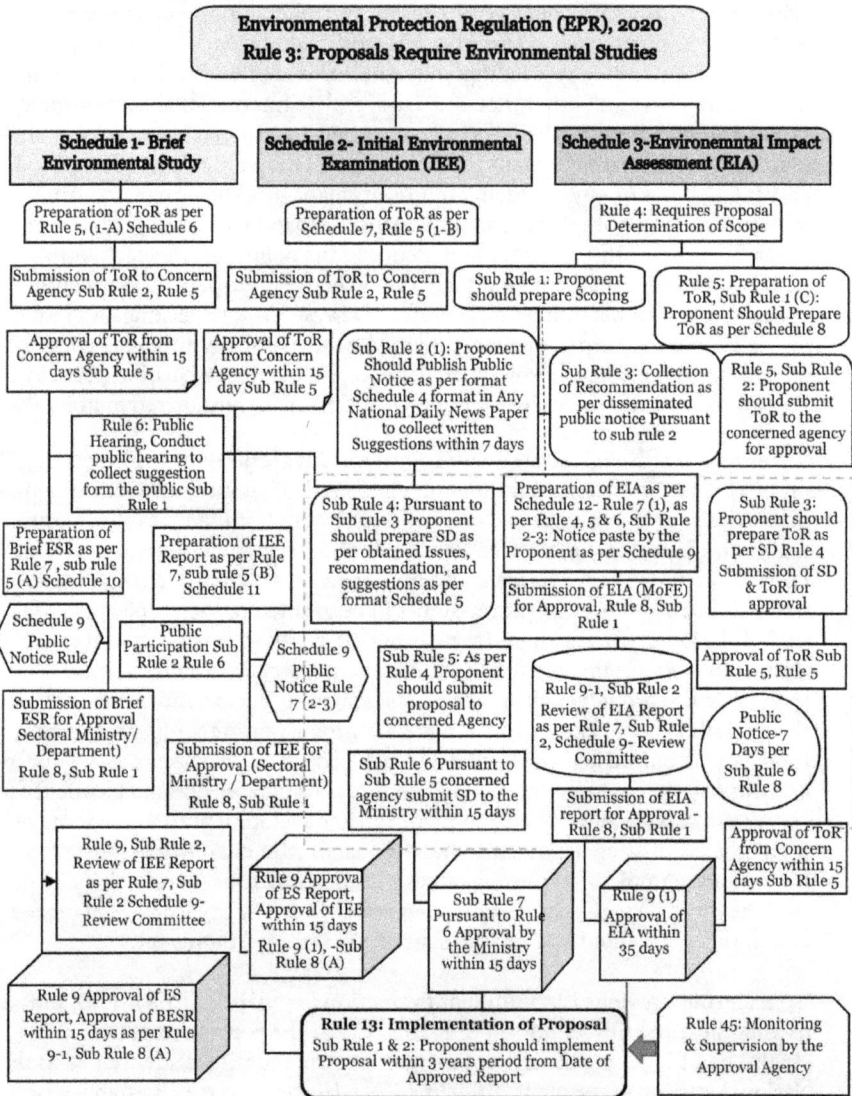

Figure 3.
Legislative framework for environmental assessment process, EPR 2020.

- The Nepal Biodiversity Strategy and Action Plan (NBSAP) 2014–2020 envision a conserved biodiversity contributing for sound and resilient ecosystems and national prosperity.

- The Forest Rules 1995, enforced as per the Forest Act 1993, has categorized number of medicinal herbs and non-timber plants or the timber species with timber use parts for legal trade. Forty-three species are listed to be licensed for their root collection; 20 species for bark; 31 species for leaves; 24 species for flower and fruits; 65 species for fruit and seeds; 12 species for whole plants; 10 species for resin, gums and lac; and other 29 herbs for whole or parts of the plants.

- Nepal became a contracting party to the Convention on International Trade in Endangered Species of Wild Fauna and Flora (CITIES) 1973 on June 18, 1975. That aims to control the trade of certain wildlife species to prevent further endangered species of their survival. CITIES classified species according to the following three criteria-Species threatened with extinction; Species, which could become endangered, and Species that are protected. As Nepal is a signatory to the convention related to species conservation, attention should be given to evaluate the impacts of the project activities on meeting their obligation. It is relevant to environmental assessment studies that species protection list could also be used to evaluate the significance of the identified and predicted impacts for practical mitigation. Plant and species of wildlife under legal protection provides a basis to purpose EMPs for their conservation and for least damaging them during project implementation.

- National Climate Change Policy (NCCP) 2019 underlined thematic areas towards developing a resilient society by reducing the risk of climate change impacts. The policy has integrated sectoral polices and strategies on agriculture and food security, forest, biodiversity and watershed conservation, water resource and energy, rural and urban habitants, industry, transport and physical infrastructure, tourism and natural and cultural heritage, health, drinking water and sanitation, disaster risk reduction and management, gender equality and social inclusion, livelihoods and good governance, awareness raising and capacity development, research, technology development and expansion and climate finance management.

- Hydropower Development Policy 2001 outlines the overall objectives and strategies for hydropower development in Nepal. In addition, National Transport Policy 2001/2002 highlights transport infrastructure development through environment-friendly green road.

- National Forest Policy 2018 defines Nepal's economic, social, and cultural prosperity through a well-managed forest and a balanced environment including production and value addition of forest-based products and services and equitable distribution of the benefits from sustainable and participatory management of the forest, protected areas, watershed, biodiversity and wildlife.

5.2 Mitigation strategy and adaptation

Climate change and risks associated towards adverse impacts on ecosystems and biodiversity, water resources, food production and infrastructure with global warming correlates to adapt reducing or avoiding adverse impacts and maximize positive consequences towards the risks. The adaptation process is complex towards risks posed by climate change and variability as Moser & Ekstrom 2010, stated adaptation process is constant of awareness and understanding, planning, implementation, and monitoring and review [71]. The main services provided by the forests is carbon removal from the atmosphere (carbon sequestration) and the long-term storage of this carbon in biomass, dead organic matter, and soil carbon pools. An estimated 55% (471 Pg C) stored in tropical forests out of the global forest carbon stocks of which more than half is stored in biomass [78]. To mitigate or avoid the projected impacts of climate change adaptation, urgently needed towards extreme changes and impacts that may occur [72].

According to Article of UNFCCC, evaluation of risk associate with climate change refers "dangerous anthropogenic interference with the climate system" [79]. The Convention on Biological Diversity (CBD) has defined ecosystem-based adaptation (EbA) in 2009 as use of biodiversity and ecosystem services as part of an overall strategy to help people adapt to the adverse effects of climate change and EbA projects have proliferated in 2015. As the EBA uses a range of opportunities for the sustainable management, conservation, and restoration of ecosystems to provide services that enable people to adapt to the impacts of climate change, Nepal had adopted Ecosystem-Based Approach (EBA) for priority ecosystems of major ecological regions. To build resilience using an ecosystem management approach in vulnerable developing countries by increasing institutional capacity, mobilizing knowledge and transferring appropriate best-practice adaptation technologies (**Figure 4**).

The developmental activities are the main drivers of biodiversity loss, proper implementation of appropriate sustainable mitigation measure against adverse impacts can support to maintain ecosystems and biodiversity loss. Although,

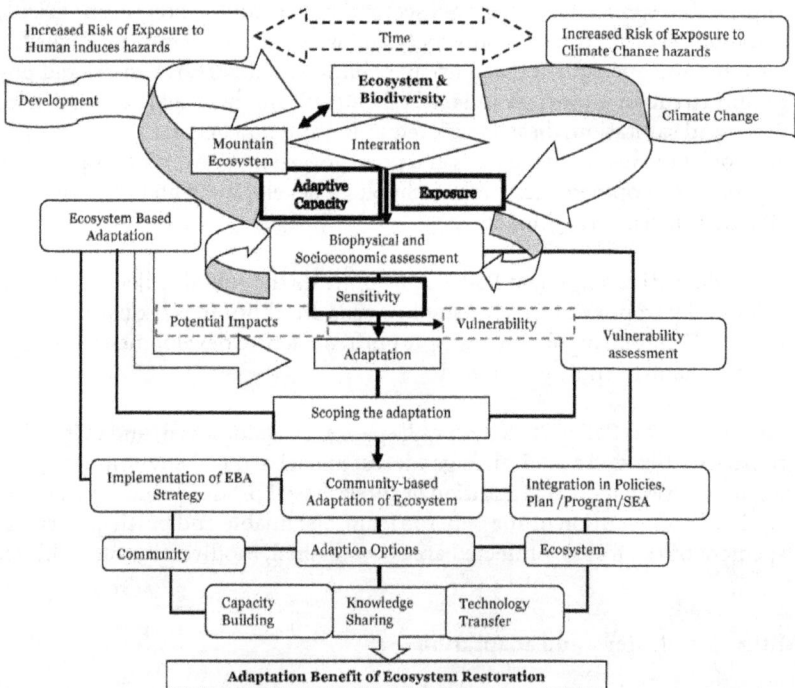

Figure 4.
Schematic diagram of ecosystem based adaptation.

Climate change influences on biodiversity and ecosystem services but not all are negative, with some species either thriving or adapting. Although, environmental assessment a planning application for the development activities generally propose mitigation measures as obligations within the legal framework for implementation. However, the assess¬ment strategy to address the likely environmental impacts and mitigation of the planned sectoral projects is limited including an effective imple-mentation of EMP. The major mitigation measures for developmental projects are presented in the **Table 2**.

Likely impacts	Mitigation measures
a. *Air Quality*	
1. Depletion of ozone layer and climatic change due to emission of some gases (SO_2, CO_2, NO_2, Fluoride, CO, CFCS etc.) to the Atmosphere	• Control the emission of SO_x, NO_x, Co and other applicable chemicals by scrubbing with water or alkaline solutions, incineration or absorption by other catalytic processes • Recycle wastes to reduce the amount of pollutants released to the atmosphere choose environmentally friendly processes, technologies, or raw materials • Treat effluent gases to reduce the amount of pollutants. • establish treatment plant
2. Reduction of air quality due to dust	• Control particular matters by scrubbers, fabric filter collectors or electrostatic precipitators • choice of environmentally friendly processes, technologies or raw materials reduce the amount and significance of pollutants watering of the area form which dust is generated
b. *Ecosystem and Biodiversity*	
Loss of flora and fauna	• locate projects far away from sensitive areas • carry out necessary rehabilitation measures when phasing out a project
Habitat Fragmentation -stability and health of an ecosystem	• plant with native species in vicinity of a project and adjacent areas to wildlife to provide additional habitats and migration routes/corridors for local animals • fence wildlife areas to avoid people interference
Direct killing of animals like collisions with vehicles	• At important areas use of tunnels/bridges reduces interference and collision rates fencing or plant barriers can reduce the interference of human beings and traffics to wildlife • Take measures, like speed break on roads to reduce the speed of vehicles where road crosses protected areas
Disturbance of ecosystem because of extraction of sand, boulders, gravel or rock	• Avoid extraction of sand, gravel boulders from riverbed, bottom/water bodies • use alternative site to exploit the resources avoid the use of dynamite/explosive in water bodies • avoid construction materials during breeding seasons in both water and terrestrial ecosystems
Exploitation of natural resources (flora and fauna) because of immigrants to project area	• before the establishment of projects, planting appropriate tree species, which can be used for different purposes, to minimize burden on the sitting natural resources • use alternative energy resources and construction materials, use proper waste management technology • Make clear demarcation between the resource and project area.

Likely impacts	Mitigation measures
Flora and fauna in wetlands are affected	• avoid the excessive clearance of vegetation from stream banks locate projects as much as far as possible from wetlands • avoid the releaser or minimize the use of hazardous chemicals in the catchments of vulnerable wetlands • If possible, the project should not modify water flow/course use soil and water conservation measures in the catchments to reduce siltation
Introduction of new species or change of cultivation may cause for development of pests, diseases, or weeds.	• Research on invasive exotic species should be carried out in enclosed areas • avoid the use of invasive exotic species for landscaping, reforestation or for other purposes • control the importation of uncertified seed or germ plasm to the region to avoid import of plant pests/disease
Direct or indirect killing of aquatic and terrestrial animals spreading of pesticide/insecticide for different purposes.	• Use integrated pest management to avoid mass killing of animals the concentration and length of time to chemicals should meet standard criteria • use appropriate and trained man power for application of chemicals avoid the use of very poisonous pesticides in particular, on fields sloping down to watercourses during rain seasons with heavy precipitation apply pesticide, when a number of fauna are at the side
Contamination or use of polluted water may affect wild life and nearby communities to the project area.	• use of chemicals or disposal of wastes in a proper way reduce the impact, handling of unused/used poisonous chemicals until they are treated and disposed properly be sure that effluents are treated to the standard before joining water bodies • Avoid the use of very poisonous pesticides in particular on fields sloping down to watercourses during seasons with heavy precipitation • Proper disposal of expired chemicals prevents the potential impacts on flora and fauna • proper disposal of wastes reduces siltation and pollution of water
Improper use of modern biotechnology or introduction of genetically modified varieties to the region may lead to genetic erosion	• Regulate/control importation of varieties to avoid genetic erosion • Regulate import of species to avoid the spoiling of the natural means of existence of existing fauna • Avoid the use of invasive exotic species for landscaping, reforestation, research or for other purposes • Care has to be taken in activities related to modern biotechnology to reduce/avoid the impacts on indigenous species or genetic erosion
Change of the living condition of fish when its migration route is blocked by constructions e.g. dams/reservoirs	• design carefully diversion wears, dams/reservoirs etc. to allow aquatic species to swim against the current • Use filters not to get away fishes to irrigation canals construct ladders so that the fishes jump and migrate against the flow of the water
Water logging may affect the flora (especially deep-rooted plants) and fauna of the area	• digging of canals to lower the water table planting high water consuming species minimizing over irrigation
Exhaustion of resources	• restrict or limit the optimum amount to be exploited/harvested according to the management plan done for the specific resource use recycling methods

Likely impacts	Mitigation measures
c. Water Resources/bodies	
Flooding, channel modification, river canal siltation	• leaving sufficient enough buffer zones of undisturbed vegetation between the site of the project and water bodies • use water flow speed reduction measures e.g. soil conservation measures • Plan carefully to avoid the change/modification of the previous channel flow/natural flow of water
Reduction/lowering of surface or groundwater table	• locate those water-consuming projects, if possible, in areas where availability of ground or surface water is not a problem • choose the most appropriate techniques to minimize the amount of water consumed ensure that the utilization of groundwater is within the capacity of natural system to replenish itself re-use the recycled wastewater
Excess increment of nutrients in water bodies (eutrophication).	• sitting projects far away from susceptible areas to erosion in order to reduce chemical pollution of water bodies • carry out soil conservation measures • leaving sufficient enough buffer zones of undisturbed vegetation between the site of the projects and water bodies • avoid direct waste disposal into or near water bodies • reduce the amount of inlet of both chemical and biological fertilizers to water bodies
Pollution of surface and groundwater through direct or indirect addition of toxic chemicals.	• sitting projects far away from susceptible areas to erosion in order to reduce chemical pollution of water bodies • leaving sufficient buffer zones of undisturbed vegetation between the site of the project and water bodies • install silting basins to reduce silt, pollutants and debris from runoff before it is discharged to adjacent water bodies • Monitoring pipeline systems and impoundments for leaks to reduce contamination of groundwater. E.g. Preparing waterproof waste water collectors • monitor sites even after the project has been closed (as necessary) • reclaiming landscapes where devastating activities have been taken place to reduce water pollution • recycling wastes to reduce water pollution • use treatment techniques especially in industrial activities • choice of the most appropriate technique, replacing processing equipment dispose safely/properly expired toxic chemicals
Increment of suspended solids (turbidity) in water bodies through soil erosion or direct release of waste from different activities.	• sitting projects far away from susceptible areas to erosion in order to reduce siltation, turbidity and chemical pollution of water bodies carry out soil conservation measures • Separation of buffer zone between project sites and water bodies for undisturbed vegetation • installing silting basins to reduce silt, pollutants and debris from runoff before it is discharged to adjacent water bodies
Increment of the amount of silt/sediment in downstream area including agricultural land, reservoirs, etc.	• Minimize the area of ground clearance; provide good vegetative cover or; control the volume and speed of water flows • Careful design/plan of projects can avoid soil erosion; carry out soil conservation measures. • Leaving sufficient buffer zones of undisturbed vegetation between the site of the project and water bodies

Likely impacts	Mitigation measures
d. *Soil*	
Soil erosion and loss of nutrients due to different activities	• Replanting right species of trees, shrubs and grasses in a right time on disturbed areas. Minimize the area of ground clearance.
	• Careful design/plan of projects carry out soil conservation and or agro-forestry measures. Reducing harvest removal.
Soil compaction due to mechanization and machineries.	• Using appropriate machineries/mechanization in appropriate time. Planting leguminous plants improve soil structure.
	• improve soil structure by planting species that improve soil structure or by adding organic matter
Salinization due to irrigation with saline water	• adding organic matter/neutralizing planting salt tolerant species
Soil acidity	• reduce the addition of artificial/organic chemical adding alkaline substance like lime appropriate use/disposal of chemicals
Imbalance of biological activities as a result of contamination of soil with toxic chemicals and loss of organic nutrients due to soil erosion	• Appropriate use of wastes/toxic chemicals take any measures that are used to minimize loss of nutrients. Adding organic matter (green maturing, compost).
	• promote cleaner production (preventing/minimizing waste)
Productive topsoil covered by proposed activities or removal of productive top soil for temporary or permanent purposes	• Collect and reuse the excavated top soil to form a superficial layer. Conversions of borrow pits and spoil dumpsites in to scenic lookouts.
	• Use vertical space than horizontal.
e. *Human Health and Safety*	
Transmission of disease between human and from plants/animals to humans	• sanitary or precaution measures can be accomplished through a comprehensive health awareness campaign curative measures should be in place
Fire, explosions, emission of toxic gases, vapors, dust, emission of toxic liquid, radiation and their cumulative effects badly affect human health in and around the project	• establishing projects far away from settlements
	• Curative measures have to be in place if accidents from different activities can happen.
	• Provide fire proofing of structures, safety buffer zones around the plant boundary, escape routes and others.
	• Store properly easily flammable/explosive gases or toxic chemicals.
	• preventive/protective instruments have to be provided
Health effects on workers due to fugitive dust, material handling, and noise, mechanical or chemical contact can be occurred.	• Prevent accidents through proper design of projects train responsible personnel how to properly handle chemicals; use protective measure, for example ear/eye masks etc.
Noise and congestion may be created and pedestrian hazards could be aggravated by heavy trucks	• Site selection can be taken as preventive measures.
Death and injuries to human beings and damages to property could be happened in factories, roads etc.	• facility should implement a safety and health program designed to identify, evaluate, monitor and control health hazards
	• Site selection can be taken as a preventive measure to minimize risk of accidents especially in road projects.
	• prevent accidents through proper design of projects use protective measure, for example ear/eye masks etc.
Extraction of sand or gravel may from unnecessary pond, which creates suitable condition for malaria and water vector borne disease	• Sanitary or precaution measures can be accomplished through a comprehensive health awareness campaign.
	• Avoid stagnating water and give consecutive awareness to reduce the occurrence of malaria and other related diseases

Likely impacts	Mitigation measures
In mining activities workers are injured when rocks/soils are collapsed	• Proper design has to be done well in such a way that rocks does not collapse. • curative measures have to be in place

Table 2.
Mitigation measures for some of the environmental impacts.

6. Conclusion

The dimensional impacts of climate change threaten species biodiversity and its functional characteristics influencing the ecosystems. This can be mitigated through implementation of mitigation and adaptation plan for actions minimizing climate change induced risks targeting to support biodiversity and entire ecosystem services. Mitigation of greenhouse gases emission through reduction in use of fossil fuel, promotion of clean energy and increase in the rate of carbon uptake through natural ecosystems is the solution. Similarly, implementation of strategic action plans, afforestation/reforestation, agroforestry, land use management, promotion of renewable energy (hydropower, biomass, wind power, solar power etc.) as well as effective implementation of plan, policy and conventions can contribute to address the global climate issue and its impacts. Likewise, minimizing negative consequences and enhancing opportunities adjusting or accommodating to climate change induced impacts can also increase the adaptive capacity of species and ecosystems. This can lead plummeting non-climatic stresses, such as pollution, over-exploitation, habitat loss and fragmentation and invasive alien species, wider adoption of conservation and sustainable use practices including through the strengthening of protected area networks and facilitating adaptive management through strengthening monitoring and evaluation systems.

Nepal has adopted several policies, strategies and EIA tool to address emerging environmental and ecological issues envisioned from developmental projects but threats on ecosystem and biodiversity and degrading rate of natural environment is constant in an alarming rate. The emerging agenda of the country like promotion of renewable energy, enhance capacity of local communities' adaptation and resilience, widen carbon storage through sustainable forest management, and reduce carbon emissions can support to the consequences of climate change issues.

To conserve biodiversity including terrestrial, freshwater and marine ecosystems, restoration of degraded ecosystems, potential impacts of developmental projects can be avoided, reduced and compensate together with pollution control. This can support restore biodiversity and sustain utilization of ecosystem services however, magnitude of climate change effect difficult to adapt all dimension of ecosystems.

Author details

Ramesh Prasad Bhatt
Institute of Ecology and Environment, Kathmandu, Nepal

*Address all correspondence to: drrameshbhatta@gmail.com

IntechOpen

References

[1] Revi, A., D.E. Satterthwaite, F. Aragón-Durand, J. Corfee-Morlot, R.B.R. Kiunsi, M. Pelling, D.C. Roberts, andW. Solecki:. *Urban areas. In: Climate Change 2014: Impacts, Adaptation, and Vulnerability. Part A: Global and Sectoral Aspects. Contribution of Working Group II to the Fifth Assessment Report of theIntergovernmental Panel on Climate Change.* United Kingdom and NewYork, NY, USA, pp. 535-612. : Cambridge University Press, Cambridge, 2014.

[2] USGCRP, Crimmins, A., J. Balbus, J.L. Gamble, C.B. Beard, J.E. Bell, D. Dodgen, R.J. Eisen, N. Fann, M.D. Hawkins, S.C. Herring, L. Jantarasami, D.M. Mills, S. Saha, M.C. Sarofim, J. Trtanj, and L. Ziska, Eds. *Impacts of Climate Change on Human Health in the United States: A Scientific Assessment.* s.l.: U.S. Global Change Research Program, Washington, DC. 312 pp. dx.doi. org/10.7930/J0R49NQX., 2016.

[3] *A climatological study of the Keetch/ Byram drought index and fire activity in the Hawaiian Islands.* Dolling, K., Chu, P.-S., Fujioka, F. s.l. : Agricultural and Forest Meteorology , 2005, Vols. 133, 17-27.

[4] *Secular temperature changes in Hawaiʻi.* Giambelluca, Thomas W., Diaz, Henry F. and Luke, Mark S. A. s.l. : Geophysical Research Letters, 35(12), n/a–n/a. doi:10.1029/2008gl034377, 2008.

[5] Frazier, A., Giambelluca, T., Diaz, H.,. Spatial rainfall patterns of ENSO and PDO in Hawaii. *In: University of Hawaii at Manoa, Department of Geography.* Poster B13B-0502. NOAA/ESRL/CIRES,University of Colorado,Boulder, 2012.

[6] IPCC. *Summary for policymakers. In: Stocker, T.F., D. Qin, G.-K. Plattner, et al. (eds) Climate change 2013: The physical science basis. Contribution of Working Group I to the fifth assessment report of*

the Intergovernmental Panel on Climate Change. Cambridge, UK : Cambridge University Press, 2013.

[7] —. *Summary of Policy Makers; In: Global Warming of 1.5°C. An IPCC Special Report on the impacts of global warming of 1.5°C above pre-industrial levels and related global greenhouse gas emission pathways, in the context of strengthening the global response.* s.l. : IPCC, 2018.

[8] *Climate change effects on organic carbon storage in agricultural soils of northeastern Spain. Agriculture.* Álvaro-Fuentes, J., Easter, M., & Paustian, K. s.l. : Ecosystems & Environment, 155, 87-94. doi:10.1016/j. agee.2012.04.001 , 2012.

[9] *Evaluating soil threats under climate change scenarios in the Andalusia region: southern Spain.* Anaya-Romero, M., Abd-Elmabod, S.K., Muñoz-Rojas, M., Castellano, G., Ceacero, C.J., Alvarez, S., Méndez, M., De la Rosa, D. s.l. : Land Degrad. Dev. 26, 441-449, 2015.

[10] Conant, R.T., Ryan, M.G., Ågren, G.I., Birge, H.E., Davidson, E.A., Eliasson, P.E., Evans, S. E., Frey, S.D., Giardina, C.P., Hopkins, F.M., Hyvonen, R., Kirschbaum, M.U.F., Lavalee, J.M., Leifield, J., Parton, W.J., Steinweg, J.M., Wallenstein, M.D., Wet. *Temperature and soil organic matter decomposition rates-synthesis of current knowledge and a way forward.* s.l. : Global chnage Biol.1,3392-3404, 2011.

[11] *Soil carbon sequestration to mitigate climate change and Advance Food Security.* Lal, R., Ronald Follet, BA Stewart. J.M. Kimble. s.l. : Geoderma 123, 1-22, 2014, Soil Science 172(12):943-946.

[12] *Muñoz-Rojas, M., Jordán, A., Zavala, Modelling soil organic carbon stocks in global change scenarios: a CarboSOIL application.* Muñoz-Rojas, M., Jordán, A., Zavala, L.M., González-Peñaloza,

F.A., De la Rosa, D., Pino-Mejias, R., Anaya-Romero, M. s.l. : Muñoz-Rojas, M., Jordán, A., Zavala, L.M., González-Peñaloza, F.A., De la Rosa, D., Pino-MejiaBiogeosciences 10, 8253-8268., 2013.

[13] IPCC. *Terrestrial and Inland Water Systems. In: Climate Change 2014: Impacts, Adaptation and Vulnerability. Part A: Global and Sectoral Aspects. Contribution of Working Group II to the Fourth Assessment Report of the Intergovernmental Panel on Climate Change.* UK and NY, USA : Cambridge University Press, Cambridge , 2014.

[14] CCSP. *The Effects of Climate Change on Agriculture, Land Resources, Water Resources, and Biodiversity in the United States. A Report by the U.S. Climate Change Science Program and the Subcommittee on Global Change Research.* s.l. : U.S. Environmental Protection Agency, Washington, DC, USA., 2008.

[15] USGCRP. Horton, R., G. Yohe, W. Easterling, R. Kates, M. Ruth, E. Sussman, A. Whelchel, D. Wolfe, and F. Lipschultz, 2014:. *Ch. 16: Northeast. Climate Change Impacts in the United States: The Third National Climate Assessment, J. M. Melillo, Terese (T.C.) Richmond, and G. W. Yohe, Eds.* s.l. : U.S. Global Change Research Program, 16-1-nn., 2014.

[16] —. Groffman, P. M., P. Kareiva, S. Carter, N. B. Grimm, J. Lawler, M. Mack, V. Matzek, and H. Tallis, 2014: Ch. 8: Ecosystems, Biodiversity, and Ecosystem Services. *Climate Change Impacts in the United States: The Third National Climate Assessment, J. M. Melillo, Terese (T.C.) Richmond, and G. W. Yohe, Eds., U.S. Global Change Research Program, 200-201.* s.l. : U.S. Global Change Research Program, 200-201. , 2014.

[17] —. Global Climate Change Impacts in the United States. *"Climate Change Impacts by Sectors: Ecosystems." Karl, T.R., J.M. Melillo, and T.C. Peterson (eds.).* s.l.:

United States Global Change Research Program. Cambridge University Press, New York, NY, 2009.

[18] ACIA. *Impacts of a Warming Arctic: Arctic Climate Impact Assessment.* s.l.: Arctic Climate Impact Assessment. Cambridge University Press, Cambridge, United Kingdom., 2004.

[19] IPCC (2014). Settele, J., R. Scholes, R. Betts, S. Bunn, P. Leadley, D. Nepstad, J.T. Overpeck, and M.A. Taboada. *Terrestrial and Inland Water Systems. In: Climate Change 2014: Impacts, Adaptation and Vulnerability. Part A: Global and Sectoral Aspects. Contribution of Working Group II to the Fourth Assessment Report of the Intergovernmental Panel on Climate Change.* Cambridge University Press, Cambridge, United Kingdom and New York, NY, USA : Contribution of Working Group II to the FouField, C.B., V.R. Barros, D.J. Dokken, K.J. Mach, M.D. Mastrandrea, T.E. Bilir, M. Chatterjee, K.L. Ebi, Y.O. Estrada, R.C. Genova, B. Girma E.S. Kissel, A.N. Levy, S. MacCracken, P.R. Mastrandrea, and L.L. White, 2014. pp. 11-13.

[20] The Kathmandu Post. "Government unveils new political map including Kalapani, Lipulekh and Limpiyadhura inside Nepal borders". 20 05 2020, p. kathmandupost.com.

[21] Dobremez, J.F. *Le Ne'pal E'colgie et Bioge'ographie (Ecology and Biogeography of Nepal).* s.l. : Centre National de la Researche Centifique, Paris, 1976.

[22] LRMP. *Land Resources Mapping Project. Survey.* Kathmandu : HMGN and Kenting Earth Sciences., 1986.

[23] H., Wang. *Guide to the World States: Nepal.* Beijing : Social Sciences Academic Press (China), 2004.

[24] *Development of 2010 national land cover database for Nepal.* Uddin K., Shrestha H.L., Murthy M.S.R.,

Bajracharya B., Shrestha B., Gilani H., Pradhan S. & Dangol B. 2014, Journal of Environmental Management, Vol. 148, pp. 82-90.

[25] DFRS. *Forest Cover Maps of Local Levels (753) of Nepal.* Kathmandu, Nepal : Department of Forest Research and Survey (DFRS), 2018.

[26] *State of Nepal's Forests. Forest Resource Assessment Nepal (FRA) Nepal.* Kathmandu, Nepal : Department of Forest Research and Survey (DFRS), 2015c.

[27] CBD. *Secretariat of the Convention on Biological Diversity.* 2010. p. 56.

[28] IPCC. *Climate change 2007: the physical basis. Summary for policy-makers. Contribution of Working Group I to the Fourth Assessment Report of the Intergovernmental Panel on Climate Change.* s.l. : Cambridge:Cambridge University Press, 2007.

[29] DFRS. *State of Nepal's Forests. Forest Resource Assessment (FRA) Nepal.* Kathmandu, Nepal : Department of Forests Research and Survey (DFRS), 2015.

[30] MoFSC. *Nepal National Biodiversity Strategy and Action Plan 2014-2020.* Kathmandu, Nepal : Ministry of Forest and Soil Conservation, Government of Nepal, 2014.

[31] GoN. *National Park and Wildlife Conservation Act 2029 (1973).* Kathmandu: Nepal Law Commission. Official English translation, 13 pp, 1973.

[32] DNPWC. *Jnawali, S.R., Baral, H.S., Lee, S., Acharya, K.P., Upadhyay, G.P., Pandey, M., Shrestha, R., Joshi, D., Laminchhane, B.R., The Status of Nepal Mammals: The National Red List Series.* Kathmandu : Department of National Parks and Wildlife Conservation, 2011.

[33] *Flora and Fauna of Nepal in CITIES Annexes.* Kathmandu : Department

of National Parks and Wildlife Conservation, 2014.

[34] Joshi, N., Dhakal, K. S., Saud, D. S. *Checklist of CITES Listed Flora of Nepal.* s.l. : Department of Plant Resources (DPR), Thapathali, Kathmandu, Nepal, 2017.

[35] Maskey, Tirtha M. Biodiversity: Conservation Challenges in Nepal. *Biodiversität: Herausforderungen des Naturschutzes in Nepal.*

[36] *Changing Pattern of Forest Resources and Manmade Implications towards Declining Biodiversity".* Bhatt, R. 4, 2018, International Journal of Forestry and Horticulture (IJFH), Vol. 4, pp. 32-42.

[37] FAO. *Global Forest Resources Assessment 2010: Main Report.* Rome : Food and Agriculture Organization of the United Nations (FAO) , 2010.

[38] WB. *Nepal - Country Environmental Analysis: Strengthening Institutions andManagement Systems for Enhanced Environmental Governance.* s.l. : The World Bank, 2008.

[39] DoF. *Summary of forest encroachment until 2068-2069 BS (2011-2012).* Kathmandu : Department of Forests, 2012.

[40] NAP. *NAP-Agriculture Project: Integrating Agriculture in National Adaptation Plans.* s.l. : UNFA/GLO/616/UND, Nepal, 2016.

[41] MoSTE. *National Adaptation Programme of Action to Climate Change.* s.l. : Ministry of Science Technology and Environment (MOSTE), Kathmandu, 2010.

[42] *The "anthropocene".* Crutzen, P. J. 2006, Earth system science in the anthropocene, Springer: 13-18.

[43] *"Emerging approaches, challenges and opportunities in life cycle 501 assessment".*

Hellweg, S. and L. Milà i Canals. 2014, pp. Science 344(6188): 1109-1113.

[44] *Hydro power development in Nepal.* Sharma, R.H. and Awal, R. 2013, Sustain. Energy Rev. 2013, 21, 684-695 .

[45] *Current Status of Renewable Energy in Nepal: Opportunities and Challenges.* al., K. Surendra et. 2011, Renewable and Sustainable Energy Reviews. Volume 15. pp. 4107-4117.

[46] *The Potential of a Renewable Energy Technology for Rural Electrification in Nepal: A Case Study from Tangting.* al., Gurung et. 2011, Renewable Energy. Volume 36. pp. 3203-3210.

[47] NEA. *A year in Review; Fiscal Year 2019/2020.* Kathmandu : Nepal Electricity Authority, Annual Report 2077, 2020.

[48] *Incorporating socio-environmental considerations into project assessment models using multi-criteria analysis: A case study of Sri Lankan hydro power projects.* Morimoto, R. 2013, Energy Policy, Vol. 59, pp. 643-653.

[49] Goldsmith, E. and Hildyard, N. *The Social and Environmental Effects of Large Dams.* s.l. : Sierra Club: San Fransisco, CA, USA, 1984, 1984.

[50] GCA/UNCC. React to Zero Dialogue. *Outcome Document; Cities and Regions.* 18 November 2020, p. 7.

[51] ADB. Hydropower Development and Economic Growth in Nepal . *ADB South Asia Working Paper Series, Herath Gunatilake, Priyantha Wijayatunga, and David Roland-Holst No. 70.* 06 2020, p. 15.

[52] R.P., Bhatt. Hydropower Development in Nepal- Climate Change, Impacts and Implications. [book auth.] INTEC (eds.) Basel I. Ismail. *In: "Renewable Hydropower Technologies".* s.l.: INTEC, 2017, pp. Chapter 5-pp. 75-98.

[53] *The role of hydropower in climate change mitigation and adaptation: a review.* Berga, L. 3, 2016, Engineering, Vol. 2, pp. 313-318.

[54] NPC. 14th Three Year Plan (2016/2017-2018/2019) . *Approach Paper, National Planning Commission.* 2016.

[55] *South Asia, Fast Growing Regions _ 2016-17.* UNESCAP. UN Web Site. Retrieved 12 01, 2016, from : s.n., 2016. UN web Site, Retrieved 12 01, 2016, from http://www.unescap.org/announcement/south-asia-expected-world.

[56] WB. *South Asia Economic Focus: A Review of Economic Developments in South Asian Countries- Creating Fiscal Space through Revenue Mobilization.* Washington, D.C : The World Bank, 2012.

[57] WEF. *The Global Competitiveness Report.* Geneva : World Economic Forum, 2016.

[58] NPC. *Annual Development Programs.* Kathmandu : National Planning Commission, 2016.

[59] ADB. *Economics of Climate Change in East Asia Michael Westphal Gordon Hughes Jörn Brömmelhörster (Editors).* Philipppins: Asian Development Bank, 2013.

[60] *Global Climate Indix 2020.* David Eckstein, Vera Künzel, Laura Schäfer, Maik Winges. Office Bonn : s.n., 2018. Germanwatch e.V.

[61] GoN/DoHM. *Observed Climate Trend Analysis of Nepal (1971-2014)".* Kathmandu : Department of Hydrology and Meteorology, 2017.

[62] *Karki, Ramchandra; Hasson, Shabeh ul; SchickhofRising Precipitation Extremes across Nepal.* Karki, Ramchandra, et al. 2017, Climate, p. 5 (1): 4. doi:10.3390/cli5010004.

[63] GoN/MoFE. *Nepal's National Adaptation Plan (NAP) Process:*

Reflecting on Lesson Learned and The Way Forward. Kathmandu : Ministry of Forests and Environment, Government of Nepal, 2018.

[64] IPBES. *Summary for policymakers of the global assessment report on biodiversity and ecosystem services of the Intergovernmental Science-Policy Platform on Biodiversity and Ecosystem Services.* s.l.: IPBES secretariat, Bonn, Germany. 56 pages, 2019.

[65] *A Summary of Climate Change Effects on Watershed Hydrology.* Pike, R.G., D.L. Spittlehouse, K.E. Bennett, V.N. Egginton, P.J. Tschaplinski, T.Q. Murdock and A.T. Werner. 2008, B.C. Min. For. Range, Res. Br., Victoria, B.C. Exten. Note 87, pp. 3-4.

[66] Pitelka, Jay R. Malcolm and Louis F. *Ecosystems and Global Climate Change: A review of potential Impacts on U.S. Terrestrial Ecosystems and Biodiversity.* s.l. : Arlington, VA : Pew Center on Global Climate Change, pp 34, 2000.

[67] MoPE. *Initial National Communication to the United Nations Framework Convention on Climate Change.* s.l. : Ministry of Population and Environment, Government of Nepal, Kathmandu, 2004.

[68] WB. *Vulnerability, Risk Reduction, and Adaptation to Climate Change Nepal.* NW Washington, DC : The World Bank Group, 2011.

[69] Agrawala, S., et al. *Development and climate change in Nepal: Focus on water resources and hydropower.* s.l. : Environment directorate, Development Co-operation, Directorate, Organization for Economic Co-operation and Development, 2003.

[70] MoSTE. *Second National Communication Report to UNFCC.* Kathmandu : Ministry of Science Techenology and Environment, 2014.

[71] *A framework to diagnose barriers to climate change adaptation.* Moser SC, Ekstrom JA. s.l. : PNAS. http://www.pnas.org/content/107/51/22026, 2010, Proc Natl Acad Sci U S A. 2010 Dec 21;107(51):22026-31.

[72] WB. *Turn down the heat: why a 4oC warmer world must be avoided.* s.l. : Washington DC: World Bank, 2012a.

[73] NPC. *The Seventh Plan (1985-1990).* Kathmandu, Nepal : National Planning Commission, 1985.

[74] —. *The Eighth Plan (1992-1997).* Nepal : National Planning Commission. Kathmandu, 1993.

[75] GoN. *National Environmental Assessment Guidelines.* s.l. : Nepal Gazette (Rajpatra), 3(5). Kathmandu, Nepal, 1993.

[76] GoN/MoLJPA. *Environmental Protection Act 1997 and Environmental Protection Regulation 1997:* s.l. : Ministry of Law, Justice, and Parliamentary Affairs. Law Books Management Board. Kathmandu, Nepal., 1997.

[77] *Environmental impact assessment system and process: A study on policy and legal instruments in Nepal.* Bhatt, Ramesh Prasad and Khanal SN. 2010, African Journal of Environmental Science and Technology, Vols. 4, 9, pp. 586-594.

[78] *"A Large and Persistent Carbon Sink in the World'S Forests".* Pan, Y., R.A. Birdsey, J. Fang, R. Houghton, P.E. Kauppi, W.A. Kurz, O.L. Phillips, B.R. Scheffers, et al. 2011, Science Express, 333: 988-993.

[79] UNFCC. "United Nations Framework Convention on Climate Change (UNFCCC)". *World Health Organization (WHO).* 22 22 October 2020.

www.ingramcontent.com/pod-product-compliance
Lightning Source LLC
Chambersburg PA
CBHW081531190326
41458CB00015B/5522

9781839626296